Methods in Molecular Biology

Series Editor
John M. Walker
School of Life Sciences
University of Hertfordshire
Hatfield, Hertfordshire, AL10 9AB, UK

For further volumes:
http://www.springer.com/series/7651

Plant Endosomes

Methods and Protocols

Edited by

Marisa S. Otegui

Departments of Botany and Genetics, University of Wisconsin-Madison, Madison, WI, USA

Editor
Marisa S. Otegui
Departments of Botany and Genetics
University of Wisconsin-Madison
Madison, WI, USA

ISSN 1064-3745 ISSN 1940-6029 (electronic)
ISBN 978-1-4939-4586-3 ISBN 978-1-4939-1420-3 (eBook)
DOI 10.1007/978-1-4939-1420-3
Springer New York Heidelberg Dordrecht London

Printed on acid-free paper

Humana Press is a brand of Springer
Springer is part of Springer Science+Business Media (www.springer.com)

Preface

The composition of the plasma membrane is tightly controlled by cells through vesicular trafficking. Cells can internalize plasma membrane transporters, enzymes, receptors, and other key signaling molecules through the formation of vesicles in a process called endocytosis. Endocytosed material is delivered to early endosomes where it can be recycled back to the plasma membrane or be further sorted into endosomal intraluminal vesicles for degradation in vacuoles/lysosomes. In plant cells, the *Trans* Golgi Network (TGN) and TGN-derived compartments have been shown to act as early endosomes whereas multivesicular bodies (MVBs) sort proteins for degradation. Both TGN and MVBs also traffic cargo material that has been synthesized in the endoplasmic reticulum and Golgi and is destined to the vacuole. In recent years, many significant contributions have dramatically changed our understanding of the plant endosomal system. The identification of key components in the regulation of plasma membrane endocytosis, recycling, and degradation supports both the unique and conserved aspects of the plant vesicular trafficking machinery. In addition, plant endocytic and endosomal trafficking plays a central role in mediating responses to biotic and abiotic stimuli and in general plant development and cell differentiation. However, the analysis of plant endosomal trafficking pathways can be difficult and entails several challenges: (1) The endosomal system comprises a dynamic set of organelles that are in continuous flux. (2) Not all plasma membrane cargoes are trafficked equally. (3) Posttranslational modifications regulate the fate of plasma membrane proteins. (4) Endosomes act as platforms for the traffic of cargo from different pathways. (5) Key endosomal subdomains, such as intraluminal vesicles of MVBs, are too small to be visualized directly by light microscopy techniques.

This book contains a collection of protocols and techniques to analyze in vivo trafficking of endocytic/endosomal cargo, including lipids, fluids, proteins, and ligands, ultrastructural features of endosomes by high-pressure freezing/freeze-substitution and electron tomography, protein-protein interactions in the endosomal and endomembrane system, sorting defects in the transport of vacuolar storage proteins, function conservation of plant endosomal proteins, endosomal trafficking during plant responses to pathogens, protein composition of endosomes and endocytic vesicles, ubiquitination of endosomal cargo proteins, and identification of novel endosomal components by chemical genomics and proteomics. I hope that these contributions from many leading and emerging plant membrane trafficking researchers from all over the world will promote and facilitate novel studies and ideas in this field. Finally, I would like to thank all the authors and colleagues that have contributed chapters and ideas to this book and the National Science Foundation for supporting research on endosomal trafficking in my laboratory.

Madison, WI, USA *Marisa S. Otegui*

Contents

Contributors

BEN AUGUST • *Electron Microscope Facility, School of Public Health, University of Wisconsin -Madison, Madison, WI, USA*

VERA BANDMANN • *Plant Cell Biology, Department of Biology, INM-Leibniz-Institute for New Materials, Germany*

SEBASTIAN Y. BEDNAREK • *Department of Biochemistry, University of Wisconsin-Madison, Madison, WI, USA*

STANLEY W. BOTCHWAY • *Rutherford Appleton Laboratory, Central Laser Facility, Research Complex at Harwell, Science and Technology Facilities Council, Didcot, UK*

YOHANN BOUTTÉ • *Membrane Biogenesis Laboratory, UMR 5200, CNRS-Université Bordeaux Segalen, INRA Bordeaux Aquitaine, Villenave d'Ornon Cédex, France*

YI CAI • *School of Life Sciences, Centre for Cell and Developmental Biology and State Key Laboratory of Agrobiotechnology, The Chinese University of Hong Kong, Shatin, NT, Hong Kong, China*

ALEXANDRA CHANOCA • *Department of Botany, University of Wisconsin, Madison, WI, USA*

GEORGIA DRAKAKAKI • *Department of Plant Sciences, University of California Davis, CA, USA*

TORU FUJIWARA • *Graduate School of Agricultural and Life Sciences, The University of Tokyo, Tokyo, Japan*

CAIJI GAO • *School of Life Sciences, Centre for Cell and Developmental Biology and State Key Laboratory of Agrobiotechnology, The Chinese University of Hong Kong, Shatin, NT, Hong Kong, China; CUHK Shenzhen Research Institute, The Chinese University of Hong Kong, Shenzhen, China*

MARKUS GREBE • *Department of Plant Physiology, Umeå Plant Science Centre (UPSC), Umeå University, Umeå, Sweden; Institut für Biochemie und Biologie, Pflanzenphysiologie, Universität Potsdam, Potsdam-Golm, Germany*

IKUKO HARA-NISHIMURA • *Department of Botany, Graduate School of Science, Kyoto University, Kyoto, Japan*

PETER HAUB • *DIPsystems.de, Altlussheim, Germany*

PING HE • *Department of Biochemistry and Biophysics, Institute for Plant Genomics and Biotechnology, Texas A&M University, College Station, TX, USA*

ANTJE HEESE • *Division of Biochemistry, Interdisciplinary Plant Group (IPG), University of Missouri-Columbia, Columbia, MO, USA*

GLENN R. HICKS • *Department of Botany and Plant Sciences, Center for Plant Cell Biology, University of California, Riverside, CA, USA*

NILOUFER G. IRANI • *Department of Plant Systems Biology, VIB, Ghent, Belgium; Department of Plant Biotechnology and Bioinformatics, Ghent University, Ghent, Belgium; Department of Plant Sciences, University of Oxford, Oxford, UK*

ERIKA ISONO • *Department of Plant Systems Biology, Technische Universität München, Freising, Germany*

EMI ITO • *Department of Biological Sciences, Graduate School of Science, The University of Tokyo, Tokyo, Japan*

ADRIANA JELÍNKOVÁ • *Institute of Experimental Botany, ASCR, Praha, Czech Republic*

LIWEN JIANG • *School of Life Sciences, Centre for Cell and Developmental Biology and State Key Laboratory of Agrobiotechnology, The Chinese University of Hong Kong, Shatin, NT, Hong Kong, China; CUHK Shenzhen Research Institute, The Chinese University of Hong Kong, Shenzhen, China*

KAMILA KALINOWSKA • *Department of Plant Systems Biology, Technische Universität München, Freising, Germany*

KOJI KASAI • *Graduate School of Agricultural and Life Sciences, The University of Tokyo, Tokyo, Japan*

YASUKO KOUMOTO • *Department of Botany, Graduate School of Science, Kyoto University, Kyoto, Japan*

JOHANNES LEITNER • *Department of Applied Genetics and Cell Biology, BOKU, University of Natural Resources and Life Sciences, Wien, Austria*

MICHELLE E. LESLIE • *Division of Biochemistry, Interdisciplinary Plant Group (IPG), University of Missouri-Columbia, Columbia, MO, USA*

IRENE LICHTSCHEIDL • *Core Facility of Cell Imaging and Ultrastructure Research, University of Vienna, Vienna, Austria*

CHRISTIAN LUSCHNIG • *Department of Applied Genetics and Cell Biology, BOKU, University of Natural Resources and Life Sciences, Wien, Austria*

KATEŘINA MALÍNSKÁ • *Institute of Experimental Botany, ASCR, Praha, Czech Republic*

TOBIAS MECKEL • *Membrane Dynamics, Department of Biology, Technische Universität, Darmstadt, Germany*

LORENA NORAMBUENA • *Plant Molecular Biology Laboratory, Department of Biology, Faculty of Sciences, University of Chile, Santiago, Chile*

MARISA S. OTEGUI • *Departments of Botany and Genetics, University of Wisconsin, Madison, WI, USA*

MIROSLAV OVEČKA • *Department of Cell Biology, Faculty of Science, Centre of the Region Haná for Biotechnological and Agricultural Research, Palacký University Olomouc, Olomouc, Czech Republic*

EUNSOOK PARK • *Department of Plant Sciences, University of California Davis, CA, USA*

TIBOR PECHAN • *Institute for Genomics, Biocomputing and Biotechnology, Mississippi State University, Missisipi State, MS, USA*

JAN PETRÁŠEK • *Institute of Experimental Botany, ASCR, Praha, Czech Republic*

FRANCISCA C. REYES • *Centro de Biotecnología Vegetal, Facultad Ciencias Biológicas, Universidad Andrés Bello, Santiago, Chile*

GREGORY D. REYNOLDS • *Department of Biochemistry, University of Wisconsin-Madison, Madison, WI, USA*

SIMONE DI RUBBO • *Department of Plant Systems Biology, VIB, Ghent, Belgium; Department of Biology, University of Washington-HHMI, University of Washington, Seattle, WA, USA*

CARLOS RUBILAR-HERNÁNDEZ • *Plant Molecular Biology Laboratory, Department of Biology, Faculty of Sciences, University of Chile, Santiago, Chile*

EUGENIA RUSSINOVA • *Department of Plant Systems Biology, VIB, Ghent, Belgium; Department of Plant Biotechnology and Bioinformatics, Ghent University, Ghent, Belgium*

JOZEF ŠAMAJ • *Department of Cell Biology, Faculty of Science, Centre of the Region Haná for Biotechnological and Agricultural Research, Palacký University Olomouc, Olomouc, Czech Republic*

OLGA ŠAMAJOVÁ • *Department of Cell Biology, Faculty of Science, Centre of the Region Haná for Biotechnological and Agricultural Research, Palacký University Olomouc, Olomouc, Czech Republic*

JENNIFER SCHOBERER • *Department of Applied Genetics and Cell Biology, BOKU, University of Natural Resources and Life Sciences, Vienna, Austria*
LIBO SHAN • *Department of Plant Pathology and Microbiology, Institute for Plant Genomics and Biotechnology, Texas A&M University, College Station, TX, USA*
TOMOO SHIMADA • *Department of Botany, Graduate School of Science, Kyoto University, Kyoto, Japan*
THOMAS STANISLAS • *Department of Plant Physiology, Umeå Plant Science Centre (UPSC), Umeå University, Umeå, Sweden*
TOMÁŠ TAKÁČ • *Department of Cell Biology, Faculty of Science, Centre of the Region Haná for Biotechnological and Agricultural Research, Palacký University Olomouc, Olomouc, Czech Republic*
JUNPEI TAKANO • *Research Faculty of Agriculture, Hokkaido University, Sapporo, Japan*
TAKASHI UEDA • *Department of Biological Sciences, Graduate School of Science, The University of Tokyo, Tokyo, Japan; Japan Science and Technology Agency (JST), PRESTO, Saitama, Japan*
JINGGENG ZHOU • *Department of Biochemistry and Biophysics, Institute for Plant Genomics and Biotechnology, Texas A&M University, College Station, TX, USA*
XIAOHONG ZHUANG • *School of Life Sciences, Centre for Cell and Developmental Biology and State Key Laboratory of Agrobiotechnology, The Chinese University of Hong Kong, Shatin, NT, Hong Kong, China*

Chapter 1

The Use of FM Dyes to Analyze Plant Endocytosis

Kateřina Malínská, Adriana Jelínková, and Jan Petrášek

Abstract

FM (Fei-Mao) styryl dyes are compounds of amphiphilic character that are used for the fluorescence tracking of endocytosis and related processes, i.e., the internalization of membrane vesicles from the plasma membrane (PM) and dynamics of endomembranes. Staining with FM dyes and subsequent microscopical observations could be performed both on the tissue and cellular level. Here, we describe simple procedures for the effective FM dye staining and de-staining in root epidermal cells of *Arabidopsis thaliana* seedlings and suspension-cultured tobacco cells. The progression of FM dye uptake, reflected by an increased amount of the dye in the endosomal compartments, is monitored under the fluorescence microscope in a time-lapse manner. The data obtained can be used for the characterization of the rate of endocytosis and the function of components of endosomal recycling machinery.

Key words FM styryl dyes, FM 4-64, FM 1-43, Endocytosis, Plasma membrane, Endomembranes

1 Introduction

Microscopical observation of endocytosis after labeling cells with FM dyes is one of the most frequently used methods for the tracking of endocytosis in plants [1]. FM styryl dyes were originally developed to stain synaptic vesicles in vivo [2] and later optimized for the tracking of endo- and exocytosis in animal cells [3]. However, as documented in numerous studies, they contributed substantially also to our understanding of processes of endocytosis and plasma membrane recycling in various fungi, algae, and vascular plants [1, 4]. There are in principle three important characteristics that favor some FM dyes like FM 4-64 (*N*-(3-triethylammoniumpropyl)-4-(6-(4-(diethylamino) phenyl) hexatrienyl) pyridinium dibromide) and FM 1-43 (*N*-(3-triethylammoniumpropyl)-4-(4-(dibutylamino) styryl) pyridinium dibromide) to be suitable markers for studies of endocytosis [2]. Firstly, thanks to their positively charged head tail, they cannot penetrate through the PM and could be therefore internalized exclusively by endocytic processes. Secondly, they are intercalated

Marisa S. Otegui (ed.), *Plant Endosomes: Methods and Protocols*, Methods in Molecular Biology, vol. 1209,
DOI 10.1007/978-1-4939-1420-3_1, © Springer Science+Business Media New York 2014

efficiently into the outer leaflet of the PM depending on the composition of their lipophilic tail. Moreover, in contrast to water solutions, they have greatly enhanced fluorescence when intercalated into the PM, providing a very good opportunity for selective observations of membranes (Fig. 1a). And thirdly, they could have various spectral characteristics corresponding to the number of double bonds in the middle bridge region (Fig. 1a, b). Thanks to their properties, FM dyes mark all early processes of plasma membrane internalization [1], including clathrin-mediated [5] and

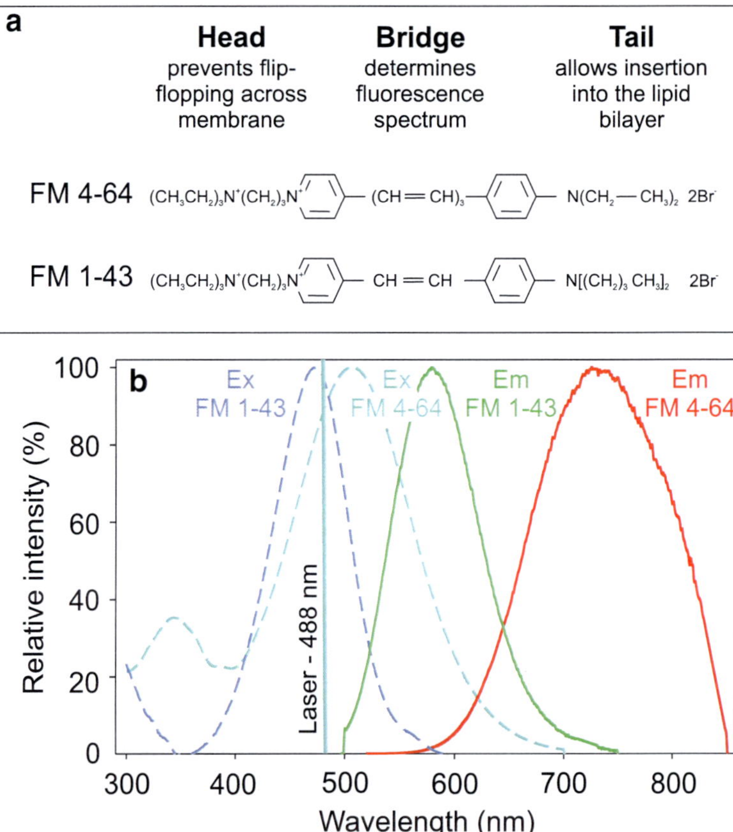

Fig. 1 Structures and spectral characteristics of FM 4-64 and FM 1-43 dyes [3]. (**a**) Comparison of head, bridge, and tail regions of FM 4-64 and FM 1-43. Positively charged head region prevents penetration through the PM (it is the same for FM 4-64 and FM 1-43). More double bonds in the bridge shift fluorescence emission of FM 4-64 to the red part of the spectrum. More carbons in the tail region of FM 1-43 increase its lipophilicity in comparison with FM 4-64. (**b**) Excitation (*dashed lines*) and emission spectra of FM 4-64 and FM 1-43. Laser line 488 nm could be used for the excitation of both FM dyes. Data for FM dyes in water solution with CHAPS are taken from Invitrogen Fluorescence Spectra Viewer. Note that both excitation and emission spectra are pH sensitive and might be shifted upon insertion of FM dyes into PM, as reported [1]

sterol endocytosis [6–8]. From early endosomes and numerous intermediate membrane vesicles, FM dyes reach several destinations that include *trans*-Golgi, prevacuolar compartments, vacuoles, as well as exocytic vesicles that are recycled back to the PM [1]. They do not stain the endoplasmic reticulum or nuclear envelope.

Here we describe simple protocols for the in vivo microscopical tracking of endocytosis with FM dyes in favorite models of *Arabidopsis thaliana* root tips and suspension-cultured cells of *Nicotiana tabacum* L., cv. Bright Yellow (BY-2) [9]. Under room temperature, the progression of FM dye internalization into these cells is observable already within several minutes after the addition of the dye. Therefore, the preincubation of seedlings or suspension-cultured cells on ice is often needed for the effective PM labeling, synchronous onset of the FM internalization process, and subsequent observation in predefined time points [7, 10]. Washing steps removing the excess of the dye applied are necessary for pulse-chase experiments [10–12], which allow to quantify the amount of internalized dye and to stain endomembranes without disturbing the staining of the PM.

2 Materials

Prepare all water solutions using standard distilled (deionized) water; there is no need of ultrapure water. Use analytical-grade chemicals and solvents. All culture media should be sterilized as well as equipment for handling cells, seeds, or seedlings (*see* **Note 1**).

2.1 Plant Material

1. Suspension-cultured cells of *Nicotiana tabacum* L., cv. Bright Yellow (BY-2) cell line (*see* **Note 2**).
2. 4- to 5-day-old *Arabidopsis thaliana* seedlings.

2.2 Culture Media, Stock, and Incubation Solutions

1. *Arabidopsis thaliana* medium (AM), also known as half-strength Murashige-Skoog (MS) medium [13]: 2.15 g/L MS basal salt mixture, 1 % (w/v) sucrose, pH 5.8. For 1 L, weigh 2.15 g MS salt mixture, transfer it into a glass or plastic beaker with around 800 mL of distilled water, and add 10 g sucrose (*see* **Note 3**). Stir thoroughly with magnetic stirrer at room temperature, adjust pH with 3 M KOH to 5.8, and bring volume to 1 L. For solidified medium, add agar (10 g/L (w/v); final concentration 1 %). Autoclave at 121 °C for 20 min under 0.1 MPa in laboratory glass bottles (*see* **Note 4**). Allow the agar medium to cool to 45–50 °C (until the container can be held with hands) and pour aseptically into sterile plastic square plates to cover approximately half of the depth of each plate. Leave the plates opened in the flow bench for 30 min to allow the agar to solidify.

2. Modified MS [13] medium for suspension-cultured tobacco cells: 4.3 g/L MS basal salt mixture, 3 % (w/v) sucrose, 100 mg/L myoinositol, 1 mg/L thiamine, 200 mg/L KH_2PO_4, and 0.2 mg/L (0.9 μM) 2,4-D, pH 5.8. For 1 L, weigh 4.3 g MS salt mixture and transfer it into a glass or plastic beaker with around 800 mL of distilled water; add 30 g sucrose, 100 mg myoinositol, 200 mg KH_2PO_4, 1 mL stock solution of thiamine, and 2 mL stock solution of 2,4-D (*see* **Note 3**). Stir thoroughly with magnetic stirrer at room temperature; adjust pH with 3 M KOH to 5.8 and bring volume to 1 L. Autoclave at 121 °C for 20 min under 0.1 MPa (*see* **Note 4**).
3. Stock solution of thiamine (3.3 mM): for 100 mL, dissolve 100 mg of thiamine in 100 mL of distilled water. Store in 2 mL aliquots in plastic tubes in −20 °C.
4. Stock solution of 2,4-D (0.45 mM): for 100 mL, dissolve 10 mg of 2,4-D in 1 mL pure ethanol. After addition of distilled water to the final volume 100 mL, heat the solution under continuous stirring for at least 4 h to dissolve 2,4-D completely.
5. Seed sterilization solution: for 10 mL, mix 5 mL of 50 % household bleach (containing 5 % sodium hypochlorite) and 5 mL of sterile water; add 5 μL 0.05 % (v/v) Tween® 20. Do not store this solution; always prepare fresh.
6. Sterile distilled (deionized) water (autoclaved at 121 °C for 20 min under 0.1 MPa).
7. Stock solutions of FM 4-64 and FM 1-43: dissolve 1 mg of lyophilized powder in DMSO or water to make 1–20 mM stock solution (*see* **Note 5**). Store aliquots at −80 °C; keep in darkness. Alternatively, FM dyes are available as 100 μg lyophilized aliquots (Molecular Probes). FM dyes are light sensitive and unstable at room temperature. Therefore, during experiments, keep aliquots on ice and ideally in darkness.

2.3 Equipment for In Vitro Cultivation, Staining Procedure, and Fluorescence Microscopy

1. Autoclave.
2. Laminar flow cabinet for handling cells and seedlings during the inoculation and sampling.
3. Sterilized standard laboratory equipment: glass Erlenmeyer flasks (250 or 100 mL) covered with aluminum foil for the incubation of cell suspensions, pipette for 1–5 mL for handling with cell suspensions (with cut tips), orbital incubator placed at 27 °C in darkness in the culture room or any air-conditioned incubator with shaker, square sterile plastic plates, and 24-well test plates, pipette for micro-volumes.
4. Microscope slides (76 × 26 × 1 mm) and cover glasses of thickness No. 1 (18 × 18 mm or 22 × 60 mm).

5. Fluorescence microscope equipped with sensitive CCD camera for acquiring weak fluorescent signals, apochromatic objectives 40× or 60× with high numerical aperture (NA 1.2 water immersion or 1.4 oil immersion), and long-pass or band-pass filters for the fluorescence excitation at 488 nm and emission at 505–550 nm (for FM 1-43 and FM 2-10) or excitation at 488 nm or 561 nm and emission >570 nm (for FM 4-64 and FM 5-95). Ideally, confocal microscope system (laser scanning or spinning disk) with photo multiplier tube detectors (laser scanning microscope, e.g., Zeiss LSM 780 or Leica SP8) or CCD cameras (laser spinning disk microscope, e.g., Yokogawa spinning disk unit-equipped systems) equipped with filters or beam splitters for the abovementioned spectral ranges of FM dye excitations and emissions. Software for image acquisition (supplied by microscope system producer) and analysis (e.g., open source image analysis software ImageJ, National Institutes of Health, http://rsb.info.nih.gov/ij).

3 Methods

3.1 Tracking Endocytosis in Root Epidermal Cells of *Arabidopsis thaliana* Seedlings

1. In the laminar flow cabinet, put seeds of *Arabidopsis thaliana* into a microcentrifuge tube (small amount is enough, not more than several hundreds), add 1 mL of freshly made sterilization solution, and incubate for 10 min with occasional manual agitation. Remove sterilization solution with 1 mL pipette (seeds settle well at the bottom of the tube placed in the stand). Add 1 mL of sterile water and incubate for 10 min; repeat this washing step three times (3 × 10 min).
2. Plant individual seeds onto the agar square plates (*see* **Note 6**). Close plates with parafilm and place them at 4 °C for 48 h for stratification (in darkness). Keep vertically oriented plates at 16/8 h light/dark photoperiod at 21/18 °C for 4–5 days.
3. Prepare a 24-well test plate with 1 mL aliquots of liquid AM medium with 2 μM FM dye by the addition of 0.1 μL of 20 mM FM dye stock solution into 1 mL AM medium (*see* **Note 7**) and 1 mL aliquots of plain AM medium for rinsing the seedlings. Place the test plate on ice for 15 min.
4. Transfer seedlings into individual wells (5–8 seedlings/well) with fine-tip tweezers (Dumont #5 or similar) and incubate them for 5 min on ice. Wash seedlings three times in liquid AM medium, still on ice (Fig. 2a).
5. Place seedlings on the microscope slide into the drop of AM medium (room temperature), close with cover glass, and start the timer (*see* **Note 8**). Shifting seedlings into room temperature triggers all processes of plasma membrane internalization, thus representing the time 0 for the FM uptake experiment.

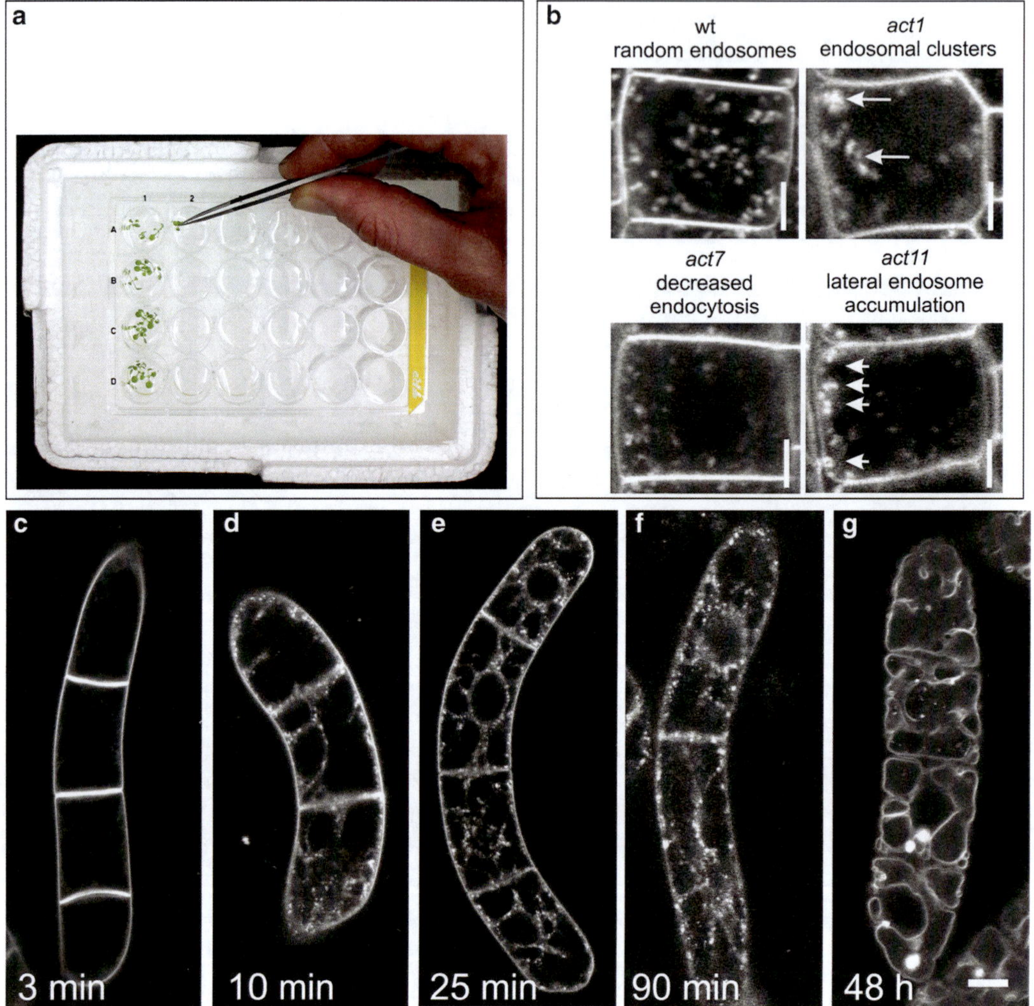

Fig. 2 FM 4-64 internalization in *Arabidopsis thaliana* root epidermal cells and tobacco suspension-cultured cells. (**a**) Multi-well test plate placed on the ice with seedlings incubated in 1 mL of 2 μM FM 4-64. (**b**) FM 4-64 internalization in root epidermal cells after 20 min at RT. Both the amount and distribution of FM dye might be used for the characterization of endocytic trafficking in individual actin mutants (*act1*, *act7*, and *act11*). In wild-type Col-0 (wt), FM 4-64-positive endosomes are distributed randomly, *act1* forms clusters of endosomes, *act7* has decreased endocytosis, and *act11* contains endosome accumulations at the periphery of cells. Scale bars, 5 μm. (**c**–**g**) FM 4-64 internalization in tobacco BY-2 cells. Staining of the PM after 3 min (**c**), early endosomes after 10 min (**d**), *trans*-Golgi and other endomembranes after 25 and 90 min (**e**, **f**) and vacuolar system after 48 h (**g**). Scale bar 20 μm

6. Observe the fluorescence at predefined time points (e.g., 5, 10, and 20 min) using an epifluorescence microscope equipped with a confocal laser scanning unit. To obtain detailed images of individual cells within the root epidermis and cortex, use 40× or 60× C-apochromatic water or oil immersion objectives (*see* **Note 8**). Fluorescence of all major FM dyes (FM 1-43,

FM 4-64, FM 2-10, and FM 5-95) could be stimulated with laser line 488 nm. Alternatively, FM 4-64 and FM 5-95 could be excited with longer wavelengths (typically with 561 nm), which might be useful for parallel observation of GFP (excited with 488 nm). The fluorescence emission is detected between 505 and 550 nm for FM 1-43 and FM 2-10 and above 575 nm for FM 4-64 and FM 2-10 dyes. Sequential scanning, i.e., using two timely isolated single excitations and emissions, can be used to avoid any crosstalk of fluorescence channels in double labeling experiments with GFP for FM 4-64 and FM 2-10. To separate FM 1-43 and GFP, the fluorescence linear unmixing method might be used on systems equipped with spectral detection unit (e.g., Zeiss LSM 780; *see* **Note 9**). FM dyes are quite photostable, which makes them suitable for long-time experiments. Both speed and distribution patterns of FM dye internalization into root epidermal cells during a 20 min uptake period might be used to characterize functions of individual elements of endocytic machinery, as documented in Fig. 2b for various mutations in actin genes.

7. Acquired images could be analyzed with image analysis (e.g., ImageJ) to determine the amount of FM dye internalized during the experimental time window. Typically, mean fluorescence intensity in the area spanning the whole cytoplasm is measured and expressed in relation to the average fluorescence at the PM [12]. FM uptake experiment should be performed in at least three biological repetitions to obtain enough data for statistical evaluation.

3.2 Tracking Endocytosis in Tobacco BY-2 Cells: Simple Labeling Protocol

1. Inoculate 1 mL of 7-day-old BY-2 cells into 30 mL of modified MS cultivation medium (*see* **Note 10**). Use pipette with tips with cut top to transfer cells under sterile conditions into the fresh medium. The laminar flow cabinet should be sterilized with UV for 20 min in advance. Cultivate cells in darkness with continuous shaking at 27 °C on an orbital incubator at 150 rpm (orbital diameter 30 mm) (*see* **Note 11**). The length of cultivation depends on the purpose of the experiment, but exponentially growing cells are the most suitable (*see* **Note 2**).
2. Pipette 1 mL aliquots of cell culture into 24-well test plate. The number of aliquots depends on the experimental setup.
3. Add 0.1 μL of 20 mM FM dye stock solution into 1 mL of cell culture (one well) to reach 2 μM final concentration (*see* **Note 7**), place test plate on orbital shaker at 26–27 °C, and start the timer.
4. Observe the fluorescence at predefined time point by following instructions described in **steps 6** and **7** of Subheading 3.1. For FM 4-64, the PM of BY-2 cells is stained within 2–3 min (Fig. 2c); endocytosed endosomes appear immediately afterwards (Fig. 2d) followed by the

gradual appearance of the dye in the *trans*-Golgi, prevacuolar compartments, and some other endomembrane structures, but never the ER and nuclear envelope (Fig. 2e, f). The dye might be applied for a longer time and used for effective staining of the tonoplast (Fig. 2g).

3.3 Tracking Endocytosis in Tobacco BY-2 Cells: Cold Pretreatment Labeling Protocol

1. Prepare cell culture according to **step 1** of Subheading 3.2.
2. Pipette 1 mL aliquots of cell culture into a 1.5 mL plastic tube. The number of aliquots (tubes) depends on the experimental setup. Place tubes on ice for 15 min.
3. Still on ice, add 0.1 μL of 20 mM FM dye stock solution into 1 mL of cell culture (i.e., into one tube) to reach 2 μM final concentration (*see* **Note 7**). Keep tubes on ice for 15 min; there is no need for shaking during this period.
4. Place tubes horizontally on orbital shaker at 26–27 °C (fix them with the tape) and start the timer.
5. Observe the fluorescence at predefined time point by following instructions described in **steps 6** and 7 of Subheading 3.1 and **step 4** of Subheading 3.2. Take one drop of cell culture with a dropper and make the preparation using a microscope slide and cover glass and start the timer (*see* **Note 8**). Initial cold pretreatment of cells reducing more than 90 % of the dye uptake [10] helps to standardize the initial steps of staining by saturating the PM with the dye.

3.4 Tracking Endocytosis in Tobacco BY-2 Cells: Pulse-Chase Labeling

1. Prepare cell culture according to **step 1** of Subheading 3.2.
2. Pipette 1 mL cell culture aliquots into 1.5 mL plastic tube. The number of aliquots (tubes) depends on the experimental setup. Place tubes on ice for 15 min.
3. Still on ice, add 0.1 μL of 20 mM FM dye stock solution into 1 mL of cell culture (i.e., into one tube) to reach 2 μM final concentration (*see* **Note 7**). Keep tubes on ice for 15 min; there is no need for shaking during this period.
4. Spin tube with cells in ice-cold BY-2 medium (1,600 × g, 4 °C, 5 min) to remove medium containing the dye. Repeat the washing step three times.
5. Place tubes horizontally on orbital shaker at 26–27 °C (fix them with the tape) and start the timer.
6. Observe the fluorescence at predefined time point by following instructions described in **step 5** of Subheading 3.3. Removal of excess dye is useful mainly for the staining of later phases of the dye uptake [14] without disturbing PM fluorescence, but it also helps to observe early phases of endocytosis.

4 Notes

1. The sterilization by autoclaving of materials and solutions for in vitro cultivation of cell cultures and Arabidopsis seedlings is necessary. Although the application of FM dyes and subsequent microscopical tracking of endocytosis is not performed aseptically, for handling and incubation of cells or seedlings, it is suggested to use sterile disposable plastic (tips, multi-well plates, etc.). This approach helps to standardize the whole technique. Note that even barely detectable contamination could interfere with the growth of cell cultures and with the progression of endocytosis.
2. For the reproducibility of results, it is absolutely necessary to work with high-quality cell suspensions. Tobacco BY-2 cell suspensions are normally yellowish, dense cultures at the end of subculture interval that normally takes 7 days. FM staining might be performed throughout this whole period, but exponentially growing cells (day 2–4) are the best.

   Avoid using cell suspensions that do not grow optimally. This could be monitored without opening the culture flask by checking cells using an inverted microscope during the first 3 days of cultivation. In general, any other suspension-grown plant cell cultures can be used for labeling with FM dyes.
3. MS salts are hygroscopic; the bottle should not be left open unless necessary. For the storage of MS salts at 4 °C, seal the cap carefully with parafilm.
4. When autoclaving liquids, always use the program containing slow release of the pressure at the end of the run. Autoclaves not equipped with this option are not suitable for this purpose. Always keep the lid loose to allow pressure equalization between the bottle and autoclave chamber. Avoid repetitive autoclaving of any media.
5. All FM dyes are easily soluble in DMSO. For most applications, 20 mM stock solution is appropriate. With limited solubility, they could also be dissolved in water as reported for 17 mM water stock solution of FM 4-64 [1]. Since DMSO as an organic solvent interferes with the composition of biomembranes, its concentration must be kept as low as possible, lower than 0.1 % (v/v) [1].
6. Plant the seeds with the sterile toothpick onto agar plates. Alternatively, individual seeds might be planted with 1 mL pipette. Upon suction of sterilized seeds in water (or 0.1 % cooled top agar), individual seeds might be planted one by one in rows.
7. Working concentrations of FM dyes range from 1 to 50 μM. However, it is suggested to keep the concentration as low as possible (1–2 μM). Since FM dyes are very lipophilic

and easily integrated into the PM, in higher concentrations they interfere with endomembrane dynamics and induce other unwanted effects like dragging of some integral PM proteins out of the PM [1]. The most frequently used FM dyes in plant research are FM 4-64 [1] and FM 1-43 [10]. For both of them, slightly less lipophilic versions exist, such as FM 5-95 and FM 2-10 (Molecular Probes Handbook, http://probes.invitrogen.com/handbook/sections/1601.html), respectively. Less lipophilic dyes are suitable in cases when the endocytic processes are too fast and, upon the addition of the dye, the uptake is too fast. Then, the less lipophilic variant allows slower uptake, and also, washing out these dyes is easier. In addition, the dynamics of FM dye internalization is much slower when using water instead of MS medium for all staining and washing steps. Therefore, it is suggested to use MS medium for all experiments. For DMSO-dissolved FM dyes, do not forget to add DMSO into mock treatments.

8. Seedlings of *Arabidopsis thaliana* placed at the slide might be completely covered with the cover glass. However, cotyledons might also remain uncovered. For inverted microscopes with immersion objectives, it is important to attach the cover glass to the slide with paper tape. Always prepare the microscope (finding the focal plane, setting the cameras, etc.) and image acquisition software before imaging experimental sample. The progression of endocytosis is very fast and there is usually not enough time for various settings. For suspension-cultured cells, time-lapse observations might be performed in a perfusion system for open cultivation, e.g., POC cell cultivation system (http://www.pecon.biz/?page_id=283).

9. Since the emission spectra of FM 1-43 and GFP overlap, the separation of these fluorochromes might be performed on confocal systems with spectral detection unit [11]. Linear unmixing of emission spectra acquired using specimens with two fluorochromes must be performed after acquiring reference spectra of samples with only FM 1-43 or GFP.

10. The inoculation density is critical for the successful growth of suspension-cultured cells. For routine propagation of BY-2 cells with a 7-day-long subculture interval, stationary cells are diluted (at the day 7) 1:50 (2 mL/100 mL of medium in 250 mL Erlenmeyer flask) or 1:30 (1 mL/30 mL of medium in 100 mL Erlenmeyer flask). Never fill flasks with more medium than specified above.

11. The temperature during cultivation of cells should not exceed 27–28 °C for BY-2 cells. Cells are also very sensitive to prolonged standing without shaking. Try to minimize this time as much as possible.

Acknowledgement

This work was supported by the Czech Science Foundation, projects GAP305/11/2476 (JP) and GPP305/11/P797 (AJ).

References

1. Bolte S, Talbot C, Boutte Y et al (2004) FM-dyes as experimental probes for dissecting vesicle trafficking in living plant cells. J Microsc 214:159–173
2. Betz WJ, Bewick GS (1992) Optical analysis of synaptic vesicle recycling at the frog neuromuscular junction. Science 255:200–203
3. Betz WJ, Mao F, Smith CB (1996) Imaging exocytosis and endocytosis. Curr Opin Neurobiol 6:365–371
4. Šamajová O, Takáč T, von Wangenheim D et al (2012) Update on methods and techniques to study endocytosis in plants. In: Šamaj J (ed) Endocytosis in plants. Springer, Berlin, pp 1–36
5. Dhonukshe P, Aniento F, Hwang I et al (2007) Clathrin-mediated constitutive endocytosis of PIN auxin efflux carriers in Arabidopsis. Curr Biol 17:520–527
6. Ueda T, Yamaguchi M, Uchimiya H et al (2001) Ara6, a plant-unique novel type Rab GTPase, functions in the endocytic pathway of Arabidopsis thaliana. EMBO J 20:4730–4741
7. Grebe M, Xu J, Mobius W et al (2003) Arabidopsis sterol endocytosis involves actin-mediated trafficking via ARA6-positive early endosomes. Curr Biol 13:1378–1387
8. Klima A, Foissner I (2008) FM dyes label sterol-rich plasma membrane domains and are internalized independently of the cytoskeleton in characean internodal cells. Plant Cell Physiol 49:1508–1521
9. Nagata T, Nemoto Y, Hasezawa S (1992) Tobacco BY-2 cell-line as the Hela-cell in the cell biology of higher-plants. Int Rev Cytol 132:1–30
10. Emans N, Zimmermann S, Fischer R (2002) Uptake of a fluorescent marker in plant cells is sensitive to brefeldin A and wortmannin. Plant Cell 14:71–86
11. Jelínková A, Malínská K, Simon S et al (2010) Probing plant membranes with FM dyes: tracking, dragging or blocking? Plant J 61: 883–892
12. Robert S, Kleine-Vehn J, Barbez E et al (2010) ABP1 mediates auxin inhibition of clathrin-dependent endocytosis in Arabidopsis. Cell 143:111–121
13. Murashige T, Skoog F (1962) A revised medium for rapid growth and bio assays with tobacco tissue cultures. Physiol Plant 15: 473–497
14. Kutsuna N, Hasezawa S (2002) Dynamic organization of vacuolar and microtubule structures during cell cycle progression in synchronized tobacco BY-2 cells. Plant Cell Physiol 43: 965–973

Chapter 2

Sterol Dynamics During Endocytic Trafficking in Arabidopsis

Thomas Stanislas, Markus Grebe, and Yohann Boutté

Abstract

Sterols are lipids found in membranes of eukaryotic cells. Functions of sterols have been demonstrated for various cellular processes including endocytic trafficking in animal, fungal, and plant cells. The ability to visualize sterols at the subcellular level is crucial to understand sterol distribution and function during endocytic trafficking. In plant cells, the polyene antibiotic filipin is the most extensively used tool for the specific detection of fluorescently labeled 3-β-hydroxysterols in situ. Filipin can to some extent be used to track sterol internalization in live cells, but this application is limited, due to the inhibitory effects filipin exerts on sterol-dependent endocytosis. Nevertheless, filipin-sterol labeling can be performed on aldehyde-fixed cells which allows for sterol detection in endocytic compartments. This approach can combine studies correlating sterol distribution with experimental manipulations of endocytic trafficking pathways. Here, we describe step-by-step protocols and troubleshooting for procedures on live and fixed cells to visualize sterols during endocytic trafficking. We also provide a detailed discussion of advantages and limitations of both methods. Moreover, we illustrate the use of the endocytic recycling inhibitor brefeldin A and a genetically modified version of one of its target molecules for studying endocytic sterol trafficking.

Key words Sterol labeling, Endocytosis, Protocols, Immunofluorescence, Confocal microscopy, *Arabidopsis* root, Endocytic trafficking mutants

1 Introduction

Sterols are membrane lipids essential to diverse cellular functions of eukaryotic cells including cytokinesis and cell polarity [1–11]. In mammalian, fungal, and plant cells, sterol function during these processes has been linked to the molecular mechanisms involved in endocytosis [9, 11–14]. In the plant model *Arabidopsis thaliana*, sterol composition plays a crucial role in endocytosis of the polarly localized PIN2 auxin efflux carrier and in endocytosis of the cytokinesis-specific KNOLLE syntaxin [9, 11, 14]. Sterols are not only needed for endocytosis but are themselves internalized through the endocytic pathway [15–18]. Endocytic sterol trafficking relies on actin and is highly sensitive to brefeldin A (BFA), an inhibitor of the plasma membrane recycling pathway in *Arabidopsis*

Marisa S. Otegui (ed.), *Plant Endosomes: Methods and Protocols*, Methods in Molecular Biology, vol. 1209, DOI 10.1007/978-1-4939-1420-3_2, © Springer Science+Business Media New York 2014

root cells [15, 19]. Within the cell, sterols predominantly localize at the plasma membrane and endocytic organelles such as the *trans*-Golgi network (TGN) and in compartments labeled by GNOM, a guanine nucleotide exchange factor (GEF) for ADP ribosylation factor (ARF) protein [11, 15, 20]. The establishment of a subcellular localization map of sterols and, in a broader sense, understanding of sterol function in cellular processes relies on the ability to visualize these lipids in cells. In mammalian cells, fluorescent sterol analogs and sterols linked to fluorescent tags are widely used to study sterol distribution in membranes [21–24]. In plant cells, the presence of pectocellulosic cell wall components decreases the accessibility of relatively big chemicals such as sterol analogs or sterols linked to fluorescent tags. In intact plant tissues, fluorescent sterol analogs accumulate in cell wall structures and are hardly loaded into membranes [25]. Successful labeling with fluorescent sterol analogs has been obtained on individual cells from which the cell wall has been chemically removed [25]. However, this is a relatively limited application, and it is not clear whether sterol analogs display similar dynamics and distribution in membranes compared to endogenous sterols in plant cells. Moreover, in contrast to animal and yeast cells, where cholesterol and ergosterol are the major sterols, respectively, plant cells contain three major sterols: sitosterol, stigmasterol, and campesterol [26, 27]. In plant cells, the fluorescent polyene antibiotic filipin III (filipin) is widely used to detect the three major forms of sterols in situ [9, 11, 15, 28–30]. Filipin is able to form complexes with sitosterol, stigmasterol, campesterol, and glycosylated sterol derivatives (steryl glucosides and acylated steryl glucosides) on dot blots [15, 31]. The specificity of filipin in probing 3-β-hydroxysterols has been previously demonstrated on dot blots and in situ using cholesterol oxidase, an enzyme that specifically catalyzes the oxidation of 3-β-hydroxysterols [15]. Hence, filipin is specific to 3-β-hydroxysterols but does not discriminate among the predominant plant sterols. A prerequisite for using filipin is the availability of a 351 nm or 364 nm laser line of a UV laser as excitation at 405 nm leads to predominant detection of unspecific tissue autofluorescence signal in *Arabidopsis* root epidermal cells [20].

Filipin can be used in live cell imaging or on fixed samples. However, even though it is possible to follow the internalization of filipin-sterol complexes in live samples, filipin also inhibits endocytosis of the endocytic tracer FM4-64 and plasma membrane proteins such as auxin efflux carriers of the PIN protein family [9, 20, 32]. Filipin can be used as a tool to inhibit sterol-dependent endocytosis, but one must be aware of the fact that the filipin-sterol interaction creates membrane deformations that may interfere with additional membrane properties. The sterol biosynthesis mutant *cpi1-1* does not display such membrane deformations but displays defects in sterol-mediated endocytosis and thus can be used as an additional tool [9]. In live samples, although possible, it is challenging to follow filipin-sterol internalization without any pharmacological treatment (Fig. 1b–e; Subheading 3.1; [15]). External application of FM4-64 (Fig. 1b) and filipin (Fig. 1c) often results in FM4-64 loading in the cell, while filipin-sterol complexes hardly enter the cell (Fig. 1b–e; Subheading 3.1). However, this limitation can be circumvented by the use of brefeldin A (BFA), an inhibitor of the endocytic recycling pathway. In live cells, treatment with BFA induces an agglomeration of TGN and other endocytic vesicles in large aggregates inside the cell that are often referred to as BFA compartments (Fig. 1f, h; Subheading 3.2). BFA inhibits the recycling activity of the GNOM ARF-GEF, most likely through binding to the Sec7 domain of the GNOM protein [19, 33, 34]. Indeed, in live cells, BFA co-agglomerates GNOM-GFP (Fig. 1k) and filipin-sterol fluorescence (Fig. 1j) into BFA compartments (Fig. 1j–l, Subheading 3.2). An ARF1 protein fused to GFP, ARF1 A1c GFP (Fig. 1n, q), also accumulates in BFA compart ments together with filipin-sterol fluorescence (Fig. 1m, p) upon BFA treatment (Fig. 1o–r, Subheading 3.2). Hence, it is possible to co-detect endosomal proteins of interest fused to a fluorescent tag with filipin in endocytic structures in live cells treated with BFA (Fig. 1j–r, Subheading 3.2). However, due to the concomitant inhibition of endocytosis caused by filipin, BFA needs to be used at high concentrations such as 100 μM (Subheading 3.2), and the method is thus of limited use only. In addition, even when applying BFA at the concentration of 100 μM, filipin labeling in live samples sometimes reveals cells without filipin-sterol labeling in BFA compartments (Fig. 1g) that, however, can be labeled with the endocytic tracer FM4-64 (Fig. 1f) and the plasma membrane marker LTI6a (Fig. 1h, Subheading 3.2). This implies a differential interference with various plasma membrane-derived cargos by filipin. Overall, the use of filipin for the imaging of endocytic events in live-cell imaging remains limited.

Experimental evidence acquired in live cells can nevertheless be correlated to filipin-sterol labeling of fixed cells where not only the plasma membrane is strongly stained but also several endomembrane compartments. Filipin-sterol staining in fixed cells expressing diverse endosomal proteins fused to fluorescent protein revealed that filipin-sterol strongly co-localized with endocytic structures also in the absence of BFA treatment (Fig. 2; Subheading 3.3; [11, 15, 20]). For example, we previously described that the TGN-localized fusion proteins ARF1-A1c-EGFP (Fig. 2a–c), VHA-a1-GFP (Fig. 2d–f), and RAB-A2a-YFP (Fig. 2g–i) display high co-localization with filipin-sterol fluorescence (Fig. 2a–i; Subheading 3.3; [11, 20]).

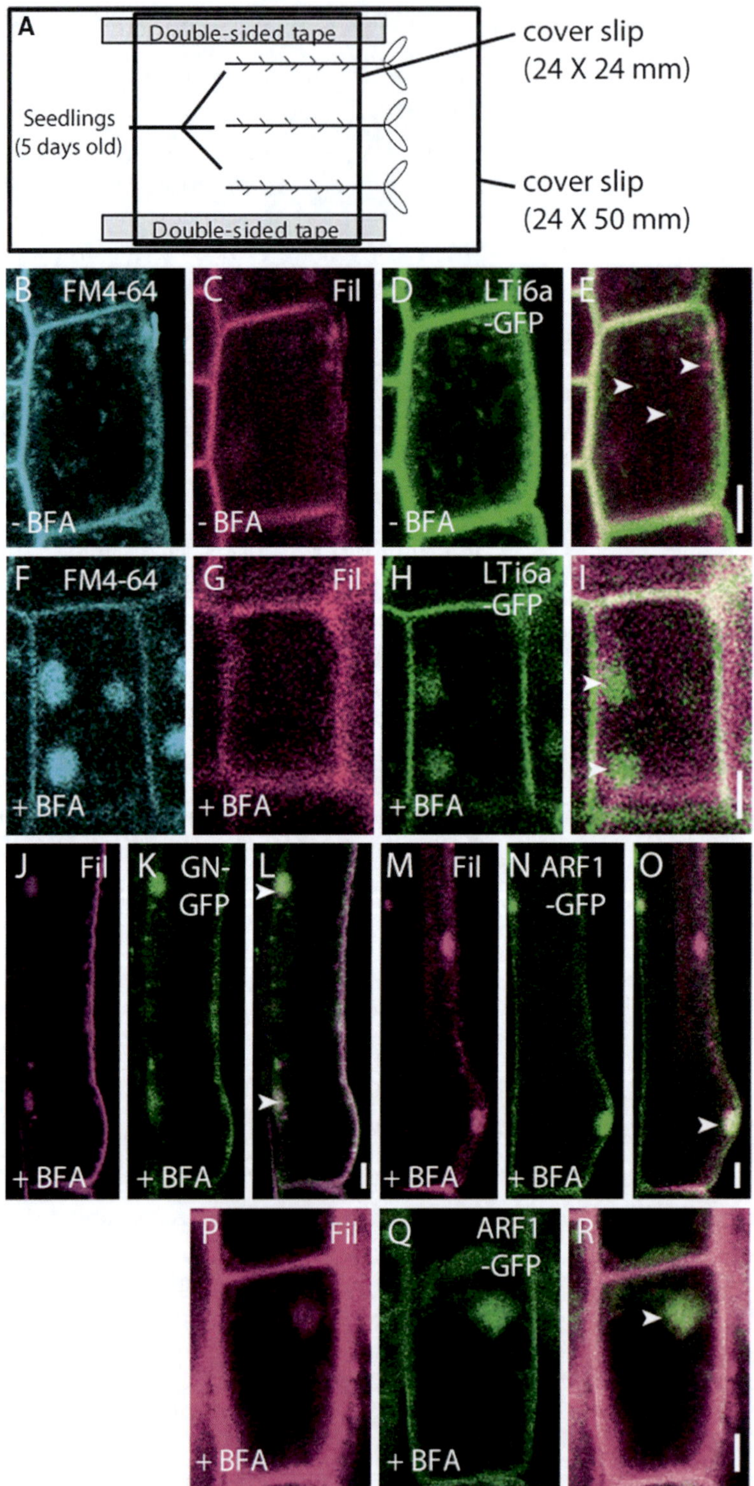

Fig. 1 Filipin-sterol labeling of live *Arabidopsis* roots cells. (**a**) Seedlings are mounted in liquid MS medium between a 24 × 50 mm coverslip and a 24 × 24 mm coverslip separated by double-sided tape placed at the edges of the 24 × 50 mm

We were also able to co-visualize filipin-sterol complexes and proteins using immunocytochemistry (Fig. 2j–o; Subheading 3.4; [11, 20]). For example, it is possible to perform filipin-sterol labeling in roots where the cell plate and TGN-early endosome-localized syntaxin KNOLLE has been immunolabeled by anti-KNOLLE antibodies (Fig. 2j–l, Subheading 3.4; [11, 20]) or where GNOM-myc has been immunodetected by anti-myc antibodies (Fig. 2m–o; Subheading 3.4; [20]). Co-labeling of filipin-sterol fluorescence with the Golgi-localized *N*-acetylglucosaminyltransferase 1 fused to EGFP (NAG1-EGFP, Fig. 2p–r) or with the multivesicular body-localized small GTPase RAB-F1 (also called ARA6) fused to EGFP (RAB-F1-EGFP, Fig. 2s–u) reveals the presence of sterols in these two organelles (Fig. 2p–u; Subheading 3.4; [11, 20]). Electron microscopic analyses suggest that the TGN and the *trans-most* Golgi cisterna carry most filipin-sterol complexes, while the *cis*- and median-Golgi cisternae are almost completely devoid of filipin-sterol complexes [11]. In BFA-treated cells, filipin-sterol complexes co-localize with KNOLLE and GNOM on BFA bodies, as demonstrated by immunofluorescence pro tein detection (Fig. 3a–f; Subheadings 3.5 and 3.6). In contrast to live cells, filipin-sterol staining of fixed samples clearly reveals co-labeling of endomembrane structures throughout the cell both in the presence and absence of BFA, because filipin penetration does not depend on endocytic trafficking of filipin-sterol complexes and the fluorophore more easily and readily enters aldehyde-fixed cells and tissues. To obtain a quantitative comparison of endocytic material accumulated in BFA compartments, BFA concentrations from 10 μM, 25 μM, or 50 μM can

Fig. 1 (continued) coverslip. (**b–e**) External application of 25 μM FM4-64 (**b**) and 30 μM filipin (**c**) in meristematic cells expressing the plasma membrane marker LTI6a-GFP (**d**). Note that in this particular cell, while endosomal structures are labeled by FM4-64 and LTI6a-GFP (co-localization indicated by *arrowheads* in the merged picture in **e**), filipin labels the plasma membrane but not endocytic structures. (**f–i**) External application of 25 μM FM4-64 (**f**) and 30 μM filipin (**g**) in meristematic cells expressing the plasma membrane marker LTi6a-GFP (**h**) followed by 100 μM BFA treatment for 60 min (Subheading 3.2). Note that in this particular cell, while FM4-64 and LTi6a-GFP (co-localization displayed by *arrowheads* in the merged picture in **i**) label endocytic structures gathered in BFA compartments, filipin does not. (**j–r**) External application of 30 μM filipin (**j**, **m**, **p**) followed by 100 μM BFA treatment for 60 min (in trichoblast cells (**j–o**) expressing GNOM-GFP (**k**) or ARF1-EGFP (**n**) and in meristematic cells (**p–r**) expressing ARF1-EGFP (**q**). Note that in these particular cells, filipin-sterol fluorescence is co-detected in BFA compartments (co-localization displayed by *arrowheads* in merged pictures in **l**, **o**, **r**) with ARF1-GFP (merged picture in **o** and **r**) or GNOM-GFP (merged picture in **l**). Scale bars are 5 μm

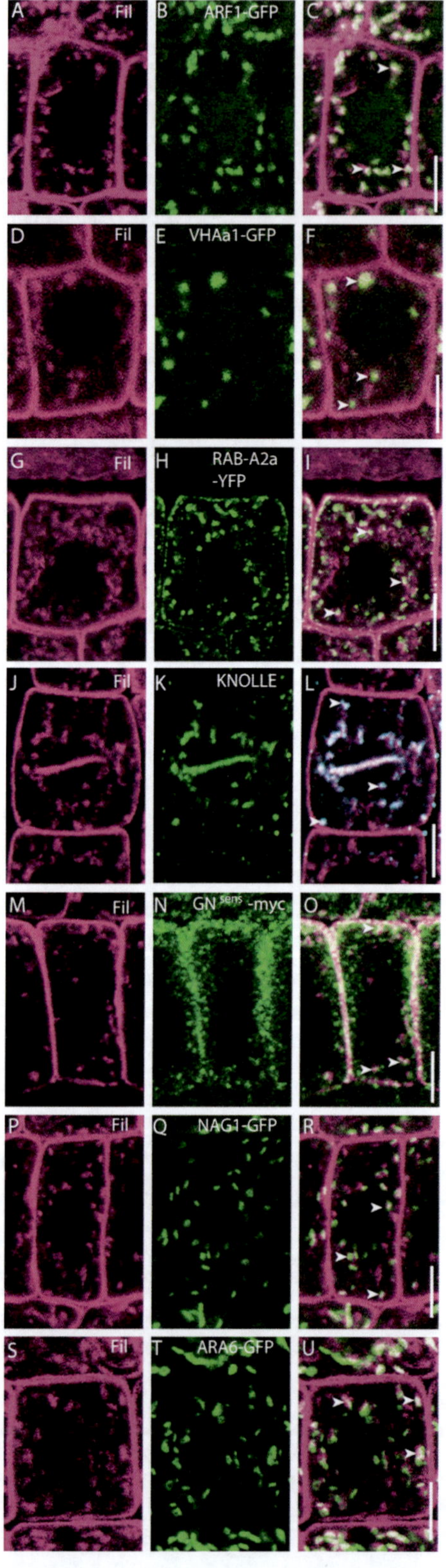

Fig. 2 Filipin-sterol labeling of fixed *Arabidopsis* root cells. (**a**–**i**) Co-detection of filipin-sterol fluorescence (**a**, **d**, **g**) with TGN-localized markers ARF1-EGFP (**b**), VHA-a1-GFP (**f**), and RabA2A-YFP (**h**) using the method described in Subheading 3.3.

be employed (Fig. 3a–f, Subheading 3.5). However, although being predominantly localized at endocytic structures, we previously demonstrated that sterols are also found in compartments of the secretory pathway such as the Golgi apparatus [11, 15]. However, in fixed cells, it is not possible to dynamically track sterols and thus to discriminate between filipin-sterol complexes derived from the endocytic pathway and filipin-sterol complexes coming from the secretory pathway. The use of mutants in the endocytic pathway can help to visualize defects in endocytic sterol trafficking. For example, in aldehyde-fixed root cells of seedlings expressing a BFA-insensitive version of the GNOM ARF-GEF [19, 33–35] and pretreated with BFA, filipin-sterol labeling of BFA bodies is much weaker than in control cells, indicating that endocytic trafficking of sterols partly depends on GNOM function (Fig. 3d–i, Subheadings 3.5 and 3.6). Hence, the BFA-resistant version of GNOM can be used as a tool to address properties of endocytic sterol trafficking.

In conclusion, filipin is a specific reagent for in situ detection of 3-β-hydroxysterols. Filipin can be used on live cells to track sterol internalization from the plasma membrane. The endocytic recycling inhibitor BFA can be used to analyze sterol agglomeration with endocytic markers. However, as filipin also inhibits endocytosis, live cell imaging results are only of limited value and need to be correlated with filipin-sterol labeling of fixed cells. In fixed samples, it is not possible to determine whether the staining observed in endomembrane compartments originates from the secretory or the endocytic pathway. Combining data obtained from filipin staining of live and fixed samples with genetic approaches, such as using the mutated BFA-resistant version of GNOM or other endocytic trafficking mutants and inhibitors, currently represents the available approaches for the analysis of endocytic sterol trafficking in plants.

Fig. 2 (continued) Note the strong co-localization between filipin-sterol fluorescence and TGN (co-localization indicated by *arrowheads* in merged pictures in **c**, **f**, **i**). (**j**–**o**) Co-detection of filipin-sterol fluorescence with the KNOLLE syntaxin (**k**) and the GNOM ARF-GEF fused to a 3xmyc tag (**n**) detected by immunocytochemistry using anti-KNOLLE and anti-myc. Note the strong co-localization between filipin-sterol fluorescence and KNOLLE or GNOM (co-localization displayed by *arrowheads* in merged pictures in **l** and **o**, respectively). (**p**–**u**) Co-detection of filipin-sterol fluorescence with the Golgi marker NAG1-EGFP (**q**) and the late endosome/multivesicular body marker ARA6/RAB-F1-EGFP (**t**). Note that structures labeled by filipin-sterol fluorescence often associate with compartments labeled by either NAG1-EGFP or ARA6-EGFP (*arrowheads* in merged pictures in **r** and **u**). Scale bars are 5 μm

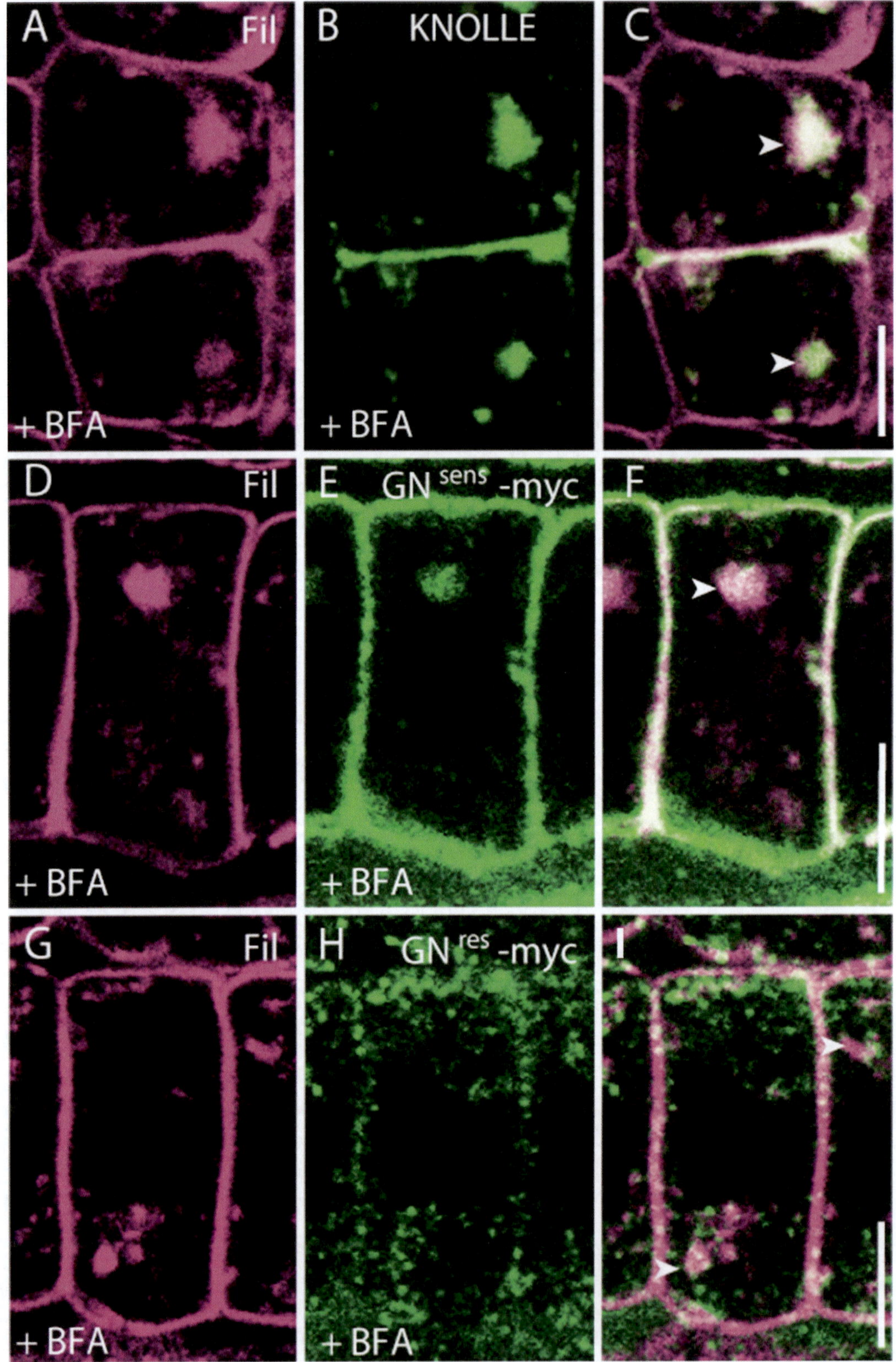

Fig. 3 Filipin labeling of fixed *Arabidopsis* root cells treated with BFA. Cells were pretreated with 50 μM BFA (**a–f**) for 60 min or 25 μM BFA (**g–i**) for 120 min prior to fixation and filipin-sterol labeling. Labeling of filipin-sterol fluorescence (**a**, **d**, **g**) in a line expressing the GNOM BFA-resistant version fused to 3x-myc (GN-myc

2 Materials

2.1 Plant Materials

Arabidopsis thaliana seedlings were grown vertically on MS-agar (*see* Subheading 2.3) square Petri dishes (12 cm × 12 cm) for 5 days. The photoperiod was set up to 16 h light/8 h dark at 22 °C. We used the *Arabidopsis* small integral membrane protein LTI6a fused to EGFP (LTI6a-GFP; [36]), the small GTPase ARF1-A1c fused to EGFP (ARF1-EGFP; [37]), the endosomal recycling GNOM ARF-GEF fused to GFP (GNOM-GFP; [34]), a GNOM BFA-sensitive version fused to myc ($GNOM^{sens}$-myc; [34]), a GNOM BFA-resistant version fused to myc ($GNOM^{res}$-myc; [34]), the small GTPase RAB-A2a fused to EYFP (RAB-A2a-EYFP; [38]), V-ATPase VHA-a1 fused to GFP (VHA-a1-GFP; [39]), *N*-acetylglucosaminyltransferase 1 fused to EGFP (NAG1-EGFP, [15]), and the small GTPase RAB-F1 fused to EGFP (RAB-F1-EGFP, [40]).

2.2 Buffers and Solutions

1. *Murashige and Skoog (MS)-agar plates:* 1× MS, 1 % sucrose, 0.8 % plant agar, and 2.5 mM MES, buffered to pH 5.8 with KOH.
2. *Liquid MS medium:* Similar to MS-agar plate medium but without plant agar.
3. 15 mM *Filipin stock solution* (10 mg/ml)*:* 1 mg of filipin III from *Streptomyces filipinensis* dissolved in 100 μl of dimethyl sulfoxide (DMSO). Be sure filipin is dissolved completely, including particles in the cap of the vial. Filipin is sensitive to freezing/thawing; resuspension of filipin is done at room temperature since resuspension on ice would lead to DMSO freezing. Filipin solution is aliquoted and can be stored in the dark at −20 °C for 2 weeks maximum.
4. 25 mM *FM4-64 stock solution:* 1 mg of FM4-64 dissolved with 65 μl of DMSO.
5. 50 mM *BFA stock solution:* 5 mg of BFA dissolved in 356 μl of DMSO.
6. *MTSB (Microtubule-stabilizing buffer):* 50 mM PIPES, 5 mM EGTA, 5 mM $MgSO_4$; adjust to pH 7 with KOH.

Fig. 3 (continued) BFA-R) (**b**) and in the GNOM BFA-sensitive version (wild-type version) fused to 3x-myc (GN-myc BFA-S) (**e**) treated with 50 μM BFA. GN-myc BFA-R and GN-myc BFA-S were detected by immunocytochemistry using anti-myc antibodies (protocol described in Subheading 3.4). Note that while filipin-sterols co-labeled strongly with GN-myc BFA-S in BFA compartments (*arrowhead* in the merged picture in **f**), filipin-sterol-labeled compartments are much less reactive to BFA in the GN-myc BFA-R line (*arrowhead* in the merged picture in **c**). (**g–i**) Co-detection of filipin-sterol (**g**) with anti-KNOLLE (**h**) in cells treated with 25 μM BFA. Note the strong co-localization of filipin-sterol and KNOLLE in BFA compartments (*arrowhead* in the merged picture in **i**). Scale bars are 5 μm

7. *4 % paraformaldehyde (PFA) solution:* Weigh 1 g of PFA and transfer it into an Erlenmeyer flask containing 25 ml of MTSB buffer. Dissolve the PFA by heating the solution to ~70 °C. Let the solution cool down to room temperature prior to use.

2.3 Material

1. 12-well multiwell plates.
2. Coverslip 24 × 24 mm.
3. Coverslip 24 × 50 mm.
4. Double-sided tape.
5. Forceps (Dumont No. 2).
6. Razorblade.
7. Microwave oven that can be adjusted to 90 W.
8. Nail varnish (clear).
9. Polylysine-coated adhesive slides for microscopy.
10. Water-vacuum pump connected to a water tap.
11. Humid chamber, i.e., a box with a humid paper towel at the bottom; a box to store microscopy slides is often chosen.
12. Confocal laser scanning microscope (CLSM). Here we used a LEICA TCS SP2 AOBS spectral system, mounted on a LEICA DM IRE2 inverted microscope (Leica Microsystems).
13. Water-corrected × 63 objective, NA = 1.2. We used an HCX PL APO 63.0 × 1.20 W BD UV (Leica Microsystems).
14. An oil-corrected × 63 objective with NA = 1.4. We used an HCX PL APO 63.0 × 1.40 OI BD UV (Leica Microsystems).
15. UV laser for 351 nm or 364 nm excitation. In this study, we used the 364 nm laser line of the Innova Technology Enterprise II ion laser (Coherent).

2.4 Antibodies

1. Mouse monoclonal anti-cMyc (clone 9E10, 1:600; Santa-Cruz Biotechnology).
2. Rabbit polyclonal anti-KNOLLE (1:4,000; [41]).
3. CY5-conjugated affinity-purified donkey anti-rabbit IgG (H + L) (1/300; Jackson ImmunoResearch).
4. Rhodamine (TRITC)-conjugated affinity-purified donkey anti-mouse IgG (H + L) (1/150; Jackson ImmunoResearch).

3 Methods

3.1 Co-visualization of Filipin-Sterol Complexes and FM4-64 Internalization in Live Arabidopsis Cells

1. Place 2 ml of cold liquid MS medium into a 12-well plate cooled on ice (*see* **Note 1**).
2. Add 4 μl from the 15 mM (10 mg/ml) filipin stock solution to obtain a working concentration of 30 μM (20 μg/ml).
3. Add 2 μl from the 25 mM (15 mg/ml) FM4-64 stock solution for a working concentration of 25 μM (15 μg/ml).

4. Submerge 12 five-day-old seedlings in liquid MS medium containing filipin and FM4-64 (*see* **Note 2**). Incubate for 10 min on ice, in the dark.
5. Wash the seedlings twice in cold liquid MS medium without filipin and FM4-64, on ice, in the dark, for 1 min per wash.
6. Transfer the seedlings to liquid MS medium equilibrated at room temperature into a well of a new 12-well plate.
7. Incubate at room temperature for 60 min (*see* **Note 3**).
8. Mount the seedlings in liquid MS medium between a 24 × 50 mm coverslip and a 24 × 24 mm coverslip separated by double-sided tape placed at the edges of the 24 × 50 mm coverslip (Fig. 1a; *see* **Note 4**).
9. Proceed with image acquisition by CLSM immediately after mounting (Subheading 3.7).

3.2 Co-visualization of Filipin-Sterol Complexes and FM4-64 Internalization in Live Arabidopsis Cells Treated with BFA

1. During the 10 min incubation time of Subheading 3.1, **step 4**, transfer 2 ml of liquid MS medium equilibrated at room temperature into a new 12-well plate, left at room temperature (*see* **Note 5**).
2. Add 4 μl from the 50 mM (14 mg/ml) BFA stock solution to reach a working concentration of 100 μM (28 μg/ml) (*see* **Note 6**).
3. After the 10 min incubation (Subheading 3.1, **step 4**), wash the seedlings twice with cold liquid MS medium without filipin and FM4-64, on ice, in the dark, for 1 min per wash.
4. Transfer the seedlings to liquid MS medium containing 100 μM BFA at room temperature.
5. Incubate at room temperature for 60 min (*see* **Note 3**).
6. Mount the seedlings in liquid MS medium containing 100 μM BFA between a 24 × 50 mm coverslip and a 24 × 24 mm coverslip separated by double-sided tape placed at the edges of the 24 × 50 mm coverslip (Fig. 1a; *see* **Note 4**).
7. Proceed with image acquisition at CLSM immediately after mounting (Subheading 3.7).

3.3 Labeling of Filipin-Sterol Complexes in Fixed Arabidopsis Root Cells

1. Prepare a fresh 4 % PFA solution in MTSB (*see* **Note 7**).
2. Pipette 300 μl of freshly prepared 4 % PFA into a 1.5 ml plastic tube.
3. Add 4.5 μl from the 15 mM (10 mg/ml) filipin stock solution to obtain a working concentration of 225 μM (150 μg/ml).
4. Transfer 12 five-day-old seedlings (*see* Subheading 2.1) into the tube containing PFA and filipin.
5. Apply six successive 30 s microwave pulses at 90 W, each separated by a pause of 60 s to avoid overheating (*see* **Note 8**).
6. Incubate at room temperature, in the dark, for 1 h.

7. Wash 3 times with 1 ml dH_2O.
8. Cut root tips (approximately 5 mm from the tip) in a drop of Citifluor mountant medium on a 24 × 50 mm coverslip.
9. Cover with a 24 × 24 mm coverslip and seal with nail varnish.
10. Proceed with image acquisition by CLSM after mounting (Subheading 3.7). Alternatively, seedlings may be kept at 4 °C and observed within 24 h (*see* **Note 9**).

3.4 Labeling of Filipin-Sterol Complexes in Fixed Cells in Combination with Immunocytochemistry

1. Proceed from Subheading 3.3, **steps 1**–7.
2. Dissect root tips (approximately 5 mm from the tip) on a coated adhesion slide in a drop of water. Root tips should be aligned horizontally on the slide (*see* **Note 10**).
3. Let the roots dry on the slide at room temperature, in the dark.
4. Cut 2 thin bands of double-sided tape (one band should be around 1 mm wide and 30 mm long) and place one band on one edge of the slide and the other band on the other edge of the slide (*see* **Note 10**).
5. Remove the protective layer of the double-sided tape and cover the dried root tips with a drop of MTSB buffer to rehydrate the samples.
6. Place a 20 × 20 mm coverslip on top of the root tips.
7. Incubate at room temperature, in the dark, for 15 min.
8. **Steps 9–13** are performed with a water-vacuum pump and in a humid chamber (*see* **Note 11**).
9. Apply 250 μl of 2 % Driselase solution in MTSB and incubate for 30 min at room temperature.
10. Wash 4 times with 250 μl of MTSB.
11. Apply 250 μl of a solution made up of 3 % Igepal CA-630 and 10 % of the 15 mM (10 mg/ml) filipin stock solution in DMSO. The final concentration is 1.5 mM filipin (1 mg/ml) and 10 % DMSO (*see* **Note 12**).
12. Incubate for 1 h at room temperature, in the dark.
13. Wash 5 times with 250 μl of MTSB.
14. Block unspecific binding sites by applying 250 μl of the blocking solution, e.g., 5 % of the serum of the animal in which secondary antibodies were produced. In our experiments, we used normal donkey serum (NDS). Alternatively, 3 % bovine serum albumin (BSA) solution can be used.
15. Incubate for 1 h at room temperature, in the dark.
16. Apply 250 μl of primary antibody diluted in the blocking solution.
17. Incubate overnight at 4 °C, in the dark.
18. Incubate for 2 h at 37 °C, in the dark.

19. Wash 5 times with MTSB.
20. Incubate 250 μl of secondary antibody diluted in the blocking solution for 3 h at 37 °C, in the dark.
21. Wash 5 times with MTSB.
22. Wash 2 times with dH_2O.
23. Starting at one side of the slide, remove the double-sided tape by pulling it off the slide by using forceps.
24. With a razorblade, lift the coverslip from the side where the double-sided tape has been removed.
25. Once the coverslip is removed, add a drop of water directly on the samples so that the root tips do not dry out.
26. Remove the second stripe of double-sided tape.
27. Remove the excess water, taking care not to let the samples dry out.
28. Apply a drop of Citifluor directly on the samples.
29. Cover with a 24 × 24 mm coverslip and seal with nail varnish.
30. Proceed with image acquisition by CLSM after mounting (Subheading 3.7). Alternatively, seedlings could be kept at 4 °C and observed within 24 h (*see* **Note 9**).

3.5 Labeling of Filipin-Sterol Complexes in Fixed Cells Pretreated with BFA

1. Transfer 2 ml of liquid MS medium equilibrated at room temperature into a 12-well plate and leave at room temperature (*see* **Note 5**).
2. Add 2 or 1 μl from the 50 mM (14 mg/ml) BFA stock solution for working concentration of 25 μM or 50 μM (7 or 14 μg/ml), respectively (*see* **Note 13**).
3. Transfer 12 seedlings to liquid MS medium containing BFA at room temperature.
4. Incubate at room temperature for 60 min with 50 μM BFA or for 120 min with 25 μM BFA (*see* **Note 3**).
5. Proceed with Subheadings 3.3 or 3.4 depending on whether immunocytochemistry is going to be performed or not.

3.6 GNOM BFA-Resistant Line in Visualization of Endocytic Sterols

1. Follow the protocol described in Subheading 3.2 to visualize sterol internalization in live samples or the protocol described in Subheading 3.5 to label sterols in fixed samples.
2. As a control for the GNOM BFA-resistant line, a GNOM BFA-sensitive line can be used (*see* **Note 14**).

3.7 Image Acquisition at CLSM

1. For live cell imaging, as the mounting medium mainly contains water (MS liquid medium), it is preferable to use a water-immersion objective (×40 or ×63, NA = 1.2). For fixed samples,

the mounting medium contains glycerol (Citifluor), and either glycerol- or oil-immersion objectives should be preferred (×63, NA = 1.4).

2. Activate the 364 nm laser line of the UV laser (*see* **Note 15**).
3. Set up the spectral acquisition window for filipin-sterol fluorescence from 400 to 484 nm. If filipin-sterol fluorescence is acquired alone, the fluorescence detection window can be extended up to 570 nm, depending on the wavelength of potential background signal (*see* **Note 16**).
4. If filipin fluorescence is acquired with other fluorochromes, sequential acquisition should be used to avoid fluorescence bleed-through from one channel into another.
5. Filipin-sterol fluorescence photobleaches. Therefore, acquisition settings should be set up in advance and then applied to the area of interest as one often only has the possibility to perform 1–3 scans on the area of interest [20].

4 Notes

1. Incubation on ice allows a homogeneous staining of the plasma membrane. Make sure the medium is well cooled on ice prior to transferring seedlings.
2. Seedlings should be completely submerged in MS liquid medium. If the seedlings are floating on the surface of the MS medium, pipette the liquid medium on top of the seedlings.
3. As filipin inhibits endocytosis, the incubation time should be at least 60 min.
4. Two thin strips of double-sided tape (1 × 30 mm) are placed longitudinally at the edges of the 24 × 50 mm coverslip (Fig. 1a). Seedling roots are then disposed longitudinally in a drop of MS liquid medium and a 24 × 24 mm coverslip is set down to stick on the double-sided tape (Fig. 1a). Cotyledons are left outside of the coverslip so that it is possible to feed the seedlings with liquid MS medium during the experiment.
5. BFA incubation is performed at room temperature.
6. Due to inhibition of endocytosis by filipin, the BFA concentration needs to be comparatively high.
7. PFA is unstable in solution. For optimal filipin staining, PFA should be prepared freshly and cooled down to room temperature prior to use. When preparing PFA, do not to overheat it.
8. After adding seedlings to the fixative, microwave pulses should preferably be applied within 10–15 min. Overheating should

be avoided when pulses are applied to samples in the microwave. Increase the pause time between each pulse if necessary.

9. Filipin staining of cells and tissues appears diffuse after 1 day, and we recommend observing samples directly or within 24 h after the staining has been performed.
10. Approximately 12 roots can be placed in the middle of one coated microscope slide. Some space should be available close to the edges of the slide to adjust thin bands (1 × 30 mm) of double-sided tape on which the 20 × 20 mm coverslip is placed.
11. A humid chamber can be any box containing a humidified paper towel at the bottom. To incubate solution, a water-vacuum pump is used to suck off solution from one end of the coverslip on the microscope slide while another solution is applied with a pipette at the other end of the coverslip.
12. This step constitutes a second filipin staining while membranes are permeabilized with Igepal and DMSO.
13. To visualize or evaluate quantitative differences, BFA should be used at a maximum concentration of 25 μM or 50 μM.
14. One experiment included both the GNOM BFA-sensitive (wild-type) and GNOM BFA-resistant line treated with BFA and, additionally, the same lines in the absence of BFA treatment.
15. A 364 nm laser line of a UV laser is employed to excite filipin-sterol complexes that display major excitation peaks at 320, 340, and 360 nm [42]. A 405 laser diode does not efficiently excite filipin in complex with sterols. As a result, fluorescence observed upon 405 nm excitation is hardly discernible from tissue background fluorescence in *Arabidopsis* root epidermal cells.
16. As a control, the background fluorescence spectrum and intensity should be checked in wild-type seedlings on which no filipin labeling has been performed but that otherwise have been treated in the same way.

Acknowledgments

We gratefully acknowledge David Ehrhardt (Stanford, USA) for making available LTI6a-GFP seeds; Gerd Jürgens (Tübingen, Germany) for providing anti-KNOLLE serum and GNOM-GFP, $GNOM^{sens}$-myc, and $GNOM^{res}$-myc seeds; Ben Scheres (Utrecht, the Netherlands) for supplying ARF1-EGFP seeds; Ian Moore (Oxford, UK) for sharing RAB-A2a-EYFP seeds; Karin Schumacher (Heidelberg, Germany) for making available VHA-a1-GFP seeds; and Takashi Ueda (Tokyo, Japan) for providing ARA6-GFP (RAB-F1-GFP) seeds.

References

1. Keller P, Simons K (1998) Cholesterol is required for surface transport of influenza virus hemagglutinin. J Cell Biol 140:1357–1367
2. Manes S, Mira E, Gomez-Mouton C et al (1999) Membrane raft microdomains mediate front-rear polarity in migrating cells. EMBO J 18:6211–6220
3. Simons K, Ikonen E (2000) How cells handle cholesterol. Science 290:1721–1726
4. Bagnat M, Simons K (2002) Lipid rafts in protein sorting and cell polarity in budding yeast *Saccharomyces cerevisiae*. Biol Chem 383: 1475–1480
5. Wachtler V, Rajagopalan S, Balasubramanian MK (2003) Sterol-rich plasma membrane domains in the fission yeast *Schizosaccharomyces pombe*. J Cell Sci 116:867–874
6. Willemsen V, Friml J, Grebe M et al (2003) Cell polarity and PIN protein positioning in *Arabidopsis* require STEROL METHYLTRANSFERASE1 function. Plant Cell 15:612–625
7. Takeda T, Kawate T, Chang F (2004) Organization of a sterol-rich membrane domain by cdc15p during cytokinesis in fission yeast. Nat Cell Biol 6:1142–1144
8. Ng MM, Chang F, Burgess DR (2005) Movement of membrane domains and requirement of membrane signaling molecules for cytokinesis. Dev Cell 9:781–790
9. Men S, Boutte Y, Ikeda Y et al (2008) Sterol-dependent endocytosis mediates post-cytokinetic acquisition of PIN2 auxin efflux carrier polarity. Nat Cell Biol 10:237–244
10. Takeshita N, Higashitsuji Y, Konzack S et al (2008) Apical sterol-rich membranes are essential for localizing cell end markers that determine growth directionality in the filamentous fungus *Aspergillus nidulans*. Mol Biol Cell 19:339–351
11. Boutte Y, Frescatada-Rosa M, Men S et al (2010) Endocytosis restricts *Arabidopsis* KNOLLE syntaxin to the cell division plane during late cytokinesis. EMBO J 29:546–558
12. Vidricaire G, Tremblay MJ (2007) A clathrin, caveolae, and dynamin-independent endocytic pathway requiring free membrane cholesterol drives HIV-1 internalization and infection in polarized trophoblastic cells. J Mol Biol 368: 1267–1283
13. Codlin S, Haines RL, Mole SE (2008) btn1 affects endocytosis, polarization of sterol-rich membrane domains and polarized growth in *Schizosaccharomyces pombe*. Traffic 9:936–950
14. Pan J, Fujioka S, Peng J et al (2009) The E3 ubiquitin ligase SCFTIR1/AFB and membrane sterols play key roles in auxin regulation of endocytosis, recycling, and plasma membrane accumulation of the auxin efflux transporter PIN2 in *Arabidopsis thaliana*. Plant Cell 21:568–580
15. Grebe M, Xu J, Mobius W et al (2003) *Arabidopsis* sterol endocytosis involves actin-mediated trafficking via ARA6-positive early endosomes. Curr Biol 13:1378–1387
16. Mukherjee S, Zha X, Tabas I et al (1998) Cholesterol distribution in living cells: fluorescence imaging using dehydroergosterol as a fluorescent cholesterol analog. Biophys J 75: 1915–1925
17. Wüstner D, Mondal M, Tabas I et al (2005) Direct observation of rapid internalization and intracellular transport of sterol by macrophage foam cells. Traffic 6:396–412
18. Wüstner D, Faergeman NJ (2008) Spatiotemporal analysis of endocytosis and membrane distribution of fluorescent sterols in living cells. Histochem Cell Biol 130: 891–908
19. Geldner N, Friml J, Stierhof YD et al (2001) Auxin transport inhibitors block PIN1 cycling and vesicle trafficking. Nature 413:425–428
20. Boutte Y, Men S, Grebe M (2011) Fluorescent in situ visualization of sterols in *Arabidopsis* roots. Nat Protocols 6:446–456
21. Wustner D (2007) Fluorescent sterols as tools in membrane biophysics and cell biology. Chem Phys Lipids 146:1–25
22. Holtta-Vuori M, Uronen RL, Repakova J et al (2008) BODIPY-cholesterol: a new tool to visualize sterol trafficking in living cells and organisms. Traffic 9:1839–1849
23. Gimpl G (2010) Cholesterol-protein interaction: methods and cholesterol reporter molecules. Subcell Biochem 51:1–45
24. Maxfield FR, Wustner D (2012) Analysis of cholesterol trafficking with fluorescent probes. Methods Cell Biol 108:367–393
25. Blachutzik JO, Demir F, Kreuzer I et al (2012) Methods of staining and visualization of sphingolipid enriched and non-enriched plasma membrane regions of *Arabidopsis thaliana* with fluorescent dyes and lipid analogues. Plant Meth 8:28
26. Benveniste P (2004) Biosynthesis and accumulation of sterols. Annu Rev Plant Biol 55:429–457
27. Hartmann M-A (2004) Sterol metabolism and functions in higher plants. In: Daum G (ed) Lipid metabolism and membrane biogenesis. Springer, Berlin, pp 183–211

28. Klima A, Foissner I (2008) FM dyes label sterol-rich plasma membrane domains and are internalized independently of the cytoskeleton in characean internodal cells. Plant Cell Physiol 49:1508–1521
29. Bonneau L, Gerbeau-Pissot P, Thomas D et al (2010) Plasma membrane sterol complexation, generated by filipin, triggers signaling responses in tobacco cells. Biochim Biophys Acta 1798: 2150–2159
30. Ovecka M, Berson T, Beck M et al (2010) Structural sterols are involved in both the initiation and tip growth of root hairs in *Arabidopsis thaliana*. Plant Cell 22:2999–3019
31. Tjellstrom H, Hellgren LI, Wieslander A et al (2010) Lipid asymmetry in plant plasma membranes: phosphate deficiency-induced phospholipid replacement is restricted to the cytosolic leaflet. FASEB J 24:1128–1138
32. Kleine-Vehn J, Dhonukshe P, Swarup R et al (2006) Subcellular trafficking of the *Arabidopsis* auxin influx carrier AUX1 uses a novel pathway distinct from PIN1. Plant Cell 18:3171–3181
33. Steinmann T, Geldner N, Grebe M et al (1999) Coordinated polar localization of auxin efflux carrier PIN1 by GNOM ARF GEF. Science 286:316–318
34. Geldner N, Anders N, Wolters H et al (2003) The *Arabidopsis* GNOM ARF-GEF mediates endosomal recycling, auxin transport, and auxin-dependent plant growth. Cell 112:219–230
35. Richter S, Anders N, Wolters H et al (2010) Role of the GNOM gene in Arabidopsis apical-basal patterning – from mutant phenotype to cellular mechanism of protein action. Eur J Cell Biol 89:138–144
36. Cutler SR, Ehrhardt DW (2002) Polarized cytokinesis in vacuolate cells of *Arabidopsis*. Proc Natl Acad Sci U S A 99:2812–2817
37. Xu J, Scheres B (2005) Dissection of *Arabidopsis* ADP-RIBOSYLATION FACTOR 1 function in epidermal cell polarity. Plant Cell 17:525–536
38. Chow CM, Neto H, Foucart C et al (2008) Rab-A2 and Rab-A3 GTPases define a trans-golgi endosomal membrane domain in *Arabidopsis* that contributes substantially to the cell plate. Plant Cell 20:101–123
39. Dettmer J, Hong-Hermesdorf A, Stierhof YD et al (2006) Vacuolar H+-ATPase activity is required for endocytic and secretory trafficking in *Arabidopsis*. Plant Cell 18:715–730
40. Goh T, Uchida W, Arakawa S et al (2007) VPS9a, the common activator for two distinct types of Rab5 GTPases, is essential for the development of *Arabidopsis thaliana*. Plant Cell 19:3504–3515
41. Lauber MH, Waizenegger I, Steinmann T et al (1997) The *Arabidopsis* KNOLLE protein is a cytokinesis-specific syntaxin. J Cell Biol 139: 1485–1493
42. Castanho MA, Prieto MJ (1992) Fluorescence study of the macrolide pentaene antibiotic filipin in aqueous solution and in a model system of membranes. Eur J Biochem 207: 125–134

Chapter 3

Live Microscopy Analysis of Endosomes and Vesicles in Tip-Growing Root Hairs

Miroslav Ovečka, Irene Lichtscheidl, and Jozef Šamaj

Abstract

Tip growth is one of the most preferable models in the study of plant cell polarity; cell wall deposition is restricted mainly to a certain area of the cell, and cell expansion at this specific area leads to the development of tubular outgrowth. Tip-growing root hairs are well-established systems for such studies, because their lateral position within the root makes them easily accessible for experimental approaches and microscopic observations. Fundamental structural and molecular processes driving tip growth are exocytosis, endocytosis, and all aspects of vesicular and endosomal dynamic trafficking, as related to targeted membrane flow. Study of vesicles and endosomes in living root hairs, however, is rather difficult, due to their small size and due to the resolution limits of conventional light microscopes. Here we present noninvasive approaches for visualizing vesicular and endosomal compartments in the tip of growing root hairs using electronic light microscopy, contrast-enhanced video light microscopy, and confocal laser scanning microscopy (CLSM). These methods allow utilizing the maximum resolution of the light microscope. Together with protocols for appropriate preparation of living plant samples, the described methods should help improve our understanding on how tiny vesicles and endosomes support the process of tip growth in root hairs.

Key words *Arabidopsis thaliana* L, Confocal laser scanning microscopy, Electronic light microscopy, Endosomes, *Medicago sativa* L, Root hairs, Tip growth, *Triticum aestivum* L, Vesicles, Video microscopy

1 Introduction

Root hairs are long tubular extensions of root epidermal cells. They greatly expand the surface area of the root, efficiently extend contacts between plant and soil, and improve the capacity of roots to take up water and mineral ions. Their tubular shape is reached and maintained by highly polarized type of elongation, tip growth, which is restricted exclusively to the tip region. Accordingly, this type of growth requires a highly polarized organization of cytoplasm, cytoskeleton, and endomembranes. Thus, root hairs, compared to other root cells, are completely different in their morphology, type, and speed of growth. Together with pollen tubes, root hairs have become a very popular model to study regulation of localized cell growth in

Marisa S. Otegui (ed.), *Plant Endosomes: Methods and Protocols*, Methods in Molecular Biology, vol. 1209,
DOI 10.1007/978-1-4939-1420-3_3, © Springer Science+Business Media New York 2014

plants, and they have provided a wealth of new information towards our understanding of plant cell polarity in general [1–5].

Growing root hairs show a characteristic zonation, including a tip region with a clear apical zone (devoid of bigger organelles and filled with small vesicles), subapical region, and vacuolated part. Fast and targeted delivery of vesicles depositing membrane and cell wall material to the expanding tip is one of the main aspects of the tip growth. Therefore, growing root hairs contain dense populations of motile vesicles in their tips, whereas larger organelles are excluded [6, 7]. Mitochondria, plastids, endoplasmic reticulum (ER), and Golgi stacks are abundant only in the subapical region, followed at some distance by the central vacuole and the nucleus [6, 8]. When tip growth of root hairs terminates, polarization of the cytoplasm gradually disappears, large organelles reach the tip region, and finally the whole tip is occupied by the large vacuole. Organelles show more uniform distribution, and they are moved by actomyosin cytoskeleton in a rotation-like streaming along the whole length of fully differentiated root hairs.

The pool of vesicles in the growing tip is permanently supplied by exocytotic vesicles delivered to the apex by anterograde cytoplasmic streaming along the cortical region of the root hair. In addition, the pool is supplied also by endocytic vesicles derived from the apical plasma membrane. The retrograde cytoplasmic streaming from the apex to the subapical region within the central area helps to maintain continuous exchange and recycling of vesicles in the tip region of growing root hairs. Evidence for the occurrence of endocytosis in the apical region of the tip is based on the localization of clathrin-coated pits and clathrin-coated vesicles in root hairs [6, 9–11]. Further proof has been provided by the internalization of fluorescent markers [12, 13]. More recent studies on sterols and flotillins clearly revealed also non-clathrin-mediated endocytosis in root hairs [14, 15].

Vesicular membrane transport is controlled and orchestrated by membrane-associated small GTPases belonging to the Rab, Arf, and Rop/Rac families. Rab GTPases, members of the Ras-related superfamily of small GTPases, spatially and temporally organize vesicular trafficking, exocytosis, endocytosis, and membrane recycling within the clear zone [3, 16, 17]. Actin-dependent transport of both vesicles and highly motile endosomes is indispensable for tip growth. Reliable tip-focused vesicular markers such as GFP-RabA1d show colocalization with the endocytic marker dye FM4-64 [14]. This provides a useful tool for the study of TGN/endocytic membrane trafficking and vesicle recycling at the growing tip of root hairs. Transgenic plants stably expressing late endosomal markers confirmed the presence of endosomes in root hairs. Although overexpression of FYVE-GFP, a highly specific molecular marker of phosphatidylinositol-3-phosphate (PtdIns(3)P), affects root growth through inhibition of signal transduction downstream of PtdIns(3)P [18], endosomes labeled with this marker are distributed in the entire root hair, with a higher abundance in the subapical zone. In addition, late endosomes visualized by GFP-RabF2a move actively also into apical and subapical regions of growing root hairs [14, 19]. All available data indicate that dynamic vesicular trafficking events, namely, polar delivery of membrane and cell wall materials to the growing cell domain, endocytosis, and vesicular recycling operate under strict spatial and temporal control. The model of cell polarization by endosomal/vesicular recycling is presently recognized as one of the key components of the tip-growth mechanism [3, 4, 20, 21].

The presence of large numbers of vesicles in the apical part of the root hair was documented by electron microscopy [6, 11, 22–25]. The size of these vesicles, however, is below the resolution limit of conventional light microscopy. Accordingly, dynamic studies on these vesicles are hindered by their small size. However, approaches such as electronic light microscopy [26] taking advantage of video- and computer-based techniques go beyond the resolution limit of the light microscope (approximately 250 nm). Thereby sub-resolution particles in living cells can be observed [27–29]. Different methods of electronic light microscopy, such as contrast-enhanced video microscopy and ultraviolet (UV) microscopy, enhance the image quality and resolution. In video-enhanced contrast microscopy, the images are captured with a high-resolution video camera, utilizing the maximum resolution of the light microscope. Digital image processing by video computers further improves the quality of the image. In UV microscopy, light of 310–360 nm is used in order to increase the resolution of the conventional bright field. Different organelles absorb UV light to a certain degree. Their contrast in the image is thus correlated to the intensity of light absorption [28]. These images are recorded by UV-sensitive video cameras and also further processed by a computer [30]. In both cases, these techniques take full advantage of live cell imaging and detection of fluorescent markers, and it makes possible the observation of the vesicle pools at the tip of living root hairs. Using these approaches, we have analyzed motility of vesicles and endosomes in the apical region of growing root hairs of *Arabidopsis thaliana*, *Medicago sativa*, and *Triticum aestivum* [12, 14, 31–33].

In this chapter, we provide detailed protocols for image acquisition and analysis of dynamic processes in growing tips of root hairs, as related to vesicular trafficking, endocytosis, and the nature of endosomal compartments. This approach integrates the identification of structural components and their dynamic behavior. We also provide details on how to manipulate plants in liquid medium without disrupting root hair growth for microscopy observations.

2 Materials

The protocol requires standard laboratory equipment, *pro analysis*-grade purity chemicals, and solutions prepared in ultrapure water (MilliQ or equivalent).

2.1 Plant Material, Culture Media, and Solutions

1. Seeds of *Arabidopsis thaliana* L. (ecotype Columbia), *Medicago sativa* L. (cv. "Europe"), and *Triticum aestivum* (e.g., cv. "Josef" and "Ludwig"). Other cultivars and transgenic lines usually need similar culture conditions.
2. Commercially available Murashige and Skoog medium in a half-strength concentration (1/2 MS) with vitamins, supplemented with 1 % (w/v) sucrose, pH 5.8 (*see* **Note 1**). Medium is used as liquid, or it can be solidified by the addition of 0.8 % (w/v) plant agar, 0.6–0.8 % (w/v) phytoagar, or 0.4–0.6 % (w/v) phytagel.
3. Modified Murashige and Skoog medium with vitamins, supplemented with 1 % (w/v) sucrose, pH 5.8. Concentration of individual components: (a) macroelements, $CaCl_2$ 111 mg/L, KH_2PO_4 340 mg/L, KNO_3 506 mg/L, $MgSO_4 \times 7H_2O$ 493 mg/L, $Ca(NO_3)_2 \times 4H_2O$ 472 mg/L; (b) microelements, $CoCl_2 \times 6H_2O$ 0.0025 mg/L, $CuSO_4 \times 5H_2O$ 0.13 mg/L, H_3BO_3 4.3 mg/L, $MnCl_2 \times 4H_2O$ 2.8 mg/L, $Na_2MoO_4 \times 2H_2O$ 0.05 mg/L, $ZnSO_4 \times 7H_2O$ 0.29 mg/L, NaCl 0.58 mg/L, Na_2-EDTA 3.722 mg/L, $FeSO_4 \times 7H_2O$ 1.115 mg/L; (c) vitamins, myoinositol 100 mg/L, nicotinic acid 1 mg/L, pyridoxine 1 mg/L, thiamin 1 mg/L, calcium pantothenate 1 mg/L, biotin 0.01 mg/L (*see* **Note 2**). Medium is used as liquid or solidified by addition of the same gelling agent concentrations as in Murashige and Skoog.
4. Fahraeus liquid culture medium: 1.36 mM KH_2PO_4, 1.12 mM Na_2PO_4, 1.36 mM $CaCl_2$, 0.97 mM $MgSO_4$, 20 μM Fe-citrate, pH 6.5 [34], filter sterilized.
5. Phosphate buffer (10 mM, pH 6.5): stock solutions of 0.2 M $Na_2HPO_4 \times 2H_2O$ and 0.2 M NaH_2PO_4 mixed at ratio 1:2.8 (v/v) and diluted 20× by MilliQ water to final volume.
6. The styryl dyes FM1-43 (*N*-(3-triethylammoniumpropyl)-4-(4-[dibutylamino] styryl)pyridinium dibromide) and FM4-64 (*N*-(3-triethylammoniumpropyl)-4-(8-(4-(diethylamino)-phenyl)hexatrienyl)pyridinium dibromide). Stock solution 1.6 mM prepared in DMSO, stored in aliquots at −20 °C. Working solution 4 μM, prepared in liquid MS culture medium (*see* **Note 3**).
7. Stock solution of $DiOC_6(3)$ (3,3′-dihexyloxacarbocyanine iodide): 100 mg/mL $DiOC_6(3)$ in DMSO. Working solution

at a concentration of 5 μg/mL prepared in liquid MS culture medium. This is the membrane dye used to stain endoplasmic reticulum (ER) in root hairs.

8. Transgenic lines of *A. thaliana*, stably expressing (under constitutive promoters) the GFP- or YFP-tagged endosomal markers, e.g., FYVE-GFP as PI3P marker [19], GFP-tagged Rab GTPase RabF2a [19], the TGN/early endosomal markers Wave line 13 (VTI12) fused with YFP [35, 36], and RabA1d (a member of the RabA subfamily of small Rab GTPases) fused with GFP [14, 37].
9. Filipin III stock solution: 25 mg/mL filipin III in DMSO. Use it in a working solution of 0.1–10 μg/mL in liquid MS culture medium. Filipin III is a polyene antibiotic fluorochrome for plant 3-β-hydroxysterols [38].
10. Latrunculin B stock solution: 10 mM latrunculin B in water. Use it in working solution of 0.1–1 μM in liquid MS culture medium. Latrunculin B is an inhibitor of actin microfilament polymerization [12, 31].
11. Brefeldin A stock solution: 150 mM brefeldin A in DMSO. Use it in a working solution of 3.5–35 μM in liquid MS culture medium. Brefeldin A is a fungal toxin inhibiting a subset of guanine nucleotide exchange factors for ADP-ribosylation factor, small GTPases which are involved in vesicular recycling [31, 39].

2.2 Sample Preparation

1. Microscope slides and coverslips. They must be clean and, for experiments with *A. thaliana* and *M. sativa*, also sterile (*see* **Notes 4** and **5**).
2. Staining jars, such as used for staining of microscope slides in histology. For cultivation of *A. thaliana*, we suggest a rectangular staining dish with cover, made of soda lime glass, for staining of horizontally oriented 75 × 25 mm glass slides. For cultivation of *M. sativa* and *T. aestivum*, we use a vertical staining jar with cover, made of soda lime glass, for staining vertically oriented 75 × 25 mm glass slides. A rectangular staining dish accommodates maximum liquid volume of 150 mL; a vertical staining jar has a volume of 50 mL. The number of required slides depends on the size of the seedlings; for *A. thaliana* seedlings, one slide can be accommodated into each slot, whereas *M. sativa* and *T. aestivum* seedlings usually need more space so that only four to max six slides can be stored in the staining dishes. This glassware must be clean and sterile (*see* **Notes 4** and **5**).

2.3 Microscopy

1. Bright-field microscope. For video-enhanced microscopy, a high-end microscope with powerful light source and high magnification is needed. We use a Univar microscope (Reichert-Leica, Vienna, Austria) equipped for bright field, differential

interference contrast, and fluorescence with an HBO 200 mercury arc lamp, 40× Plan Apochromat oil immersion objective (numerical aperture of 1.00), 100× Plan Apochromat oil immersion objective (numerical aperture of 1.32), and additional 1.6× or 2.5× magnification. For UV microscopy, all lenses in the beam path must be made of quartz. We use a Zeiss ACM stand equipped with HBO 100 mercury arc lamp or XBO 75 Xenon lamp and monochromator for selection of appropriate wavelengths.

2. High-resolution video camera. It is needed for a high-contrast acquisition of structures at the limit of light microscopy resolution and with low refraction index. We have been successful with the high-resolution camera Chalnicon C 1000.1 from Hamamatsu (Hamamatsu Photonics, Germany). For digital recording, a back-illuminated CCD camera (e.g., C8000-30, Hamamatsu Photonics, Germany) yields excellent results. Both cameras can be also used for image acquisition in the UV microscope.
3. Digital image processor. It allows for digital processing of the image with functions like image addition, background subtraction, and tracking (*see* **Note 6**). We use a digital Polyprocessor frame image memory unit DVS 3000 (Hamamatsu Photonics, Germany) or the Hokawo software from Hamamatsu, which can be operated on any standard computer.
4. Video recording. We acquire images from the analogue video system (Hamamatsu DVS 3000) on mini-DV video tapes (Sony and JVC). Videotape recording of images with digitally improved gain and contrast utilizing the maximum resolution of the microscope is a basic principle of the video-enhanced microscopy. With digital cameras and software, images are directly saved to the computer.
5. Monitor screen for real-time observation of acquired and recorded signal.
6. Computer workstation for video microscopy data analysis.
7. Confocal laser scanning microscope for fluorescence detection of fluorescent markers (e.g., FM1-43, FM4-64, $DiOC_6(3)$, filipin) and GFP/YFP-tagged markers of endosomal compartments in living root hairs.

3 Methods

3.1 Plant Material

1. Seeds of *Arabidopsis thaliana* L. are surface sterilized in 70 % (v/v) ethanol for 2 min and in 1 % (v/v) sodium hypochlorite with 0.05 % Tween 20 (v/v) for 8 min and rinsed with sterile distilled water. Seeds are then plated on solid ½ MS or solid

modified MS medium supplemented with vitamins, 1 % sucrose with pH adjusted to 5.8. Seeds germinate in culture chamber at 22 °C, 50 % humidity, and 16/8 h light/dark regime (*see* **Note 7**).

2. Seeds of *Medicago sativa* L. are surface sterilized with 1 % (v/v) sodium hypochlorite and 70 % (v/v) ethanol for 3 min, thoroughly washed with sterile water, and imbibed in sterile water at 4 °C for 4–6 h for breaking dormancy (this is also improving and synchronizing seed germination). Seeds are then placed on moist filter paper in Petri dishes for germination (*see* **Note 8**). Leave them in culture chamber in darkness at 25 °C.
3. Seeds of *Triticum aestivum* are soaked in water at 4 °C for 24 h on filter paper and then germinated in Petri dishes with wet filter paper at 22–24 °C under continuous light.

3.2 Transfer, Stabilization, and Processing of Plants

1. Select germinated plantlets with well-developed roots: *A. thaliana* seedlings 2–3 days after germination, seedlings of *M. sativa* 3 days old, and seedlings of *T. aestivum* 3 to 4 days old.
2. Prepare micro-chambers for cultivation of individual plants in liquid media. Micro-chambers are made from a microscope slide and a coverslip [12]. Stick small strips of Parafilm "M," 20 mm long, sideways in the middle position of cleaned slide, perpendicularly or parallel to its long axis. Place another one in the same way in a distance, fitting the size of the coverslip you will use (*see* **Note 9**). Parafilm strips act as spacers between slide and coverslip, defining the width, and together with the size of a coverslip also the volume of a micro-chamber (*see* **Note 10**).
3. Mount seedlings into micro-chambers. Place a drop of culture medium on slide between Parafilm spacebars. Transfer seedling carefully from the plate using a forceps, and put it on the drop of medium in the way that cotyledons are in upper position, 2–3 mm below the upper edge of the slide. Place very carefully a coverslip on the top of Parafilm spacebars, covering with it the root and hypocotyl of the seedling, but not the cotyledons. In the case of *T. aestivum*, coverslip should cover the roots below the grain (*see* **Notes 11** and **12**).
4. Prepare sterile glass jar containing liquid medium (*see* **Note 13**), and transfer slides with plants there. Add the medium to the jar so that it reaches the open lower edge of the micro-chamber after the transfer of slides to the jar. This allows, first, complete filling of the micro-chamber with the medium and, second, free exchange of medium between micro-chamber and jar. Let seedlings grow in vertical position in closed jars under continuous light (or in darkness for *M. sativa*) for 12–24 h. During this period, roots of seedlings grow stably and develop new root hairs (*see* **Note 14**).

3.3 Video-Enhanced Microscopy of Growing Root Hairs

1. After 12–24 h period of stabilization, check seedlings in micro-chambers under the stereomicroscope. Select only morphologically normal seedlings with regular and constant growth.
2. After placing the micro-chamber under the microscope, find the root hair formation zone and select a young growing root hair, which is showing typical zonation with a "clear zone" and cytoplasm at the tip.
3. Check the position of a selected root hair in the microscope and on the monitor screen. Record root hair at low magnification, and measure the root hair tip-growth rate for at least 15 min without any movement or mechanical perturbation (*see* **Note 15**).
4. Measure the rate of the root hair elongation directly from the monitor screen; mark the position of the tip at the start of the 15-min period, and record new positions of the tip every 5 min from time-lapse images. When the hair grows out of the recorded camera field, reposition the hair apex and mark its new position at the meantime of two recording time slots.

3.4 Detection of Endosomes and Analysis of Their Motility by Video-Enhanced Light Microscopy

1. Select root hairs showing normal average speed of tip growth (*see* **Note 16**). Record the tip region of growing root hair at high magnification for at least 60 s. Make sure that the hardware is set up correctly for good visibility of the resolved structures on the monitor screen.
2. Record video sequences representing the velocity of vesicles and endosomes in the tip of growing root hairs in a longitudinal view. Analyze motility of endosomes and their trajectories from video recordings. Data can be visualized on the monitor screen using the frame-by-frame playback function.
3. For documentation of endosomes with UV microscopy, mount plants in the quartz slides. Transfer them to the UV microscope. Set appropriate wavelength (generally within the range of 280–360 nm) by the monochromator, and acquire images using UV-sensitive camera. Record the root hairs for short periods of time, to prevent harmful effects of the UV light (*see* **Note 17**).
4. The video computer offers tracking functions for the analysis of endosomal movements. Every second, third, or fourth frame can be displayed on the monitor screen, thus forming an integrated picture. Short displacements of single organelles in various directions are leveled out, straight or curvilinear tracks result in rows of organelle locations, and static organelles are displayed by strong signals in their positions.
5. Import videos to workstation, convert videos to a digital format, and process them as .avi files (or equivalent format) and .tif images using appropriate software, e.g., Adobe Premiere (http://www.adobe.com/) and Pinnacle Studio (Pinnacle Systems).

3.5 Detection of Endosomes by CLSM

Maintenance of normal growth conditions and careful checking of growth characteristics are crucial for any conclusions to be drawn from experiments, since root hairs are extremely sensitive to mechanical and chemical changes.

1. Document endosomal compartments in growing root hairs by their labeling with the styryl dyes FM1-43 and FM4-64. Prepare staining solution of dyes with a final concentration of 4 μM in the culture medium, and infiltrate this solution into the micro-chamber by perfusion. Remove the culture medium from the micro-chamber with a small strip of filter paper that is placed between slide and coverslip. Simultaneously, slowly and continuously apply the new medium containing FM1-63 or FM4-64 on the other side of the micro-chamber with a micropipette at a loading speed of 10 μL/min. Loading for 5–10 min without washing is enough to get satisfactory staining of the plasma membrane in root cells (*see* **Note 18**). Subsequently, in a stepwise manner, you can follow FM dye internalization and sequential staining of endosomal compartments in the microscope. Set up excitation laser line 488 and 514 nm for FM1-43 and FM4-64, respectively, with emission spectrum interval above 520 nm for FM1-43 and above 650 nm for FM4-64.
2. You can simultaneously stain endosomes and use of the membrane dye $DiOC_6(3)$ for ER staining in root hairs. Treat root hairs by pulse loading with 4 μM FM4-64 for 5 min, then wash out the dye with culture medium. After 10–30 min of cultivation, stain root hairs with working solution of $DiOC_6(3)$ for 5 min by perfusion, wash it out with culture medium, and image. Alternatively, apply both dyes simultaneously for 10 min and wash out before observation.
3. Use transgenic lines of *A. thaliana*, expressing different GFP- or YFP-tagged endosomal molecular markers for visualization of endosomes and analysis of their motility. You can colocalize double FYVE-GFP [19], RabF2a-GFP [19], VTI12-YFP [35, 36], or RabA1d-GFP [14, 37] with FM4-64 and with $DiOC_6(3)$.
4. In addition to special particle tracking software, you can also use the maximum intensity projection function of the CLSM from time series. It will display an integrated picture, summing all positions of the endosomal compartments within the studied area, referring to intracellular distribution and velocities of each individual endosomal particle.

3.6 Experimental Interference with Root Hair Tip Growth and Motility of Endosomes

1. Select a young root hair with typical cytoarchitecture in the microscope, and record its growth for 15 min by taking a short video sequence every 5 min. Make sure that light in the microscope is turned on only during the recording time.
2. Perfuse the micro-chamber gently with plain culture medium directly on the microscope stage. The speed of perfusion

should be 10 μL/min (*see* **Note 19**). Check the morphology and shape of the root hair apex every 5 min during perfusion. If necessary, you can use this checking also for repositioning of the root hair tip before further recording. After perfusion, measure the tip growth of the same root hair during additional 15 min. This is to ensure that no artifacts in the morphology and tip growth of root hairs are induced by the medium exchange. Only root hairs that show normal average speed of the tip growth after control perfusion should be used for further experiments.

3. As a next step, perfuse the micro-chamber directly on the microscope stage with culture medium containing drugs of interest (e.g., latrunculin B, brefeldin A, filipin III). After perfusion (with or without washing), record tip growth of the same root hair for additional time, according to the experimental design.
4. Repeat the same approach at high magnification for the analysis of drug-induced changes in vesicular trafficking and endosomal motility.
5. Collect sufficient amount of data for well-supported quantitative interpretation of experiments. As minimum, three independent biological replicas should be performed with five different root hairs from at least three independent seedling roots.

4 Notes

1. MS medium can be buffered by MES (2-(*N*-morpholino)ethanesulfonic acid), added at 10 mM concentration, and further titrated to desired pH 5.8 with KOH. In our experiments, however, *A. thaliana* plants germinate and grow only for 2–3 days on solidified media. After transfer to liquid media and subsequent stabilization in the micro-chambers, they are observed in the range of hours. Excessive pH stabilization therefore is not necessary, and accordingly, media are buffered with KOH only without the addition of MES.
2. Macroelements, microelements, and vitamins of modified MS medium are prepared separately as stock solutions (concentrated 10×, 100×, and 100×, respectively). Stock solution of macroelements is kept at 4 °C, and stock solutions of microelements and vitamins are filter sterilized, aliquoted, and kept at −20 °C. Macroelements, sucrose, and gelling agents are mixed in MilliQ water to final concentrations before pH adjustment and sterilized in autoclave. Microelements and vitamins are added in sterile box only when medium after autoclaving cools down to 60–70 °C. Solution must be mixed well before pouring to plates.

3. The styryl dyes FM1-43 and FM4-64 are light sensitive. Keep them in darkness.
4. Glass material must be clean and free of fat and mechanical impurities. It should be washed thoroughly in water with detergent first. After washing, place it in 96 % ethanol (v/v) for 2–3 min and wash by deionized water. After drying, it should be used directly or stored at clean and dry place.
5. Because glass staining dishes should not be used at temperatures above 80 °C, they are not autoclavable or microwaveable. Sterilization of staining dishes (including slides and coverslips inside) is thus made by filling them with 96 % ethanol (v/v) for 5–10 min, inside of the sterile box. After removing ethanol, let them dry inside of the sterile box. Glass can be used only after thorough and complete drying. Do this sterilization only before the use of glass for experiments.
6. Analogue enhancement of contrast and gain is possible by the video cameras. It is further improved digitally by the image processor. Background subtraction function utilizes either background subtraction, using a reference image of out-of-focus details and blurs, or an in-focus subtraction, using a reference image of the undesired parts of the cell itself, like stationary thin details. As a result, small structures and fine details, barely visible in the unprocessed picture, could be visualized in plant cells.
7. Prior to transferring to the culture chamber, plates with *A. thaliana* seeds are placed at 4 °C for stratification taking 2–3 days. This helps to obtain uniform germination of seeds.
8. To ensure sterile germination and growth of *M. sativa* plants, place sterile filter paper into sterile Petri dishes, pour sterile culture solution, and place seeds on it. For sterilization of the filter paper, put pack of cut filter paper at desired size into glass Petri dish, wrap it in aluminum foil, and autoclave it using dry sterilization program at 121 °C for 20 min.
9. To perform undisturbed observation under the microscope and effective exchange of any solutions, we put only one plant into the micro-chamber. For this reason, we use coverslips with size 20 × 20 mm for *A. thaliana* (for horizontally oriented slide) and 32 × 24 mm for *M. sativa* and *T. aestivum* (for vertically oriented slide).
10. Width of the micro-chambers is defined by strips of Parafilm. To fit it well to the thickness of cultivated roots, micro-chambers for *A. thaliana* contain one layer, for *M. sativa* three layers, and for *T. aestivum* six layers of Parafilm as spacers.
11. Correctly positioned coverslip should be distanced 2 mm from the lower edge of the slide in case of horizontally mounted

seedling (*A. thaliana*) and 5–10 mm from the lower edge of the slide in case of vertically mounted seedling (*M. sativa*, *T. aestivum*).

12. Work in sterile box either with gloves or without them, but with fingers surface sterilized by common aseptic solutions.
13. Liqiud medium to be used in micro-chambers and in jars for *A. thaliana* is the same medium as for plant germination, but without gelling agents. For *M. sativa*, use Fahraeus liquid culture medium, and for *T. aestivum*, use 10 mM phosphate buffer at pH 6.5.
14. Growing root hairs are very sensitive to different stresses. To prevent damage or unintentional changes in tip growth, it is very important to protect them before experiments. Stabilization of plants in custom-made micro-chambers enables placing the plants to the microscope stage and analyzing them without any further transfer or manipulation. In addition, this method ensures that the roots and root hairs could adapt to the liquid growth medium. Samples are thus transferred to the microscope in the stage of active undisturbed growth.
15. To avoid external disturbances, tip growth should be recorded in a time-lapse mode. For the first 15 min, root hairs are recorded only every 5 min, and only for short 5-s periods. This short recording time is needed to prevent harmful effects of the intense microscope illumination at high magnifications.
16. Normal average speed of tip growth in liquid media in micro-chambers is within the range of 1.2–1.4 μm/min in *A. thaliana* [14], 0.4–0.6 μm/min in *M. sativa* [31], and 0.6–1.2 μm/min in *T. aestivum* [33].
17. Utilization of short wavelengths of the light (UV range) increases the resolution in the microscope. UV light, however, has harmful effects and may cause serious damage in the cells. Shorter wavelengths are obviously more destructive. Some cells, like onion inner epidermis, can be observed in the UV light for reasonable long time without visible damage [30]. Root hairs, however, are so sensitive that the observation times should be minimized to a minimum.
18. You can observe root cells either immediately after FM dye loading without washing, or, for pulse-labeling, you can remove FM dye after 5-min incubation in the micro-chamber by perfusion of the micro-chamber with plain culture medium [12]. Cell membranes are already stained and only excess of soluble dye (not incorporated into plasma membrane) is removed by washing.
19. Using coverslip with the size of 20 × 20 mm (for *A. thaliana*), the volume of the micro-chamber is app. 50 μL. Use double amount of medium for perfusion (100 μL) to ensure a complete

replacement of the internal volume by fresh solution. Since slow application of 10 μL takes app. 1 min, it gives a total time period of 10 min for application of 100 μL. In micro-chambers with a coverslip size of 32 × 24 mm (for *M. sativa* and *T. aestivum*), apply ten times 20 μL.

Acknowledgements

This work was supported by grant LO1204 from the National Program of Sustainability I to the Centre of the Region Haná for Biotechnological and Agricultural Research in Olomouc and by the Austrian ŒAD/appear funded project BIOREM (Appear 43).

References

1. Baluška F, Salaj J, Mathur J, Braun M, Jasper F, Šamaj J, Chua NH, Barlow PW, Volkmann D (2000) Root hair formation: F-actin-dependent tip growth is initiated by local assembly of profilin-supported F-actin meshworks accumulated within expansin-enriched bulges. Dev Biol 227:618–632
2. Hepler PK, Vidali L, Cheung AY (2001) Polarized cell growth in higher plants. Annu Rev Cell Dev Biol 17:159–187
3. Šamaj J, Müller J, Beck M, Böhm N, Menzel D (2006) Vesicular trafficking, cytoskeleton and signalling in root hairs and pollen tubes. Trends Plant Sci 11:594–600
4. Campanoni P, Blatt MR (2007) Membrane trafficking and polar growth in root hairs and pollen tubes. J Exp Bot 58:65–74
5. Ketelaar T, Galway ME, Mulder BM, Emons AMC (2008) Rates of exocytosis and endocytosis in Arabidopsis root hairs and pollen tubes. J Microsc 231:265–273
6. Galway ME, Heckman JW, Schiefelbein JW (1997) Growth and ultrastructure of Arabidopsis root hairs: the *rhd3* mutation alters vacuole enlargement and tip growth. Planta 201:209–218
7. Galway ME (2000) Root hair ultrastructure and tip growth. In: Ridge RW, Emons AMC (eds) Root hairs: cell and molecular biology. Springer, Tokyo, pp 1–17
8. Ketelaar T, Faivre-Moskalenko C, Esseling JJ, de Ruijter NC, Grierson CS, Dogterom M, Emons AMC (2002) Positioning of nuclei in Arabidopsis root hairs: an actin-regulated process of tip growth. Plant Cell 14:2941–2955
9. Robertson JG, Lyttleton P (1982) Coated and smooth vesicles in the biogenesis of cell walls, plasma membranes, infection threads and peribacteroid membranes in root hairs and nodules of white clover. J Cell Sci 58:63–78
10. Emons EMC, Traas JA (1986) Coated pits and coated vesicles on the plasma membrane of plant cells. Eur J Cell Biol 41:57–64
11. Ridge RW (1995) Micro-vesicles, pyriform vesicles and macrovesicles associated with the plasma membrane in the root hairs of *Vicia hirsuta* after freeze-substitution. J Plant Res 108:363–368
12. Ovečka M, Lang I, Baluška F, Ismail A, Illeš P, Lichtscheidl IK (2005) Endocytosis and vesicle trafficking during tip growth of root hairs. Protoplasma 226:39–54
13. Yoo C-M, Quan L, Cannon AE, Wen J, Blancaflor EB (2012) AGD1, a class 1 ARF-GAP, acts in common signaling pathways with phosphoinositide metabolism and the actin cytoskeleton in controlling Arabidopsis root hair polarity. Plant J 69:1064–1076
14. Ovečka M, Berson T, Beck M, Derksen J, Šamaj J, Baluška F, Lichtscheidl IK (2010) Structural sterols are involved in both the initiation and tip growth of root hairs in *Arabidopsis thaliana*. Plant Cell 22:2999–3019
15. Li R, Liu P, Wan Y, Chen T, Wang Q, Mettbach U, Baluška F, Šamaj J, Fang X, Lucas WJ, Lin J (2012) A membrane microdomain-associated protein, Arabidopsis Flot1, is involved in a clathrin-independent endocytic pathway and is required for seedling development. Plant Cell 24:2105–2122
16. Nielsen E, Cheung AY, Ueda T (2008) The regulatory RAB and ARF GTPases for vesicular trafficking. Plant Physiol 147:1516–1526
17. Stenmark H (2009) Rab GTPases as coordinators of vesicle traffic. Nat Rev Mol Cell Biol 10:513–525

18. Lee Y, Bak G, Choi Y, Chuang W-I, Cho H-T, Lee Y (2008) Roles of phosphatidylinositol 3-kinase in root hair growth. Plant Physiol 147:624–635
19. Voigt B, Timmers ACJ, Šamaj J, Hlavačka A, Ueda T, Preuss M, Nielsen E, Mathur J, Emans N, Stenmark H, Nakano A, Baluška F, Menzel D (2005) Actin-propelled motility of endosomes is tightly linked to polar tip-growth of root hairs. Eur J Cell Biol 84:609–621
20. Žárský V, Cvrčková F, Potocký M, Hála M (2009) Exocytosis and cell polarity in plants – exocyst and recycling domains. New Phytol 183:255–272
21. Richter S, Müller LM, Stierhof Y-D, Mayer U, Takada N, Kost B, Vieten A, Geldner N, Koncz C, Jürgens G (2012) Polarized cell growth in Arabidopsis requires endosomal recycling mediated by GBF1-related ARF exchange factors. Nat Cell Biol 14:80–87
22. Bonnett HT, Newcomb EH (1966) Coated vesicles and other cytoplasmic components of growing root hairs of radish. Protoplasma 62: 59–75
23. Emons AMC (1987) The cytoskeleton and secretory vesicles in root hairs of *Equisetum* and *Limnobium* and cytoplasmic streaming in root hairs of *Equisetum*. Ann Bot 60:625–632
24. Galway ME, Lane DC, Schiefelbein JW (1999) Defective control of growth rate and cell diameter in tip-growing root hairs of the rhd4 mutant of *Arabidopsis thaliana*. Can J Bot 77:494–507
25. Ridge RW (1988) Freeze-substitution improves the ultrastructural preservation of legume root hairs. Bot Mag Tokyo 101:427–441
26. Shotton DM (1988) Video-enhanced light microscopy and its application in cell biology. J Cell Sci 89:129–150
27. Allen RD, Allen NS, Travis JL (1981) Video-enhanced contrast, differential interference contrast (AVEC-DIC) microscopy: a new method capable of analyzing microtubule-related motility in the reticulopodial network of *Allogromia laticollaris*. Cell Motil 1:291–302
28. Lichtscheidl IK (1995) An introduction to the video microscopy of plant cells: principles of modern light microscopical techniques and their application for the study of plant cells. Wiss Film (Wien) 47:11–40
29. Lichtscheidl IK, Foissner I (1996) Video microscopy of dynamic plant cell organelles: principles of the technique and practical application. J Microsc 181:117–128
30. Lichtscheidl IK, Url WG (1987) Investigation of the protoplasm of *Allium cepa* inner epidermal cells using ultraviolet microscopy. Eur J Cell Biol 43:93–97
31. Šamaj J, Ovečka M, Hlavačka A, Lecourieux F, Meskiene I, Lichtscheidl I, Lenart P, Salaj J, Volkmann D, Bögre L, Baluška F, Hirt H (2002) Involvement of the mitogen-activated protein kinase SIMK in regulation of root hair tip-growth. EMBO J 21:3296–3306
32. Ovečka M, Baluška F, Lichtscheidl IK (2008) Non-invasive microscopy of tip growing root hairs as a tool for study of dynamic, cytoskeleton-based processes. Cell Biol Int 32:549–553
33. Volgger M, Lang I, Ovečka M, Lichtscheidl IK (2010) Plasmolysis and cell wall deposition in wheat root hairs under osmotic stress. Protoplasma 243:51–62
34. Fahraeus G (1957) The infection of clover root hairs by nodule bacteria studied by a simple glass slide technique. J Gen Microbiol 16:374–381
35. Sanderfoot AA, Assaad FF, Raikhel NV (2000) The Arabidopsis genome. An abundance of soluble *N*-ethylmaleimide-sensitive factor adaptor protein receptors. Plant Phys 124:1558–1569
36. Geldner N, Dénervaud-Tendon V, Hyman DL, Mayer U, Stierhof Y-D, Chory J (2009) Rapid, combinatorial analysis of membrane compartments in intact plants with a multicolor marker set. Plant J 59:169–178
37. Takáč T, Pechan T, Šamajová O, Ovečka M, Richter H, Eck C, Niehaus K, Šamaj J (2012) Wortmannin treatment induces changes in Arabidopsis root proteome and post-Golgi compartments. J Proteome Res 11:3127–3142
38. Grebe M, Xu J, Möbius W, Ueda T, Nakano A, Geuze HJ, Rook MB, Scheres B (2003) Arabidopsis sterol endocytosis involves actin-mediated trafficking via ARA6-positive early endosomes. Curr Biol 13:1378–1387
39. Geldner N, Anders N, Wolters H, Keicher J, Kornberger W, Muller P, Delbarre A, Ueda T, Nakano A, Jürgens G (2003) The Arabidopsis GNOM ARF-GEF mediates endosomal recycling, auxin transport, and auxin-dependent plant growth. Cell 112:219–230

Chapter 4

Analysis of Fluid-Phase Endocytosis in (Intact) Plant Cells

Vera Bandmann, Peter Haub, and Tobias Meckel

Abstract

Endocytosis is a continuous process at the plasma membrane at least of all eukaryotic cells. Regardless of the molecular machinery, which drives the formation and uptake of endocytic vesicles, it is reasonable to assume that this process inevitably collects external fluid. Hence, at least for the majority of apoplastic solutes, the endocytosis of the fluid phase is likely to be an inevitable process. Due to its independence from the molecular machinery and low selectivity with respect to the cargo, it is thus perfectly suited to be used as a tracer to follow the activity of all endocytic events. Here we describe simple protocols based on fluorescence microscopy, which yield quantitative information about endocytic vesicle sizes—with sub-diffraction accuracy—as well as the size exclusion limits for these uptake routes.

Key words Fluid-phase endocytosis, Fluorescence, Microscopy, Confocal, Single-molecule, Diffraction limit, Fluorescence intensity, Size exclusion limit

1 Introduction

The composition, size, and integrity of the plasma membrane is maintained by a continuous traffic of membranous vesicles. To deliver lipids, membrane proteins, and compounds destined for secretion to the apoplast, membranous vesicles fuse with the plasma membrane in a process known as exocytosis. In reverse, these components are collected and internalized via endocytosis, a process that starts with an invagination of the plasma membrane followed by separation and intracellular release of a vesicle. Hence, the dynamic properties of the plasma membrane are tightly linked to the dynamics of these trafficking processes.

In this protocol we will not distinguish between the various forms of endocytosis, such as clathrin, caveolin, and flotilin dependent forms, or macropinocytosis [1, 2]. Rather, we focus on how the uptake of the fluid phase by *any* endocytic processes can be measured and quantitatively analyzed. In general, fluid-phase endocytosis can conceptually—much less than experimentally—be separated into two distinct processes: (1) Uptake of the fluid phase may be considered

Marisa S. Otegui (ed.), *Plant Endosomes: Methods and Protocols*, Methods in Molecular Biology, vol. 1209,
DOI 10.1007/978-1-4939-1420-3_4,

an inevitable secondary effect of endocytosis. For example, there is evidence that receptor mediated endocytosis is accompanied by uptake of extracellular fluid. In addition, there is evidence that (2) fluid-phase endocytosis is also adeliberate and regulated process. In presence of excess glucose levels in the culture media, tobacco BY-2 protoplasts are able to internalize glucose via endocytosis [3]. Notably, however, glucose is not internalized by default as a random cargo during normal membrane turnover. Rather, glucose uptake is stimulated even if clathrin-dependent endocytosis is blocked. Fluid-phase endocytosis of glucose in this case appears to be more a physiologically controlled mechanism than a secondary effect. Whether regulated fluid-phase endocytic is a widespread phenomenon for other soluble extracellular compounds is not known. Nevertheless, fluid-phase endocytosis is known to be required for many tightly controlled processes, such as cell wall homeostasis, pathogen defense, and heavy metal sequestration. Clearly, fluid-phase endocytosis is more than a simple side effect and therefore, warrants specific considerations for the study of membrane turnover and fate of all compounds taken up along with it.

The investigation of fluid-phase uptake by plants using a fluorescent aqueous fluid has been performed both on intact plant cells and protoplasts [4–6]. Reports on uptake of the popular fluid-phase marker Lucifer Yellow by carriers or anion channels [7, 8] and misleading interpretations of some results, e.g., that endocytic vesicles may be too small for a successful detection [5], led to a somewhat confusing and unclear view on fluid-phase endocytosis in plants. A suitable fluid-phase endocytosis tracer should not leak through intact lipid bilayers, be transported by channel proteins, or bind to extracellular proteins- or sugar domains, thereby initiating some form of stimulated uptake. Among the best fluorescent markers to follow endocytosis are FITC-Dextrans and Alexa 488 Hydrazide for the fluid-phase uptake [9, 10], FM4-64 for the membrane internalization [11–13], and fluorescent beads or quantum dots [14], as tracers to probe for size exclusion.

The small size of the endocytic vesicles, which generally fall well below the diffraction limit of light microscopy (~250 nm), in combination with the speed of their formation and movement, provides a challenging task to obtain reliable quantitative information of the process.

Here we describe a protocol for fluorescent imaging and quantitative analysis of fluid-phase endocytosis in *Vicia faba* guard cells and Tobacco Bright-Yellow 2 (BY-2) cells. We describe how to prepare appropriate test samples to quantify the sensitivity and resolution of a confocal microscope, describe how reliable quantitative information can be obtained from recorded images, and provide software to aid in these tasks. Instead of using elaborate preparations of liposomes with a narrow size distribution to determine the molecular brightness of a single dyes molecule on a commercial confocal setup [10], we now determine the intensity ratio of fluorescent beads and individual fluorophores using a single molecule setup [15–17]. This technique is applicable to the study of fluid-phase endocytosis in any intact plant cell or tissue which can be observed with a fluorescence microscope.

It is important to note that protoplasts of all plant cells are amenable to an alternative, quantitative method, namely, patch clamp capacitance recordings, to measure for example vesicle size, endocytic events per time, or pore sizes [18–21]. The clear advantage of the technique lies in its high spatiotemporal resolution, which not only allows for precise measurements of the vesicle size but also kinetic information of single fusion and fission events. Even formation and size of fusion pores can be monitored in real time [21]. The technique therefore provides detailed information on exocytic and endocytic processes, which should always be considered complementary to the procedures described herein.

2 Material

2.1 Plant Material

2.1.1 Preparation for Imaging Tobacco BY-2 Cells

1. Tobacco BY-2 cells.
2. BY-2 culture medium: Linsmaier and Skoog medium supplemented with 30 g/L sucrose, 1 mg/mL 2,4-dichlorophenoxyacetic acid, and 1 mg/mL thiamine.
3. Imaging buffer: Linsmaier and Skoog medium without 2,4-dichlorophenoxyacetic acid and thiamine.
4. Imaging Spacer (e.g., Secure Seal™ SSX13 from Grace Bio-Labs, Oregon, USA). While double sided tape with a thickness of 100–120 μm may work as well, organic solvents may leak from such tapes and impair cell viability.
5. Coverslips, #1 or #1.5.
6. Microscope slides.

2.1.2 Preparation for the Microscopy of Vicia faba L. guard Cells

1. Seeds of *Vicia faba* L. cv. Bunyan
2. Imaging buffer: 10 mM 2-morpholinoethanesulfonic acid (MES, pH 6.1/KOH), 20 mM KCl, and 100 mM CaCl.
3. 35 mm petri dish.
4. Wire-meshes. Stopping screens used with particle delivery systems work well (e.g., stopping screens, Bio-Rad, Hercules, CA, USA).
5. Two sharp and bent tweezers.
6. Coverslip bottom dishes.

2.1.3 Protoplast Production

1. BY-2 cell suspension at exponential growth phase.
2. BY-2 digestion medium: 3 % cellulase R10, 0.2 % macerozyme, 0.1 % pectolyase, 8 mM $CaCl_2$, 25 mM MES/KOH, pH 5.6, adjusted to 540 mOsmol/kg using sucrose (*see* **Note 1**).
3. BY-2 wash solution: 30 mM $CaCl_2$, 30 mM KCl, 20 mM MES, pH 6.5, adjusted to 430 mOsmol/kg using sorbitol.
4. Guard cells digestion medium: 1.8–2.5 % of Cellulase Onozuka RS (Yacult Honsha, Japan), 2 % Cellulysin (Calbiochem, Behring Diagnostics, La Jolla), 0.026 % Pectolyase Y-23, 0.26 % (BSA), 1 m M $CaCl_2$ pH 5.56 adjust the osmolarity to 520 mOsmol/kg with sorbitol (*see* **Note 1**).
5. Guard cell wash solution 200 mM $CaCl_2$, 10 mM Mes/KOH, pH 5.6 adjusted to 520 mOsmol/kg with sorbitol.
6. 100 μm nylon mesh.

2.2 Fluorescent Dyes and Beads

1. Alexa Fluor® 488 hydrazide, Sodium Salt, 1 mg (Invitrogen, Carlsbad, CA, USA).
2. FluoSpheres® Size Kit #2, Carboxylate-modified Microspheres, Yellow-Green Fluorescent (505/515), 2 % solids, 6 bead diameters (0.02, 0.1, 0.2, 0.5, 1.0, and 2.0 μm) (Invitrogen, Carlsbad, CA, USA).
3. Phosphate buffered saline (PBS) (pH 7.4) containing 1 % (w/v) poly(vinyl alcohol) (PVA, M_w 89000-98000, 99+% hydrolyzed).

2.3 Microscopy and Image Analysis

2.3.1 Confocal Laser Scanning Microscopy

1. Confocal Laser Scanning Microscope, e.g., Leica TCS SP5 (Leica Microsystems, Mannheim, Germany), 488 nm excitation line; emission settings for GFP.
2. Objective with high numerical aperture, preferably a water immersion type (e.g., 63×, NA 1.2), but oil immersion objectives with NA of 1.3 or better will work as well (40×, NA 1.3/60×, NA 1.4/100×, NA 1.44, etc.).

2.3.2 Single Molecule Microscopy

1. 488 nm Laser with sufficient power to achieve an illumination intensity of ~1 kW/cm², i.e., if the collimated beam exiting the objective has a diameter of ~30 μm (FWHM, full width at half maximum peak height) the power measured at the objective should at least reach 7 mW.
2. Objective with numerical aperture of 1.3 or better.
3. Dichroic mirrors, bandpass emission filters for fluorescein or GFP detection.
4. Back illuminated, frame transfer electron-multiplying charge coupled device (EMCCD) camera, such as Andor iXonEM⁺ (Andor, Belfast, UK). Back illuminated EMCCD cameras capture more than 90 % photons and amplify signals to overcome instrument noise.

2.3.3 Software for Image Acquisition and Analysis

1. Image acquisition software, e.g., Micromanager [22]
2. Fiji (http://fiji.sc/Fiji), a distribution of ImageJ [23].
3. GaussFitOnSpot, a plugin for Fiji or ImageJ (http://rsb.info.nih.gov/ij/plugins/gauss-fit-spot/) to perform two-dimensional Gaussian fits on spots in an image.
4. Matlab (Mathworks, Nattick, MA), to run the attached script GaussFitOnSpot.m (*see* **Note 14**).
5. Program for regression analysis, i.e., fitting of a Gaussian to 1D data (*see* **Note 2**).

3 Method

3.1 Plant Material

3.1.1 Preparation of BY-2 Cells for Imaging

1. Culture BY-2 cells in Linsmaier and Skoog medium under rapid shaking (100 rpm).
2. Collect the cells at exponential growth phase. To reduce shear forces while pipetting BY-2 cells, cut the pipette tip with a sharp razorblade. Take the cells directly from shaker to keep them in good condition during the experiment.
3. For imaging, dilute BY-2 cells suspension 1:10 in LS medium (e.g., add 20 μL cells to 180 μL medium).
4. Place 15–20 μL of the cell dilution on a glass slide with an imaging spacer and put a cover slide on top.

Preparation of BY-2 Protoplasts

1. For protoplast isolation, collect the cells at exponential growth phase.
2. Place culture in a 50 mL plastic tube and let the cells settle down to the bottom of the tube.
3. After sedimentation of the cells, discard excess BY-2 culture medium and add digestion medium.
4. Add the same volume of digestion medium as cell volume after sedimentation (e.g., use 5 mL of sedimented cells with 5 mL digestion medium) and incubate for 3–4 h at 37 °C under slow shaking (50 rpm). Living protoplasts will move to the surface of the digestion medium while cell debris will slide to the bottom and build a fluffy pellet.
5. After digestion, collect BY-2 protoplasts by directing a long-tip Pasteur pipette through the band of living protoplasts to the bottom of the plastic tube. Remove cell debris and digestion medium until the band of living protoplasts reach the bottom.
6. Wash the BY-2 protoplasts twice with wash solution by centrifugation at 100 × *g* for 10 min (*see* **Note 3**).
7. Wash the cell pellet twice with BY-2 wash solution (*see* **Note 4**).

8. For imaging, dilute BY-2 protoplasts 1:10 in BY-2 wash solution, place them on glass slide with spacers and put a cover slide on top.
9. Turn the slide around and wait for BY-2 protoplasts to settle down to the bottom of the cover slide (5 min).

3.1.2 Preparation of Epidermal Peels of Vicia faba L. for Imaging

1. Grow *Vicia faba* L. cv. Bunyan 14 h at 300 mE, 20 °C, 60 % relative humidity, and 10 h in the dark, at 17 °C, and 70 % relative humidity cycles.
2. Harvest 2–3 three week old leafs.
3. Fill a 35 mm dish with 2 mL of imaging buffer.
4. Submerge a wire screen in the dish.
5. Use one sharp and bent tweezer to lift an epidermal peel from the abaxial side of the leaf. Leave the end of the strip attached to the leaf and use a razor blade to cut it from the remaining epidermis.
6. Use the tweezers as a fork-lift, to transfer the peel to the surface of the buffer in the petri dish. The cuticle should face the air.
7. Grab the wire screen with the tweezers, slowly lift it towards the floating peel thereby removing both from the buffer.
8. Put the wire screen with the peel in a coverslip bottom dish with the peel facing the coverslip and slowly add buffer solution. The wire screen safely keeps the peel submerged and in place for microscopic investigation.
9. Check guard cells for cytoplasmic streaming as a viability test.

Preparation of *Vicia faba* Protoplasts

1. Fill a 35 mm dish with 2 mL of guard cell digestion medium.
2. Prepare epidermal peels as described in Subheading 3.1.2.
3. Let epidermal peels float on digestion medium. The cuticle should face the air. Incubate epidermal stripes on guard cell digestion medium for 2–3 h at 28 °C under gently shaking (50 rpm) (*see* **Note 5**). Incubate epidermal stripes on guard cell digestion medium for 2–3 h at 28 °C under gently shaking (50 rpm) (*see* **Note 5**).
4. Purify released protoplasts by passage through a 100 μm nylon mesh.
5. Wash the nylon mesh with guard cell wash solution to release further protoplasts from the tissue.
6. Incubate purified protoplasts on ice for 2–3 min.
7. Collect protoplasts by centrifugation at 100 × *g* for 10 min (*see* **Note 3**).
8. Wash protoplast pellet twice with guard cell wash solution.
9. Keep isolated protoplasts at 4 °C in fresh wash solution.

3.2 Determination of the Spatial Resolution and Sensitivity of the Confocal Setup

3.2.1 Determination of the Spatial Resolution of the CLSM in Presence of the Specimen

1. Dilute the 100 nm FluoSpheres stock solution (2 % solids) 10^5 times and add between 10 and 100 μL to the dish containing the epidermal peel and mix slowly by pipetting up and down while gently holding the wire screen down with the pipette tip.
2. On a confocal setup, focus on the parenchymal side of a guard cell and identify an individual, immobile fluorescent bead.
3. For calibration, select beads located at the center of the field of view, i.e., in the optical axis of the objective, unless you deliberately try to characterize the objectives' off-axis performance (*see* **Note 6**).
4. Record a 3D-stack of at least 20 beads with an 80 nm lateral (xy) and 200 nm axial (z) pixel spacing.
5. Load the image stacks into Fiji.
6. Select the center of the bead with a square ROI (region of interest) of 5 × 5 pixels.
7. Select the command Image > Stacks > Plot *Z*-axis Profile.
8. Copy the data using the "Copy" button under the plot and paste in your preferred software (e.g., SciDaVis, *see* **Note 2**) to perform nonlinear curve fitting.
9. Fit using a Gaussian function of the form

$$(x) = y_0 + \frac{2}{\text{FWHM}}\sqrt{\frac{\ln 2}{\pi}}Ae^{-\left(\left(\frac{x-x_c}{\text{FWHM}}\right)^2 4\ln 2\right)}$$

Use the following notation of the function, to create a user-defined fitting function in the regression analysis software of your choice:

```
f(x)=y0+((2/FWHM)*sqrt(ln(2)/PI)*A*exp(-((x-xc)/FWHM)^2*4*ln(2)))
```

The fitting parameters are y0 (base or background), FWHM (full with at half maximum), A (integral or intensity), and xc (center). The resulting FWHM corresponds to the axial resolution of you microscope setup (*see* **Note 7**).

10. The location of the maximum (x_c) tells the frame number of the image stack, where the beads axial center has been recorded. Save this image frame to determine the lateral resolution in the next step.
11. Copy the Matlab-Skipt "GaussFitOnSpot.m" (*see* **Note 14**) in the same folder as the image and start it. Select the image saved in **step 10**.
12. The fitting process is started by clicking on the spot-like image of the bead. Once finished, the fitting results will be presented in a new window. A dialog window asks whether you want to keep or discard the fitting results. Depending on your choice, they will/will not be saved in the Matlab workspace. You will be

returned to the image, where you can select another bead. The FWHM tells the lateral resolution of you setup (*see* **Note 7**).

13. Alternatively, open the image in Fiji and start the plugin "GaussFitOnSpot" (Subheading 2.3.3).

3.2.2 Determination of the Intensity Ratio Between Fluorescent Beads and Individual Alexa 488 Hydrazide Fluorophores on Single Molecule Setup

1. Clean coverslips carefully. Both acid and plasma cleaning approaches work well.
2. Prepare a solution of Alexa 488 hydrazide (1×10^{-10} M) in PBS with 1 % PVA.
3. Dilute the 100 nm FluoSpheres stock solution (2 % solids) 10^5 times in PBS. Sonicate at every dilution step (*see* **Note 8**).
4. Mix a small part of the above dilutions 10:1 (Alexa 488–FluoSpheres) to obtain a combined calibration sample.
5. Embed samples in PVA films on cleaned glass slides by spin-coating 50 μL of the solution on a spin coater following a two-step protocol: 10 s at 300 rpm and 1 min at 2,000 rpm.
6. Keep coverslips with the now immobilized Alexa 488 hydrazide molecules, FluoSpheres, as well as the combined sample in the dark at all times.
7. Record images of all samples at various positions on a setup, capable to detect signals from single Alexa 488 hydrazide molecules (Subheading 2.3.2).
8. Analyze the diffraction limited spots corresponding to the beads and single Alexa 488 hydrazide molecules using the Matlab or Fiji GaussFitOnSpot routines (Subheading 3.2.1).
9. Calculate the intensity ratio (i.e., fit parameter "A") between 100 nm FluoBeads and Alexa 488 hydrazide molecules. A value of around 100 is expected (*see* **Note 9**).

3.3 Imaging of Fluid-Phase Endocytosis

Fluid-phase endocytosis, i.e., the uptake of soluble compounds or molecules from fluid-phase surrounding a cell can be measured using two approaches. In the first case, the fluid phase is labeled with bright fluorescent dye molecules, which lack any apparent affinity to bind to the cell wall or plasma membrane and are small enough to be enclosed by a forming vesicle. The number of molecules within the final vesicle can be used to estimate its size.

In the second case, the fluid phase is loaded with fluorescent beads. Again, in the absence of an apparent affinity to bind and immobilize on cell wall or plasma membrane components, beads can be used to probe for the dimensions of soluble components, which are taken up cells.

3.3.1 Endocytosis of Alexa 488 Hydrazide

1. To record fluid-phase endocytosis on a confocal setup in intact *Vicia faba* guard cells, exchange the standard buffer with buffer containing 3–5 mM Alexa 488 hydrazide.

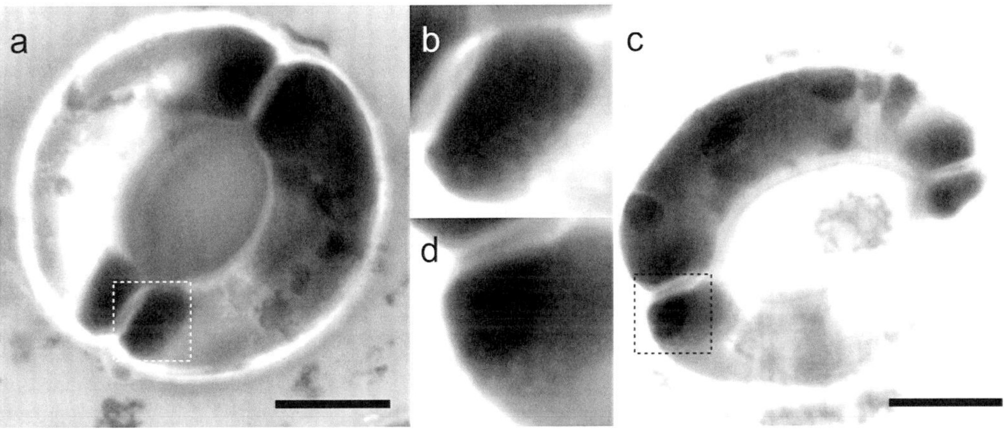

Fig. 1 Fluid-phase endocytosis in guard cells. (**a**, **c**) Alexa 488 hydrazide-stained endocytic vesicles are visible in the cortical cytoplasm of *Vicia faba* guard cells. Scale bars = 20 μm. (**b**, **d**) Higher magnification views of the areas marked by *boxes* in (**a**) and (**c**), respectively; single endocytic structures are clearly visible. Note that the brightest signal of Alexa 488 hydrazide remains close to the plasma membrane, i.e., within the cell wall, regardless whether cells have been washed with dye free buffer (as in **a**) or not (as in **b**)

2. Choose appropriate filter settings. The dye should be excited at 488 nm and fluorescence emission should be detected at 500–550 nm. Settings for the detection of FITC or GFP work well.
3. Endocytic vesicles should appear within 15 min. On the confocal setup, record a single image, not a *z*-stack, with pixel sizes of around 80 nm (*see* Fig. 1).
4. Repeat the measurement on different epidermal peels using decreasing concentrations of Alexa 488 hydrazide.
5. The lowest concentration able to load endocytic vesicles with a sufficient amount of dye to produce contrast in your recordings is needed for further calculations (*see* Subheading 3.4.1).
6. Record a sample of fluorescent beads (*see* Subheading 3.3.2) with the same imaging settings as applied in the last step.

3.3.2 Endocytosis of Fluorescent Beads

1. Use fluorescent beads as a 1:1,000 v/v dilution (*see* **Note 10**). For each sample type (guard cells, protoplasts), use corresponding image buffer to prepare the bead dilution.
2. Sonicate the bead solution for at least 10 min at room temperature directly before imaging to avoid agglomeration.
3. Add the fluorescent bead dilution to plant specimen of interest with a final concentration of 1:10,000 (*see* **Note 11**). Place on slides with imaging spacers.
4. Use a 488 nm line of a 25 mW argon laser for fluorescent beads excitation.

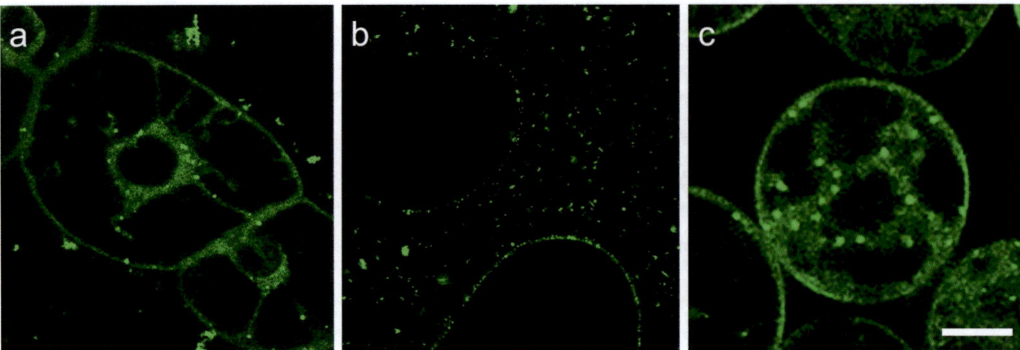

Fig. 2 Endocytosis of fluorescent beads by intact tobacco BY-2 cells and protoplasts. (**a**) Fluorescent images of BY-2 cells 15 min after addition of 40 nm beads. (**b**) Fluorescent images of BY-2 cells 15 min after addition of 100 nm beads. Fluorescent beads accumulate on the cell wall but are excluded from endocytic uptake. (**c**) Fluorescent images of BY-2 protoplasts 15 min after addition of 200 nm beads. Protoplasts are able to internalize beads up to 1,000 nm in diameter. Scale bar = 10 μm

5. Detect fluorescent beads at 505–580 nm.
6. For time series of fluorescent beads internalization, take an image every 5–10 min over a time period of 60 min.
7. Internalization of fluorescent beads should be clearly visible after 15 min (*see* Fig. 2) and the fluorescence should be detectable in bright spots and as a diffuse distribution in the cytosol.
8. For labeling of fluid-phase endocytosis in intact cells use fluorescent beads with a diameter smaller than 100 nm (*see* Fig. 2b); bigger beads will not pass through the cell wall [24].
9. For the detection of fluid-phase endocytosis in protoplasts bigger beads can be used (*see* Fig. 2c). BY-2 protoplasts are able to take up beads up to 1,000 nm in diameter by endocytosis [24]. Note that in protoplasts the bead diameter also reflects the pore size of endocytic vesicle.

3.4 Image Analysis

3.4.1 Analysis of Vesicle Intensities

1. Analyze the intensities of the fluorescent beads using the Matlab or Fiji GaussFitOnSpot routines (Subheading 3.2.1, Fig. 3)
2. Analyze the intensities of the vesicles stained with Alexa 488 hydrazide. Due to the high and uneven background in these recordings, we recommend using the Gaussian function with the tilted base (i.e., type GaussFitOnSpot('tiltedbase') when using the Matlab Script.
3. Calculate the number of Alexa 488 hydrazide molecules internalized by a given vesicle, by first relating the vesicle brightness analyzed in **step 2** to the brightness of the fluorescent beads quantified in **step 1**. Then, multiply the result with the intensity

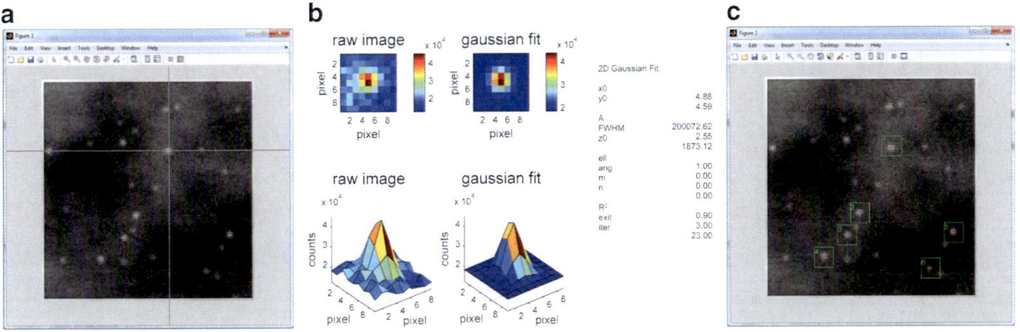

Fig. 3 Screenshots of the analysis of fluorescent spots with the Matlab-Script GaussFitOnSpot.m (**a**) After opening the image, a spot can be selected with a cross-hair cursor. (**b**) At the position selected in (**a**), a Gaussian fit is initiated. The raw image (*upper left*), the Gaussian function resulting from the fit (*upper right*), as well as the fit parameters are shown (text field to the *right*). In addition, 2D surface graphs of the raw image (*lower left*) and the Gaussian function (*lower right*) are shown. (**c**) After each round of spot selection and fitting, fit areas and fitted centers are displayed for each performed fit

ratio between FluoBeads and single Alexa 488 hydrazide molecules from Subheading 3.2.2.

4. Under the assumption that vesicles are perfect spheres and the thickness of a vesicles membrane is $p=5$ nm (i.e., $5e{-}9$ m), the outer diameter of the vesicle (d) can be calculated from the number of Alexa 488 hydrazide molecules per vesicle (N) and the concentration of the dye solution (c):

$$d = 2\sqrt[3]{\frac{N}{1000cN_A\frac{4}{3}\pi}}$$

Example: A vesicle, which contains 432 molecules of a 4 mM dye solution has an outer diameter of 80 nm (*see* **Note 12**).

3.4.2 Quantification of Total Fluid-Phase Uptake

For quantification of total uptake the ratio of intracellular versus whole cell fluorescence can be estimated and plotted over time. This analysis provides an easy tool to compare different conditions (e.g., nano bead uptake in the presence of endocytic inhibitors in comparison to control conditions.). For better comparability the presentation of the ratio per area is a good choice.

1. Open Picture in Fiji.
2. Select > Analyze > Set Measurements set checkmarks on Area and Integrated density.
3. Use polygon selection to set the first ROI (Fig. 4a). Mark the outline of the cell to measure the whole cell fluorescence (*see* **Note 13**).

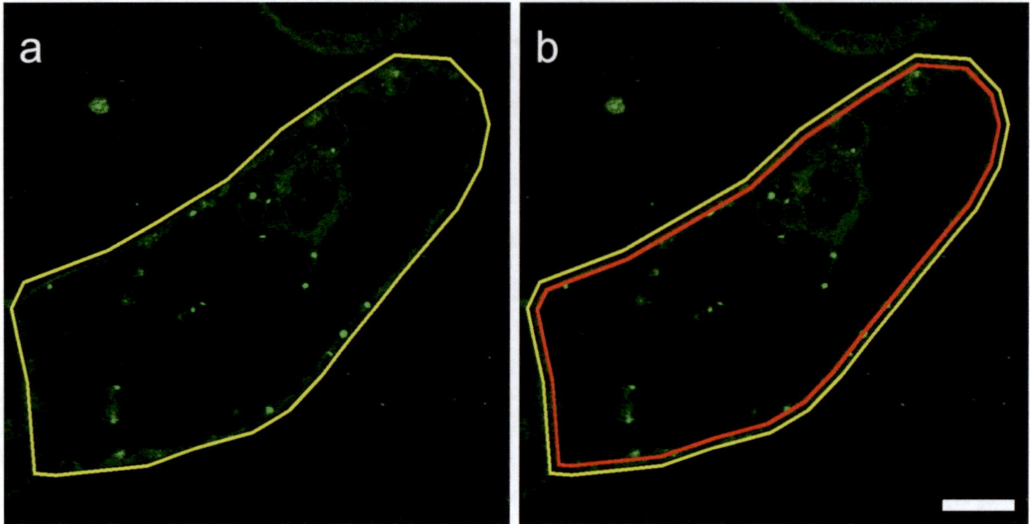

Fig. 4 Quantification of total uptake of 40 nm fluorescent beads using subroutine of Fiji. (**a**) ROI 1 (*yellow*) surrounds the outline of the cell. (**b**) ROI 2 (*red*) surrounds the cytosol of cell just beneath the plasma membrane. Scale bar = 10 μm

4. Select Analyze > Measure or use the shortcut *Str m*.
5. The outcome for ROI 1 appear in an new window (Results).
6. Select Edit > Selection > Enlarge. Type in negative number (e.g., −15 Pixel) to decrease ROI which than should mark the cell just beneath the plasma membrane (ROI 2, Fig. 4b).
7. Repeat **step 4**.
8. Copy the data from the result window and paste in a spreadsheet. Calculate the ratio of relative intracellular fluorescence intensity (integrated density of ROI2) versus relative whole cell fluorescence intensity (integrated density of ROI2) per area.
9. For time series experiments repeat the image analysis for each time point and plot the ratio of intracellular versus whole cell fluorescence against time.

4 Notes

1. Incubate digestion medium for 10 min at 40 °C to minimize proteolytic activity.
2. We recommend SciDaVis, a powerful, free, open source, and cross platform software for scientific data analysis and visualization (http://sourceforge.net/projects/scidavis/).

3. For centrifugation of protoplasts use a swing out rotor centrifuge and reduce break strength to a minimum if possible.
4. Protoplasts can be stored in the fridge for 24 h. After that time period fresh protoplasts have to be isolated from intact cells.
5. Force the release of guard cell protoplasts by lightly tapping the 35 mm dish on the table.
6. An objectives off-axis image quality is degraded due to chromatic and spheric aberrations.
7. The value reflects the lateral or axial resolution, respectively, under the conditions the specimen was in at the time the images were recorded. It includes the contributions of *all* factors influencing the final resolution and image quality, i.e., all irregularities of the objective used, the refractive properties of the specimen, the settings of the confocal pinhole and its (mis-)alignment, to just name the most important factors.
8. This method is also compatible with isolated fluorescent proteins, which allows to estimate their brightness in relation to fluorescent beads.
9. As different batches of FluoSpheres may vary in brightness, we cannot give an exact value here.
10. Do not store fluorescent beads in the freezer as this will promote aggregates and leakage of fluorescent dye from the beads.
11. Fluorescent beads should be freshly prepared every 2 weeks.
12. The size of a small fluorescent structure, which is determined by the Gaussian fit (i.e., the FWHM) reports the lateral resolution of the microscope, which is limited by diffraction. Signals from fluorescent beads with diameters of 200 nm, single molecules and most endocytic vesicles should cause fluorescent signals with the same diffraction limited size. The size calculation via the intensity, however, is not limited by diffraction, as it exploits the information provided by the intensity, i.e., the number of molecules, taken up by a given endocytic vesicle from a solution with a known and constant dye concentration.
13. Quantification of total uptake is easier to handle if the plasma membrane is clearly visible. To label plasma membranes add 10 μM FM4-64 to your cells (Excitation of FM4-64 with 488 nm line of a 25 mW argon laser detection at 630–700 nm).
14. Matlab Script GaussFitOnSpot.m:

```
function spots = GaussFitOnSpot(option,sis)
%% -------------------------------------------------
%  Matlab Skript to manually select spots in images
%  and fit them with a 2D Gaussian with a tilted base
%  to account for strong backround variations
%
%  1. run the skript with the following options
%       GaussFitOnSpot('standard')     --> circular Gaussfit with flat base
%       GaussFitOnSpot('ellipticity') --> elliptic Gaussfit with flat base
%       GaussFitOnSpot('tiltedbase')  --> circular Gaussfit with tilted base
%       GaussFitOnSpot('both')         --> elliptic Gaussfit with tilted base
%  2. choose tif image file
%  3. click the spot you want to fit
%  4. keep or discard the fitresults or end the session
%  5. find the fit results under "spots" in the workspace
%
%  The output "spots" contains the results of the fit parameters:
%       p(1)=x0    p(2)=y0    p(3)=A     p(4)=FWHM   p(5)=z0
%       p(6)=ell   p(7)=ang   p(8)=m     p(9)=n
%
%  6. Optionally, the size if the subimage, on which the gauss fit is
%     performed, can be chosen. Note however, that only uneven subimage
%     sizes are supported. Even values will be increased by 1
%     automatically:
%     EXAMPLE: GaussFitOnSpot('tiltedbase',11)
%
% 19 July 2013
% PD Dr. Tobias Meckel
% Membrane Dynamics (www.bio.tu-darmstadt.de/meckel)
% Department of Biology, Technische Universität Darmstadt
% -------------------------------------------------

%% checks inputs and set defaults
if nargin < 1                option = 'standard';
elseif nargin > 0 && strcmp(option,'standard') == 1;
elseif nargin > 0 && strcmp(option,'ellipticity') == 1;
elseif nargin > 0 && strcmp(option,'tiltedbase') == 1;
elseif nargin > 0 && strcmp(option,'both') == 1;
else error('Check input!'); end

if exist('sis','var') == 0; sis = 9; end
if rem(sis,2) == 0; sis = sis+1; end
%% get, read, and display image file
[FileName,PathName] = uigetfile('*.tif','select image');
img = imread(fullfile(PathName, FileName));

figure(1);
warning('off','images:initSize:adjustingMag');
imshow(img,[], 'InitialMagnification',1000);

%% cycle trough spots
choice = 'x';
n      = 0;
while strcmp(choice,'end') == 0;
    try close(2); catch end;

    figure(1);
    [xc,yc] = ginput(1);
    hold on;

    %% create subimage around selected spot with a size accorsing to subimgsize
    imgspot = img(round(yc-floor(sis/2)):round(yc+floor(sis/2)),...
        round(xc-floor(sis/2)):round(xc+floor(sis/2)));

    %% 2D gaussian fit
    options = optimset('Maxfunevals',1e5,'TolFun',1e-12, 'TolX',1e-10,'Display','off');

    % params          x0      y0     A    w     z0  ell   ang       m       n
    paramsLB   = [    1,      1,    10,  1,     1  0.5,    0, -1000, -1000];
    paramsInit = [sis/2, sis/2, 1000,  7, 3000,   1,    0,     0,     0];
    paramsUB   = [  sis,    sis,   1e7, 20, 2^16, 2.0,   90,  1000,  1000];

    [params,resnorm,residual,exitflag,output,lambda,jacobian] = ...
        lsqcurvefit(@gauss2Dfit_plane,paramsInit,double(imgspot),...
        double(imgspot),paramsLB,paramsUB,options);
    gaussfit = gauss2Dfit_plane(params,double(imgspot));

    SSE = norm(residual)^2;        % sum of squared errors
    SST = norm(double(imgspot) - mean(mean(imgspot)))^2; % total sum of squares;
    Rsq = 1 - SSE ./ SST;          % see regstats for more statistics

    %% plot fit result
    titlefontsize = 16; axeslabelfontsize  = 12; axisnumberfontsize = 10;

    figure(2);
    xmin = 1; xmax = size(imgspot,1);
    ymin = 1; ymax = size(imgspot,2);
    zmin = min(min([imgspot; gaussfit]));
    zmax = max(max([imgspot; gaussfit]));
```

```
    subplot(2,3,1)
    imagesc(imgspot);
    colormap('jet'); colorbar;
    axis('equal','tight');
    title('raw image','FontSize',titlefontsize);
    xlabel('pixel','FontSize',axeslabelfontsize);
    ylabel('pixel','FontSize',axeslabelfontsize);
    set(gca,'FontSize',axisnumberfontsize);
    hold on;

    subplot(2,3,2)
    imagesc(gaussfit); colorbar;
    axis('equal','tight');
    title('gaussian fit','FontSize',titlefontsize);
    xlabel('pixel','FontSize',axeslabelfontsize);
    ylabel('pixel','FontSize',axeslabelfontsize);
    set(gca,'FontSize',axisnumberfontsize);

    subplot(2,3,4)
    surf(double(imgspot))
    axis([xmin xmax ymin ymax zmin zmax]);
    set(gca,'YDir','reverse');
    title('raw image','FontSize',titlefontsize);
    xlabel('pixel','FontSize',axeslabelfontsize);
    ylabel('pixel','FontSize',axeslabelfontsize);
    zlabel('counts','FontSize',axeslabelfontsize);
    set(gca,'FontSize',axisnumberfontsize);
    h = axis;

    subplot(2,3,5)
    surf(gaussfit)
    axis([xmin xmax ymin ymax zmin zmax]);
    axis(h); set(gca,'YDir','reverse');
    title('gaussian fit','FontSize',titlefontsize);
    xlabel('pixel','FontSize',axeslabelfontsize);
    ylabel('pixel','FontSize',axeslabelfontsize);
    zlabel('counts','FontSize',axeslabelfontsize);
    set(gca,'FontSize',axisnumberfontsize);

    subplot(2,3,[3;6])
    info = {...
        sprintf('2D Gaussian Fit: \n\n')...
        ,sprintf('x0   \t\t %0.2f    ',params(1))...
        ,sprintf('y0   \t\t %0.2f \n ',params(2))...
        ,sprintf('A    \t\t %0.2f    ',params(3))...
        ,sprintf('w    \t\t %0.2f    ',params(4))...
        ,sprintf('z0   \t\t %0.2f \n ',params(5))...
        ,sprintf('ell  \t\t %0.2f    ',params(6))...
        ,sprintf('ang  \t\t %0.2f    ',params(7))...
        ,sprintf('m    \t\t %0.2f    ',params(8))...
        ,sprintf('n    \t\t %0.2f \n ',params(9))...
        ,sprintf('R^2  \t\t %0.2f    ',Rsq)...
        ,sprintf('exit \t\t %0.2f    ',exitflag)...
        ,sprintf('iter \t\t %0.2f \n ',output.iterations)...
        ,sprintf(output.message)};
    annotation('textbox',get(gca,'Position'),'String',info,'LineStyle','none');
    axis off;
    hold off;

    %% dialog
    choice = questdlg('What do you want to do with these results?', ...
        'Options', ...
        'keep','discard','end','keep');
    switch choice
        case 'keep'
            % write fit results to the output matrix 'spots'
            n = n + 1;
            spots(n,1)      = round(xc)-sis/2 -0.5 + params(1);
            spots(n,2)      = round(yc)-sis/2 -0.5 + params(2);
            spots(n,3:9)    = params(3:9);
            spots(n,10)     = Rsq;
            spots(n,11)     = exitflag;
            spots(n,12)     = output.iterations;

            % display fit result and position of subimage on choosen spot(s)
            figure(1)
            plot(spots(n,1), spots(n,2), 'r.');
            rectangle('Position',[round(xc)-sis/2,round(yc)-sis/2,sis,sis],...
                'LineWidth',1,'EdgeColor','g')
            text(round(xc)-sis/2+0.2,round(yc)-sis/2+0.2,num2str(n),...
                'FontSize',12, 'VerticalAlignment','top', 'Color','green');
            hold on;
            set(gca,'ydir','reverse');
        case 'end'
            close(2)
            break
    end
end
```

```
%% GaussianFit function
    function f = gauss2Dfit_plane(p,mat)
        if     strcmp(option, 'standard') == 1;
            p(6)=1; p(7)=0; p(8)=0; p(9)=0;
        elseif strcmp(option, 'ellipticity') == 1;
            p(8)=0; p(9)=0;
        elseif strcmp(option, 'tiltedbase') == 1;
            p(6)=1; p(7)=0;
        else
        end
        [maxX, maxY] = size(mat);
        p(7) = ((2*pi)/360)*p(7); % change degree to radians
        efac = 4*log(2)/p(4)^2;
        pfac = p(3)*efac/pi;
        xv = ones(maxY,1)*[1-p(1):maxX-p(1)];
        yv = [1-p(2):maxY-p(2)]'*ones(1,maxX);
        x  = -efac*((cos(p(7))*xv+yv*sin(p(7)))/p(6)).^2;
        y  = -efac*((cos(p(7))*yv-xv*sin(p(7)))*p(6)).^2;
        f  = pfac*exp(y+x)+p(5)+(p(8).*xv + p(9).*yv);
    end
end
```

References

1. Li R, Liu P, Wan Y et al (2012) A membrane microdomain-associated protein, Arabidopsis Flot1, is involved in a clathrin-independent endocytic pathway and is required for seedling development. Plant Cell 24:2105–2122
2. Doherty GJ, McMahon HT (2009) Mechanisms of endocytosis. Annu Rev Biochem 78: 857–902
3. Bandmann V, Homann U (2012) Clathrin-independent endocytosis contributes to uptake of glucose into BY-2 protoplasts. Plant J 70: 578–584
4. Wartenberg M, Hamann J, Pratsch I et al (1992) Osmotically induced fluid-phase uptake of fluorescent markers by protoplasts of Chenopodium album. Protoplasma 166:61–66
5. Diekmann W, Hedrich R, Raschke K et al (1993) Osmocytosis and vacuolar fragmentation in guard-cell protoplasts – their relevance to osmotically-induced volume changes in guard-cells. J Exp Bot 44:1569–1577
6. Low PS, Chandra S (1994) Endocytosis in plants. Annu Rev Plant Physiol Plant Mol Biol 45:609–631
7. Ballatori N, Hager DN, Nundy S et al (1999) Carrier-mediated uptake of lucifer yellow in skate and rat hepatocytes: a fluid-phase marker revisited. Am J Physiol Gastrointest Liver Physiol 277:G896–G904
8. Oparka K, Murant E, Wright K et al (1991) The drug probenecid inhibits the vacuolar accumulation of fluorescent anions in onion epidermal cells. J Cell Sci 99:557–563
9. Horn MA, Heinstein PF, Low PS (1992) Characterization of parameters influencing receptor-mediated endocytosis in cultured soybean cells. Plant Physiol 98:673–679
10. Gall L, Stan RC, Kress A et al (2010) Fluorescent detection of fluid phase endocytosis allows for in vivo estimation of endocytic vesicle sizes in plant cells with sub-diffraction accuracy. Traffic 11:548–559
11. Meckel T, Hurst AC, Thiel G et al (2004) Endocytosis against high turgor: intact guard cells of Vicia faba constitutively endocytose fluorescently labelled plasma membrane and GFP-tagged K+-channel KAT1. Plant J 39:182–193
12. Meckel T, Hurst AC, Thiel G et al (2005) Guard cells undergo constitutive and pressure-driven membrane turnover. Protoplasma 226:23–29
13. Bolte S, Talbot C, Boutte Y et al (2004) FM-dyes as experimental probes for dissecting vesicle trafficking in living plant cells. J Microsc 214:159–173
14. Etxeberria E, Gonzalez P, Baroja-Fernandez E et al (2006) Fluid phase endocytic uptake of artificial nano-spheres and fluorescent quantum dots by sycamore cultured cells: evidence for the distribution of solutes to different intracellular compartments. Plant Signal Behav 4:196–200
15. Lindhout BI, Meckel T, van der Zaal BJ (2010) Zinc finger-mediated live cell imaging in Arabidopsis roots. Methods Mol Biol 649: 383–398
16. Lindhout BI, Fransz P, Tessadori F, Meckel T et al (2007) Live cell imaging of repetitive DNA sequences via GFP-tagged polydactyl zinc finger proteins. Nucleic Acids Res 35(16): e107
17. Langhans M, Meckel T, Kress A et al (2012) ERES (ER exit sites) and the "secretory unit concept". J Microsc 247:48–59
18. Homann U, Thiel G (1999) Unitary exocytotic and endocytotic events in guard-cell

protoplasts during osmotically driven volume changes. FEBS Lett 460:495–499

19. Thiel G, Kreft M, Zorec R (2008) Unitary exocytotic and endocytotic events in Zea mays L. coleoptile protoplasts. Plant J 13:117–120
20. Weise R, Kreft M, Homann U et al (2000) Transient and permanent fusion of vesicles in Zea mays coleoptile protoplasts measured in the cell-attached configuration. J Membr Biol 174:15–20
21. Bandmann V, Kreft M, Homann U (2010) Modes of exocytotic and endocytotic events in tobacco BY-2 protoplasts. Mol Plant 4:241–251
22. Edelstein A, Amodaj N, Hoover K et al (2010) Computer control of microscopes using µManager. Curr Protoc Mol Biol Chapter: Unit14.20
23. Schindelin J, Arganda-Carreras I, Frise E et al (2012) Fiji: an open-source platform for biological-image analysis. Nat Methods 9:676–682
24. Bandmann V, Müller JD, Köhler T (2012) Uptake of fluorescent nano beads into BY2-cells involves clathrin-dependent and clathrin-independent endocytosis. FEBS Lett 586:3626–3632

Chapter 5

Immunogold Labeling and Electron Tomography of Plant Endosomes

Alexandra Chanoca and Marisa S. Otegui

Abstract

High-resolution imaging of endosomal compartments and associated organelles can be achieved using state-of-the-art electron microscopy techniques, such as the combination of cryofixation/freeze-substitution for sample processing and electron tomography for three-dimensional (3D) analysis. This chapter deals with the main steps associated with these imaging techniques: selection of samples suitable for studying plant endosomes, sample preparation by high-pressure freezing/freeze-substitution, and electron tomography of plastic sections. In addition, immunogold approaches for identification of subcellular localization of endosomal and cargo proteins are also discussed.

Key words Electron tomography, Endosomes, *Trans*-Golgi network, Multivesicular body, Cryofixation, Immunolabeling

1 Introduction

Endosomes are dynamic organelles that traffic both biosynthetic and endocytic cargo to the vacuole/lysosomes. In the context of their function in the endocytic pathway, animal endosomes are generally classified as early, recycling, and late endosomes (also called multivesicular bodies or MVBs). Early and recycling endosomes present tubule-vesicular profiles and sort and recycle endocytosed membrane proteins back to the plasma membrane and vacuolar cargo receptors back to the *Trans*-Golgi Network (TGN). MVBs sort membrane proteins into intraluminal vesicles to be degraded in vacuoles/lysosomes. In addition, MVBs also carry newly synthesized proteins from the Golgi to lysosomes/vacuoles.

Although plants can perform all these trafficking functions, few morphologically distinct endosomes have been clearly identified and characterized in plants: the TGN/TGN-derived compartments that act as early/recycling endosomes and the MVBs, also called prevacuolar compartments [1]. The typical tubulo-vesicular recycling endosomes found in animals have not been identified in plants.

Marisa S. Otegui (ed.), *Plant Endosomes: Methods and Protocols*, Methods in Molecular Biology, vol. 1209,
DOI 10.1007/978-1-4939-1420-3_5,

The plant TGN receives and recycle cargo from the endocytic pathway and sorts biosynthetic cargo destined either to the plasma membrane/cell wall/cell plate (glycoproteins, cell wall polysaccharides) [2–6] or to the vacuole (tonoplast and soluble vacuolar proteins) [7, 8]. The TGN derives from the trans-most Golgi cisterna by cisternal maturation and is eventually released from the Golgi stack as a free organelle. The TGN is easily recognized in samples prepared for transmission electron microscopy (TEM) by its abundant budding profiles and clathrin-coated forming vesicles [5].

Plasma membrane proteins internalized by endocytosis and targeted for vacuolar degradation are usually modified by ubiquitination. At the late endosomes or MVBs, ubiquitinated cargo proteins are recognized by the ESCRT (Endosomal Sorting Complex Required for Transport) machinery and sorted into intraluminal vesicles [1, 9]. Once endosomes produce intraluminal vesicles, they can be recognized morphologically as MVBs by transmission electron microscopy. Plant MVBs commonly range between 270 and 380 nm and their intraluminal vesicles are typically 35–37 nm in diameter [10].

The combination of cryofixation/freeze-substitution methods with electron tomography provides essential information about sorting events on endosomes that are beyond the resolution of conventional light microscopy. By combining superb cellular preservation, immunogold labeling, and three-dimensional (3D), high-resolution (7 nm axial resolution) imaging, it has been possible to analyze the maturation of MVBs in embryo cells [7] and define subdomains within the TGN [5]. TEM techniques have also made possible the quantitative analysis of structural changes in endosomes upon drug treatments or mutations in endosomal trafficking components [4, 10–12] and to estimate the efficiency in cargo sorting within intraluminal vesicles of endosomes [11]. In addition, the detection of vacuolar enzymes and their processed substrates with the MVB lumen in Arabidopsis cells have demonstrated that MVB are enzymatically active and not just passive prevacuolar carriers [7].

To obtain reliable 3D electron tomographic data, it is very important to work with very well preserved biological samples. Therefore, this chapter not only deals with the calculation and segmentation of electron tomograms but also with plant sample preparation by the best preservation method available, high-pressure freezing and freeze-substitution. Since it is also very important to be able to correlate structure with composition, a protocol for immunogold detection of proteins is also included.

2 Materials

2.1 Material for Plant Growth

1. Deionized autoclaved water.
2. Solution of 10 % commercial bleach in sterile water.
3. 70 % ethanol.
4. Sterile 1 mL pipette tips and automatic pipette.
5. *Arabidopsis thaliana* seeds.
6. *Arabidopsis thaliana* growth plates: 1 % agar plates containing 1/2 strength Murashige and Skoog (MS) basal medium (2.2 g of powder per L of medium).
7. Laminar flow hood.

2.2 High-Pressure Freezing

1. High pressure freezer (Leica EM HPM100 or Bal-tec/RMC/ABRA Fluid AG HPM 010).
2. Freezing brass planchettes ("hats") type B if a Bal-tec/RMC/ABRA Fluid AG HPM 010 is used.
3. Cryoprotectant: 0.1 M sucrose.
4. 2.0 mL Cryovials.
5. Fine point tweezers.
6. Stereomicroscope
7. Liquid nitrogen.

2.3 Freeze-Substitution and Resin Embedding

1. Freeze-substitution media for structural analysis: 2 % OsO_4 in anhydrous acetone. Prepare 1.5 mL aliquots in cryovials in hood using gloves. Can be stored in liquid nitrogen) *see* **Note 1**.
2. Freeze-substitution media for immunogold labeling: 1.5 mL 0.2 % glutaraldehyde plus 0.2 % uranyl acetate in anhydrous acetone. Prepare 1.5 mL aliquots in cryovials in hood using gloves. Aliquotes can be stored in liquid nitrogen *see* **Note 1**.
3. Automated freeze-substitution and low-temperature resin embedding/polymerization system (for example, AFS from Leica with or without a freeze-substitution processor).
4. Aluminum block with holes to fit cryovials.
5. Dry ice and Styrofoam box.
6. Epoxy-based resin. For example, Eponate 12 kit or Embed 812 kit.
7. Methacrylate-based resin: Lowicryl HM20 resin kit.
8. Glass Pasteur pipettes.
9. Silicone rubber flat embedding molds.
10. Coverwell® silicone imaging chambers (2.8 mm deep; 20 mm diameter).

11. Glass slides and coverslip.
12. Hypodermic needles.
13. Jeweler saw and blade.
14. Super glue.
15. Plastic mounting cylinders; 8 mm diameter × 13 mm length.

2.4 Preparation of Sections for Electron Tomography

1. Copper/rhodium slot grids coated with 0.7–1 % (w/v) formvar in ethylene dichloride.
2. Ultramicrotome.
3. Glass knife maker.
4. Ultramicrotomy and histology glass strips for making glass knifes.
5. Ultra 45° diamond knife.
6. 2 % uranyl acetate on 70 % methanol.
7. Reynold's lead citrate (2.6 % lead nitrate and 3.5 % sodium citrate, pH 12).
8. 10- or 15-nm colloidal gold particles (store at 4 °C).
9. High vacuum carbon evaporator (for instance DV-502 High Vacuum Evaporation System, Denton).

2.5 Image Acquisition and Calculation of Dual-Axis Electron Tomograms

1. Intermediate-voltage (300 kV) electron microscope (for example FEI Tecnai G2 30 TWIN) equipped with high-tilt rod for tomographic image acquisition.
2. Software: SerialEM [13–15] for image acquisition and IMOD [16] package for tomogram reconstruction (http://bio3d.colorado.edu/docs/software.html) (*see* **Note 2**).

2.6 Image Segmentation

1. IMOD package [16] (http://bio3d.colorado.edu/docs/software.html) (*see* **Note 2**).

2.7 Immunolabeling

1. Nickel single slot grids coated with 0.25–0.5 % (w/v) Formvar in ethylene dichloride.
2. Non-magnetic fine tip tweezers.
3. 10× phosphate-buffered saline (PBS) stock solution: 1.76 g of NaH_2PO_4; 11.49 g of Na_2HPO_4, 85 g sodium chloride in 1 L of distilled water, pH 6.8 (store at room temperature).
4. PBS-T-0.1 %: add 10 μL of Tween-20 to 10 mL 1× PBS.
5. PBS-T-0.5 %: add 0.5 mL of Tween-20 to 1 L 1× PBS.
6. Blocking buffer: 5 % (w/v) nonfat milk in PBS-T-0.1 %.
7. Primary antibody solution: dilutions between 1:10 and 1:50 in blocking buffer.
8. Secondary antibody conjugated to gold particles (5, 10, or 15 nm in diameter) diluted (1:10) in blocking buffer.

3 Methods

3.1 Plant Material

3.1.1 Arabidopsis Seedlings

1. Root cells show very little autofluorescence and therefore, they are often used for light microscopy imaging of endocytic dyes and endosomal markers. Root tips are also very amenable to high pressure freezing/ freeze-substitution. To obtain root tips from 5 to 7 day-old seedlings, seeds are germinated on 1 % agar plates containing 1/2 strength MS basal medium (*see* **Note 3**).
2. Place approximately 20 μL of *Arabidopsis* seeds in microcentrifuge tube. Add 10 % bleach, mix by inverting, and incubate for 5 min.
3. Spin the seeds down briefly and remove bleach solution using sterile glass Pasteur pipette or sterile 1 mL pipette tips. Rinse the seeds three times with sterile water.
4. Add 70 % ethanol, mix, and incubate for 5 min.
5. Spin the seeds down briefly and remove ethanol. Rinse three times with sterile water and place the seeds on 1 % agar plates supplemented with 1/2 strength MS.

3.1.2 Developing Embryos

Developing embryos accumulate large amounts of vacuolar storage protein during the late stages of embryo development. These storage proteins traffic through MVBs, making embryos a useful system for analyzing MVB function. Storage proteins are easily detected by electron microscopy at the late torpedo-bent cotyledon stages (approximately 11–15 days after pollination).

3.2 High-Pressure Freezing

1. 1-mm segments of root tips or excised developing embryos are loaded into a type B freezing planchette containing 0.1 M sucrose (*see* **Note 4**).
2. A second freezing planchette is placed on top, flat side down, to close the chamber. To avoid air bubbles, it is important to completely fill the chamber with 0.1 M sucrose.
3. Place freezing planchettes in the sample holder tip and high-pressure freeze them in a HPM 010 unit (*see* **Note 5**).
4. Under liquid nitrogen, pry open the two freezing planchettes with the tips of a pair of forceps that have been precooled in liquid nitrogen. The freezing planchettes containing the samples can either be stored in cryovials containing liquid nitrogen (*see* **Note 6**) or placed directly in freeze-substitution medium. At this point, the samples can either be further processed for structural analysis/electron tomography (Subheading 3.3.1) or immunogold labeling (Subheading 3.3.2). In some cases, immunolabeled samples can also be used for electron tomography analysis [17].

3.3 Freeze-Substitution and Resin Embedding

The samples are then fixed under low temperatures in cryosubstituion media and embedded in resins. The freeze-substitution medium and resin should be chosen according to the kind of analysis to be performed. Good preservation and staining of membranes is achieved using 2 % OsO_4 in acetone during freeze-substitution, followed by epoxy-based resin embedding. However, OsO_4 and epoxy resins are not suitable for most immunolabeling approaches. Cryosubstitution in acetone without fixatives or low concentrations of glutaraldehyde (0.2 %) and uranyl acetate (0.2 %) [18] followed by embedding in methacrylate-based, low-temperature, UV-curing resins, such as Lowicryl HM20, is preferred for immunogold labeling applications.

3.3.1 Freeze-Substitution in OsO_4

1. Transfer freezing planchettes with frozen samples to cryovials containing 1.5 mL of 2 % OsO_4 in acetone. Be sure to keep cryovials in liquid nitrogen during planchette transferring and to precool the tip of the tweezers in liquid nitrogen before touching the freezing planchettes.
2. Place the cryovials in an aluminum block inside a Styrofoam box containing dry ice for 3–5 days. The aluminum block should be precooled in dry ice (at −80 °C). Refill the box with dry ice if necessary.
3. Gradually bring the temperature of the samples to room-temperature in the following steps: Transfer aluminum block with cryovials to a freezer at −20 °C for 24 h, then to a fridge at 4 °C for at least 3 h and finally transfer aluminum block with cryovials to the fume hood and leave it at room temperature for 1 h (*see* **Note 7**). The remaining steps should be performed in the fume hood.
4. Discard freeze-substitution medium and rinse the samples with fresh anhydrous acetone at least 5 times, every 5 min. Be careful not to accidentally discard the samples.
5. Remove freezing planchettes (freezing planchettes can be reused after sonication in methanol or acetone).
6. Rinse the samples again with fresh acetone.
7. Prepare the epoxy-based resin mix without accelerator according to manufacturer's instructions. It can be done in advance and kept at 4 °C for several days.
8. Prepare dilutions of 10, 25, 50, and 75 % of the epoxy-based resin without accelerator in acetone to be used for sample embedding.
9. Remove and discard most of the acetone from the cryovials, being careful not to remove the samples or dry them out. Add 1.5 mL of 10 % resin mix in acetone and incubate them in an angled slow agitation rotator at room temperature.

Repeat this step replacing with solutions of increasing concentrations of resin. Incubate the samples at each resin concentration for at least 4 h.

10. Incubate the sample in 100 % resin mix without accelerator for at least 8 h.
11. Prepare the epoxy-based resin mix with accelerator according to manufacturer's instructions (*see* **Note 8**).
12. Incubate the sample in 100 % resin mix with accelerator. After 12 h, replace the resin with fresh 100 % resin mix with accelerator and incubate for another 12 h.
13. Place the samples in silicone rubber flat embedding molds filled with fresh resin with accelerator. Gently separate the individual samples with a clean toothpick to avoid clustering. Polymerize for 24 h at 60 °C (*see* **Note 9**).
14. After polymerization, analyze the resin block with the help of a stereomicroscope. The samples will be dark due to osmication. Cut out resin pieces containing samples from resin block with a jeweler saw and mount them on plastic mounting cylinders with super glue.

3.3.2 Freeze-Substitution for Immunolabeling Using the Leica AFS

Cryosubsitution for immunolabeling can also be performed using dry ice as explained in Subheading 3.3.1. However, if a low-temperature, UV-curing resin is used, it is advisable to use an automated freeze-substitution device such as the Leica AFS. This device allows for a very precise control of the temperature during freeze-substitution and resin embedding. In addition, it includes a UV lamp that can be directly attached to the sample chamber for resin polymerization. The Leica AFS can also be combined with a freeze-substitution processor, which is an automated reagent-handling system that eliminates the need to change solutions manually. In this procedure, the freeze-substitution and the resin embedding and polymerization are completely carried out inside the sample chamber of the AFS. All the tools (glass pipettes, tweezers) have to be precooled to the sample chamber temperature before getting in contact with samples or the media.

Before starting, set the program(s) in the Leica AFS; some suggested programs/steps are included in Table 1.

1. Fill the Leica AFS device with liquid nitrogen and start Program 0 (*see* Table 1). Wait for the machine to reach the initial temperature of –90 °C before proceeding.
2. In the hood and under liquid nitrogen, transfer freezing hats containing the samples into cryovials with freeze-substitution media for immunogold labeling (0.2 % glutaraldehyde plus 0.2 % uranyl acetate in anhydrous acetone) *see* **Note 5**.

Table 1
Suggested programs/steps to control temperature during freeze-substitution and resin embedding using a Leica AFS

Program/step	Temperature 1	Temperature 2	Ramp	Duration (h)
0	-90 °C	-90 °C	–	120
1	-90 °C	-60 °C	5 °C/h	54
2	-60 °C	-50 °C	5 °C/h	24
3	-50 °C/h	18 °C	5 °C/h	24

3. Place the cryovials containing the samples in the sample chamber of the Leica AFS. Let the freeze-substitution take place at -90 °C for 3–5 days *see* **Note 10**.
4. Change to Program 1 (*see* Table 1) and wait for the temperature to reach -60 °C.
5. With the sample chamber at -60 °C, precool a glass Pasteur pipette and acetone. Remove the cryosubstitution media from the vial and add precooled acetone, repeating this washing step 3 times (*see* **Note 11**).
6. Precool tweezers and carefully remove the freezing hats from the cryovials, leaving the samples free in anhydrous acetone.
7. In the hood, prepare HM20 resin mix according to manufacturer's instructions. Keep the resin in a dark bottle at -20 °C.
8. Prepare dilutions of 30 and 60 % HM20 resin mix in acetone to be used for sample embedding and keep them at -20 °C.
9. Precool the 30 % HM20 solution and a glass Pasteur pipette to -60 °C. Remove and discard the acetone from the cryovials being careful not to remove the samples. Add 30 % HM20 in acetone and incubate for at least 3 h. Repeat this step adding increasing concentrations (60 and 100 %) of precooled HM20 solutions.
10. Over the following 24 h, replace the resin at least three times with fresh, precooled 100 % resin to ensure proper infiltration.
11. Prepare the embedding chambers by attaching the Coverwell® Silicone Imaging chambers to a glass slide and label them. Precool embedding chamber and glass Pasteur pipette in the sample chamber.
12. Transfer samples from the cryovial to the embedding chambers, making sure not to leave bubbles (*see* **Note 12**). Cover the embedding chambers with a glass cover slip.
13. Connect the UV lamp attachment and start Program 2 followed by Program 3 (*see* Table 1). After 48 h under UV light, the resin should be fully polymerized.

14. Detach resin blocks with samples from embedding chamber. Label the resin blocks.
15. Analyze the resin block with the help of a stereomicroscope. Cut out samples from resin block with a jeweler saw and mount it on plastic mounting cylinders with super glue.

3.4 Preparation of Sections for Electron Tomography

In preparation to performing electron tomography, thin sections (60–70 nm thick) of samples embedded in epoxy-based resin are analyzed by regular transmission electron microscopy. It is important to assess the preservation of the sample, as well as define the areas/cells of interest. Appropriate specimens are further processed for electron tomography.

1. Using a diamond knife, section and collect 250- or 300-nm-tick sections on copper/rhodium slot grids coated with Formvar. The sections should be placed close to the center of the slot.
2. Stain the sections with 2 % uranyl acetate and Reynold's lead citrate for 10 and 5 min, respectively (*see* **Note 13**).
3. Apply 10 μL of 10- or 15-nm colloidal gold solution to each side of the sections for 5 min. Remove the excess solution by touching the grid with filter paper. Gold particles are used as fiducials, aiding in the fine alignment of the tilt images during the tomographic reconstruction.
4. Using a high vacuum carbon evaporator, carbon-coat both sides of the grids to reduce charging and drifting of the section during imaging under the electron beam.

3.5 Image Acquisition and Calculation of Dual Axis Electron Tomograms

The resolution of a tomographic reconstruction depends on different factors. Given that the quality of the tilt series (image focus and alignment) is optimal, the main factors affecting resolution are (a) the magnification at which the images are collected, (b) the angular interval, (c) the angular range, and (d) section thickness [19, 20]. Before collecting the images for calculating tomographic reconstructions, the variables involved in these four factors have to be carefully considered.

(a) *Magnification*: If large areas (several square microns) need to be analyzed, imaging at low magnification is not recommended because of the resulting loss in resolution. If a charge-couple device (CCD) camera is used to collect the images, the magnification should be high enough that each pixel in the image is $\leq$1 nm. If the area of interest cannot be imaged in a single frame, montaged images should be use to not compromise resolution [21].

(b) *The angular interval*: 1° angular interval is recommended.

(c) *Angular range*: Contrary to medical computed tomography in which it is possible to image a patient over a full 360° rotation, the angular range allowed by the conventional tilting specimen holders used for ET is much more restricted; this limitation results in a wedge of missing information between the maximal tilt angle collected and 90° [13, 22] and consequently, in distorted tomograms with anisotropic resolution [19]. To improve the isotropy in resolution, it is recommended to collect images from two orthogonal axes and combine the two resulting tomograms into a dual-axis tomogram [13].

(d) *Section thickness*: If an intermediate-voltage (200–300 kV) electron microscope is used, sections thicker than 300 nm will likely result in poor resolution images, particularly at high-tilt angles. Sections 300 nm in thickness do not usually contain whole MVBs and TGNs; larger volumes can be analyzed by obtaining serial tomograms from serial sections [23, 24]. A ribbon of serial sections is placed on the same slot grid and an area of interest is located and imaged on each relevant section in the ribbon. It is important to note that this approach suffers from a 15 to 25 nm gap of missing information between serial tomograms [25].

1. Place specimens in a high-tilt sample holder of an intermediate (300 kV) electron microscope and collect images at 1° angular intervals and over an angular range of ±60° to 70° using the free software SerialEM (*see* **Note 2**).
2. After collecting the first stack of images, rotate the grid 90° and collect images at 1° angular intervals along the second axis.
3. Process the resulting images using the *eTomo* program in the IMOD package (*see* **Note 2**).

3.6 Segmentation and Quantitative Analysis of Tomographic Reconstructions

Organelles contained in electron tomographic reconstructions can be manually segmented using the *3dmod* program of the IMOD package. Segmentation should result in graphic objects that accurately represent the 3D positions of features of interest in a tomogram. Image segmentation is the most time-consuming part of the process and it can somehow be a subjective task. The *3dmod* program allows the operator to draw on the image data, placing points, chosen shapes (circles, etc.), or sets of point (lines or curves that match the structure of interest) as "overlays" on the image data (Fig. 1). Each such representation is called a "contour," and these are generally drawn on a single tomographic slice extracted from the tomogram (Fig. 1). Contours can be either closed (if they represent the membrane that surrounds an MVB or TGN) or open (if they represent a linear structures, such as a microtubule or actin filament).

Three different types of image displays or "windows" are generally used during modeling: the Zap windows that shows tomographic slices parallel to the surface of the physical section

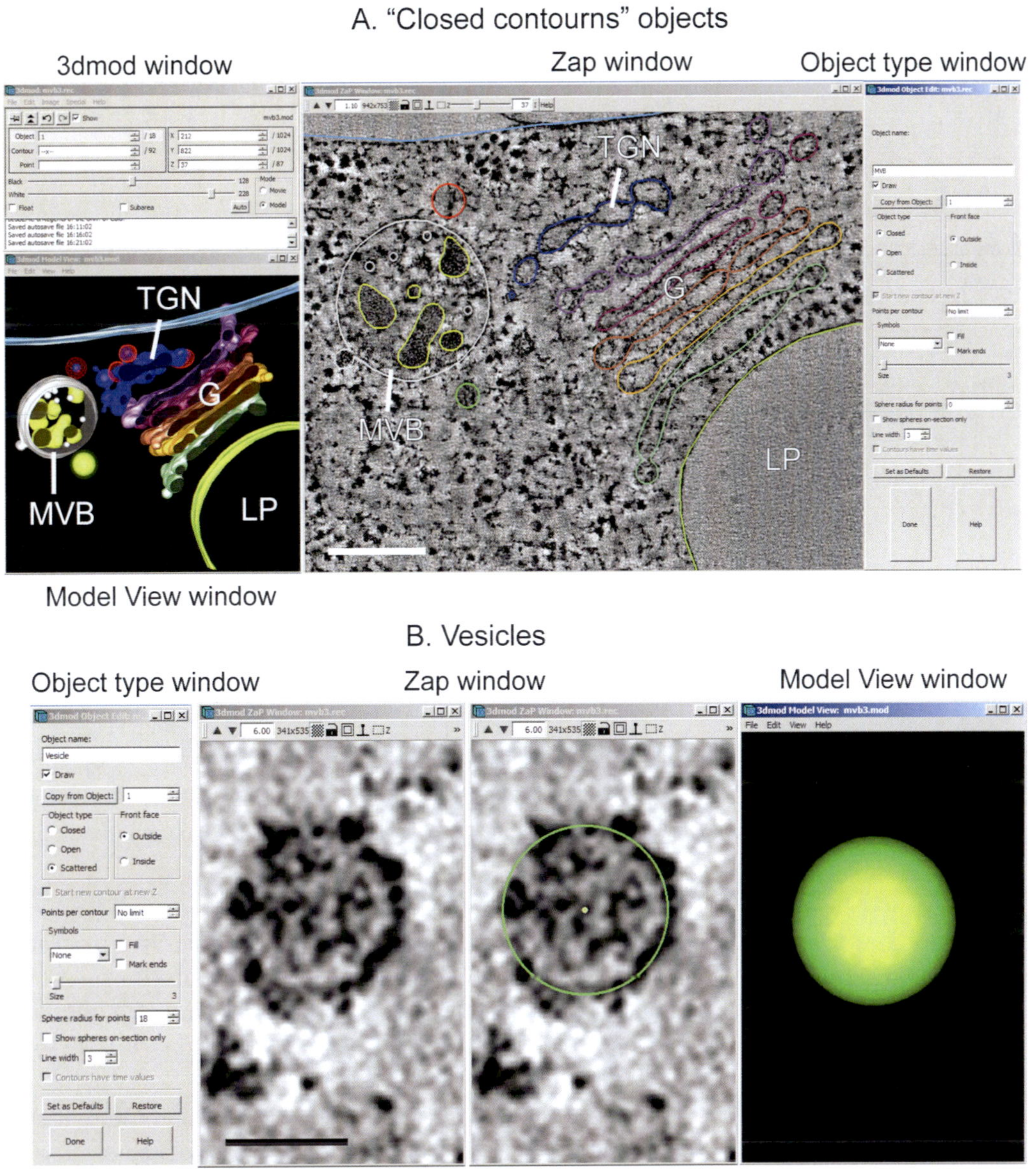

Fig. 1 Segmentation of membranous organelles and vesicles using *3dmod*. (**a**) Golgi (G), TGN, and MVB membranes are segmented as overlay contours in the Zap window using the "closed" object type option and rendered as a meshed 3D object in the Model View window. (**b**) Vesicles are rendered as spheres of variable radii. The tomographic reconstruction shown in (**a**) was obtained from Arabidopsis embryo cells; MVBs contain typical electron dense deposit of vacuolar storage proteins here segmented with yellow contours. *LB* lipid body. Scale bars = 200 nm (**a**), 50 nm (**b**)

that was reconstructed, the Slicer window that allows the operator to display a slice of a selected thickness and at an arbitrary angle through the volume, and the Model View window that shows the segmented objects and contours (Fig. 1). A general guide that provides a comprehensive description of the *3dmod* segmentation program can be found at http://bio3d.colorado.edu/IMOD.

Quantitative analysis on tomograms (see below) is based on the tomographic models that have been meshed, a process that generates 3D graphic objects with defined surfaces or "skins." It is important to keep in mind that meshes derive from contours, so it is critical to draw the contours accurately.

3.6.1 MVBs, TGN, and Membranous Organelles

Membrane-bound organelles should be segmented as separate closed objects. When segmenting membranes, it is recommended to place contours in the middle of the lipid bilayer [25, 26].

1. Create a new object under the "Edit" menu and select the "closed" option (Fig. 1a).
2. Draw in each slice (if the organelle is highly irregular in shape) or every 2–3 slices (for more regular shapes) the outline of the organelle by moving along the stack of tomographic slices (Fig. 1a).
3. Save model and run the *imodmesh* command. *imodmesh* offers a number of options for capping off objects, connecting contours in nonadjacent tomographic slices, etc. (*see* **Note 14**). For a complete reference of *imodmesh* options go to http://bio3d.colorado.edu/imod/doc/man/imodmesh.html.

3.6.2 Microtubules and Actin Filaments

In the simplest case, microtubules and actin filaments can be modeled as tubes of a given diameter.

1. Create a new object under the "Edit" menu of *3dmod* and chose the "open contours" option.
2. Move along the stack of tomographic slices using the Zap window, identify one end of the microtubule and place the first point, as close to the center of the microtubule as possible. Place more points every 5th or 6th tomographic slice along the microtubule length until reaching the other end of the microtubule.
3. It is also possible to use the "slicer" window to model microtubules [22]. Adjust the X, Y, Z sliders to find a tomographic slice that contains as long a segment as possible of the microtubule length and place points at the beginning and end of the segment. In either case, each microtubule/ filament should be considered a new contour within one object.
4. To obtain a 3D representation of the microtubule/filament, use the command *imodmesh* with the options -t (for tube), -d (the pixel diameter of the tube, 25 nm for a microtubule, 5–9 nm for an actin filament), and -E (this will "cap" the end of the tube). Complete instructions for use of the *imodmesh* command can be found at http://bio3d.colorado.edu/imod/doc/man/imodmesh.html (*see* **Note 14**). Be sure to save the model before running *imodmesh*.

3.6.3 Vesicles

A simple method to segment vesicles assumes that their shape approximately corresponds to a sphere. Vesicles can be modeled as "scattered points," where a sphere is computed from a single point placed in the center of the vesicle, and its size can be adjusted to match the diameter of the vesicle (Fig. 1b). All similar vesicles (for example, all clathrin-coated vesicles) can be contained in a single object.

1. Create a new object under the "Edit" menu and select the "scattered" option.
2. Identify the center of the vesicle by moving along the stack of tomographic slices in the Zap window.
3. Place a point in the center of the vesicle and define the radius by adjusting the "Sphere radius for points" option.
4. Spheres do not need to be meshed and can be directly displayed as 3D objects in the "Model" window (Fig. 1b).

3.6.4 Quantitative Analysis

One of the powerful advantages of ET is the possibility of performing quantitative analysis on tomographic models. One can analyze spatial relationships between organelles, membrane surface area and volume variations in endosomes, density and sizes of intraluminal vesicles in MVBs, etc. Measurements are generally expressed in terms of the pixel size described in the model header as a number in nanometers. The "thinning factor," which reports the extent to which the section thinned during image acquisition under the electron beam, is also described in the model header and is applied to the calculations. Therefore, for quantitative analysis of electron tomograms, it is very important to enter accurate values in the model header window (under the "Edit" menu).

Volume and Surface Area of Endosomal Compartments

The *imodinfo* command is used for this analysis. Information about volume and surface area of meshed, whole objects can be extracted directly by running this command with the option -s. A complete description of all options available for the *imodinfo* command can be found at http://bio3d.colorado.edu/imod/doc/man/imodinfo.html (*see* **Note 15**).

Density of Vesicles

1. All individual vesicles should be points in the same contour of the same object.
2. Extract defined size boxes/areas from the model and run the *imodinfo* command to obtain the number of points contained in the box volume.

3.7 Immunolabeling

One of the main challenges in TEM is to identify the biochemical identity/composition of cellular components. One approach to identify molecules in TEM is immunogold labeling. Immunogold protein detection on plastic sections can reveal critical information

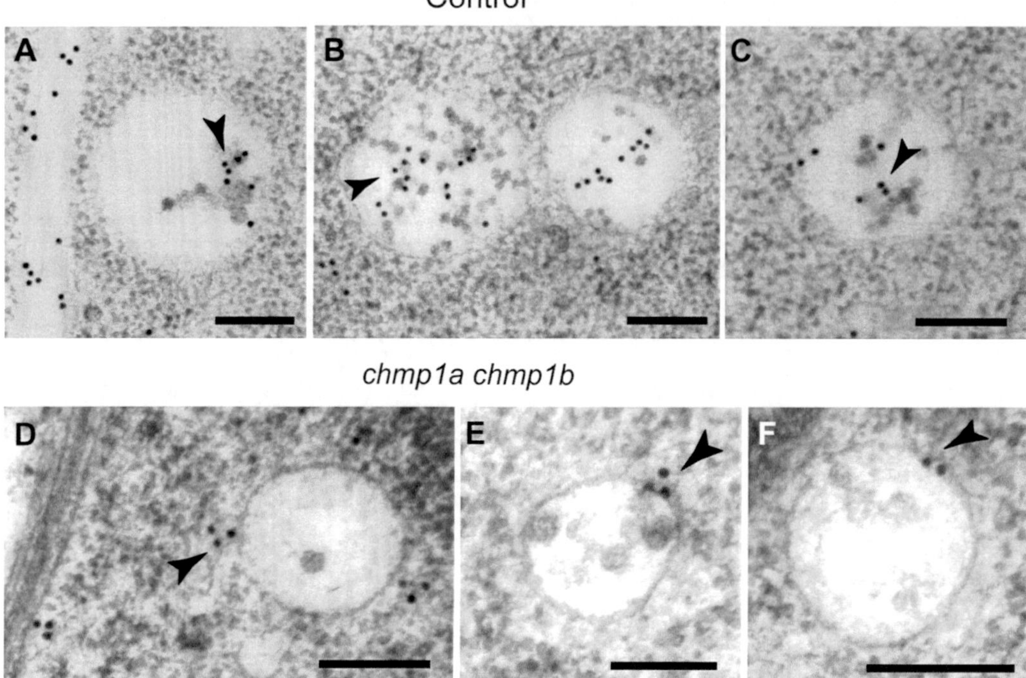

Fig. 2 Immunogold detection of PIN1-GFP using anti-GFP antibodies on HM20-embedded roots from control and *chmp1a chmp1b* (an ESCRT mutant) plants. Whereas heavy labeling (*arrowheads*) of PIN1-GFP is detected on intraluminal vesicles of control MVBs (**a–c**), the *chmp1a chmp1b* MVBs (**d–f**) fail to sort PIN1-GFP into intraluminal vesicles. Scales = 50 nm. Reproduced from [11] (Copyright © 2009 American Society of Plant Biologists)

about sorting functions of endosomes. For examples, PIN1-GFP, a plasma membrane protein sorted at MVBs for degradation, can be specifically detected on intraluminal vesicles (indicating normal MVB sorting) or only at the MVB limiting membrane (for example, in an ESCRT mutant that fails to internalize MVB cargo proteins; Fig. 2).

For electron tomography, immunogold labeling can be performed on the same section used for tomography reconstruction [17]. However, in this case, only the surface of the plastic section is labeled and no labeling information is available for the rest of the reconstructed tomographic volume.

1. Using a diamond knife, collect 60–70 nm thick sections on nickel single slot grids coated with formvar.
2. Distribute 15 μL drops of blocking buffer on a parafilm strip inside a glass petri dish. With non-magnetic fine tip tweezers, place the grids floating on top of the drops for 20 min (*see* **Note 16**).
3. Distribute 10 μL drops of primary antibody solution on the parafilm strip. Remove grids from blocking buffer; carefully

blot them with filter paper and place them on top of the primary antibody drops for 1 h (*see* **Note 17**).

4. Distribute 10 μL drops of secondary antibody solution on the parafilm strip.
5. Gently, rinse grids with a continuous stream of PBS-T-0.5 % for 1 min. Blot grids with filter paper to remove excess of liquid, and place them in the secondary antibody drops for 1 h.
6. Rinse grids with a continuous stream of PBS-T-0.5 % for 1 min followed by rinsing with distilled water.
7. Post-stained sections with uranyl acetate and Reynold's lead citrate for 10 and 5 min, respectively, and image in a transmission electron microscope.

4 Notes

1. Osmium tetroxide, glutaraldehyde, and uranyl acetate are toxic and should be handled with appropriate caution. Always manipulate these chemicals in a fume hood, read the material datasheet, and comply with the suggested personnel protective equipment.
2. SerialEM is a free software for image acquisition (http://bio3d.colorado.edu/SerialEM) that is compatible with FEI and JEOL electron microscopes. IMOD is a free software package that runs in both PC and Macintosh systems and was developed primarily by David Mastronarde, Rick Gaudette, Sue Held, and Jim Kremer at the Boulder Laboratory for 3D Electron Microscopy of Cells. It contains around 140 programs for image processing, tomogram calculation, image segmentation, display, and quantitative analysis.
3. Cell at the root tip are adequate for the structural analysis of plant endosomes by high-pressure freezing/freeze-substitution. However, the techniques described here can be used for virtually any plant tissue, with minor modifications.
4. Loading the freezing hats with samples is a delicate step. After the roots are excised, the samples have to be processed quickly but carefully to minimize tissue damage. It is advisable to practice loading the hats with samples beforehand.
5. Right after the samples are high-pressure frozen, quickly move the holder tip with the freezing hat to a styrofoam box containing liquid nitrogen. From this moment on, the samples should always be manipulated under liquid nitrogen, and all materials that will be in direct contact with them (tweezers, cryovials etc.) should always be precooled in liquid nitrogen. An increase in sample temperature could lead to the formation of ice crystals, damaging the cellular contents.

6. High-pressure frozen material can be stored in liquid nitrogen for months without suffering changes in cellular preservation.
7. OsO_4 in acetone is highly volatile. Even when the acetonic OsO_4 solution is kept in closed cryovials, osmication of objects around cryosubstitution vials can easily happen. If a freezer/fridge is used during cryosubstitution with OsO_4, it is advisable to have a freezer/fridge fully dedicated to this use to avoid contamination of other reagents and labware.
8. BDMA is recommended as the accelerator for the Eponate 12 resin mix because of its low viscosity.
9. Unpolymerized resin waste and resin-contaminated items should be placed in the oven for 24 h for polymerization before disposal.
10. During the freeze-substitution and resin-embedding process using the AFS it will be necessary to refill the liquid nitrogen tank of the machine. Be aware of the liquid nitrogen content and plan accordingly.
11. Although some automated freeze-substitution and low-temperature resin embedding systems are equipped with attached lights sources and/or stereomicroscopes, when performing washes and resin changes in the sample chamber your visibility will often be limited. To decrease the accidental loss of sample on those steps, do not remove more than half the volume of the cryovial at a time, and wait between washes so the samples can sediment to the bottom of the tube.
12. When transferring the samples from the cryovials to the embedding chambers, the samples might not be conspicuous. Since the infiltrated samples will tend to sediment to the bottom of the tube, discard the upper half of the cryovial content, take the bottom half with a glass Pasteur pipette and transfer it completely to the embedding chamber. Add more fresh resin if needed.
13. A methanolic solution of uranyl acetate is recommended because it increases the contrast of membranes more than aqueous solutions do.
14. *imodmesh* generates triangles that connect neighboring points within a contour and nearby points on adjacent contours. The resulting triangles represent a surface in space that is an excellent approximation to all the contour information, so they are used for all subsequent quantifications of area, volume, and distance. They can also be used to generate a shaded surface that provides a good visual representation of the modeled object.
15. *imodinfo* provides information about IMOD models, such as lists of objects, contours and point data, lengths and centroids of contours; and surface areas and volumes of objects or surfaces.

16. It is important that the sample section is in contact with the liquid. Float the grid on top of the antibody solution drop being sure that the grid side containing the section faces down.
17. Antibody binding can be improved with mild agitation. During the incubation with the primary and secondary antibody, carefully place the petri dish containing the grids on top of a magnetic stirrer set to low, so that the grids can gently rotate on top of the antibody solution drop.

Acknowledgements

This work was supported by NSF grant MCB1157824 to M.S.O.

References

1. Reyes FC, Buono R, Otegui MS (2011) Plant endosomal trafficking pathways. Curr Opin Plant Biol 14:666–673
2. Dettmer J, Hong-Hermesdorf A, Stierhof YD, Schumacher K (2006) Vacuolar H^+-ATPase activity is required for endocytic and secretory trafficking in *Arabidopsis*. Plant Cell 18:715–730
3. Lam SK, Siu CL, Hillmer S, Jang S, An G, Robinson DG, Jiang L (2007) Rice SCAMP1 defines clathrin-coated, trans-Golgi-located tubular-vesicular structures as an early endosome in tobacco BY-2 cells. Plant Cell 19:296–319
4. Viotti C, Bubeck J, Stierhof YD, Krebs M, Langhans M, van den Berg W, van Dongen W, Richter S, Geldner N, Takano J, Jurgens G, de Vries SC, Robinson DG, Schumacher K (2010) Endocytic and secretory traffic in *Arabidopsis* merge in the trans-Golgi network/early endosome, an independent and highly dynamic organelle. Plant Cell 22:1344–1357
5. Kang BH, Nielsen E, Preuss ML, Mastronarde D, Staehelin LA (2011) Electron tomography of RabA4b- and PI-4Kbeta1-labeled trans Golgi network compartments in Arabidopsis. Traffic 12:313–329
6. Chow C-M, Neto H, Foucart C, Moore I (2008) Rab-A2 and Rab-A3 GTPases define a trans-Golgi endosomal membrane domain in *Arabidopsis* that contributes substantially to the cell plate. Plant Cell 20:101–123
7. Otegui MS, Herder R, Schulze J, Jung R, Staehelin LA (2006) The proteolytic processing of seed storage proteins in *Arabidopsis* embryo cells starts in the multivesicular bodies. Plant Cell 18:2567–2581
8. Hanton S, Matheson L, Chatre L, Rossi M, Brandizzi F (2007) Post-Golgi protein traffic in the plant secretory pathway. Plant Cell Rep 26:1431–1438
9. Henne William M, Buchkovich Nicholas J, Emr Scott D (2011) The ESCRT pathway. Developmental Cell 21:77–91
10. Haas TJ, Sliwinski MK, Martínez DE, Preuss M, Ebine K, Ueda T, Nielsen E, Odorizzi G, Otegui MS (2007) The *Arabidopsis* AAA ATPase SKD1 is involved in multivesicular endosome function and interacts with its positive regulator LYST-INTERACTING PROTEIN5. Plant Cell 19:1295–1312
11. Spitzer C, Reyes FC, Buono R, Sliwinski MK, Haas TJ, Otegui MS (2009) The ESCRT-related CHMP1A and B proteins mediate multivesicular body sorting of auxin carriers in *Arabidopsis* and are required for plant development. Plant Cell 21:749–766
12. Scheuring D, Viotti C, Kruger F, Kunzl F, Sturm S, Bubeck J, Hillmer S, Frigerio L, Robinson DG, Pimpl P, Schumacher K (2011) Multivesicular bodies mature from the trans-Golgi network/early endosome in Arabidopsis. Plant Cell 23:3463–3481
13. Mastronarde DN (1997) Dual-axis tomography: an approach with alignment methods that preserve resolution. J Struct Biol 120:343–352
14. Mastronarde DN (2005) Automated electron microscope tomography using robust prediction of specimen movements. J Struct Biol 152:36–51
15. Mastronarde DN (2008) Correction for non-perpendicularity of beam and tilt axis in tomographic reconstructions with the IMOD package. J Microsc 230:212–217
16. Kremer JR, Mastronarde DN, McIntosh JR (1996) Computer visualization of three-dimensional image data using IMOD. J Struct Biol 116:71–76
17. Donohoe BS, Kang BH, Staehelin LA (2007) Identification and characterization of COPIa- and

COPIb-type vesicle classes associated with plant and algal Golgi. Proc Natl Acad Sci U S A 104:163–168

18. Giddings TH (2003) Freeze-substitution protocols for improved visualization of membranes in high-pressure frozen samples. J Microsc 212: 53–61
19. McEwen BF, Frank J (2001) Electron tomographic and other approaches for imaging molecular machines. Curr Opin Neurobiol 11: 594–600
20. Marsh BJ (2005) Lessons from tomographic studies of the mammalian Golgi. Biochim Biophys Acta 1744:273–292
21. Otegui MS, Austin JR II (2007) Visualization of membrane-cytoskeletal interactions during plant cytokinesis. Methods Cell Biol 79:221–240
22. O'Toole ET, Giddings TH, JR., Dutcher SK (2007) Understanding microtubule organizing centers by comparing mutant and wild-type structures with electron tomography. In: McIntosh JR (ed) Cellular Electron Tomography, vol 79. Methods Cell Biol, 2007/03/01 edn. Elsevier, pp 125–143
23. Otegui MS, Staehelin LA (2004) Electron tomographic analysis of post-meiotic cytokinesis during pollen development in *Arabidopsis thaliana*. Planta 218:501–515
24. Segui-Simarro JM, Austin JR II, White EA, Staehelin LA (2004) Electron tomographic analysis of somatic cell plate formation in meristematic cells of *Arabidopsis* preserved by high-pressure freezing. Plant Cell 16:836–856
25. Ladinsky MS, Mastronarde DN, McIntosh JR, Howell KE, Staehelin LA (1999) Golgi structure in three dimensions: functional insights from the normal rat kidney cell. J Cell Biol 144: 1135–1149
26. Otegui MS, Mastronarde DN, Kang BH, Bednarek SY, Staehelin LA (2001) Three-dimensional analysis of syncytial-type cell plates during endosperm cellularization visualized by high resolution electron tomography. Plant Cell 13:2033–2051

Chapter 6

Investigating Protein–Protein Interactions in the Plant Endomembrane System Using Multiphoton-Induced FRET-FLIM

Jennifer Schoberer and Stanley W. Botchway

Abstract

Real-time noninvasive fluorescence-based protein assays enable a direct access to study interactions in their natural environment and hence overcome the limitations of other methods that rely on invasive cell disruption techniques. The determination of Förster resonance energy transfer (FRET) by means of fluorescence lifetime imaging microscopy (FLIM) is currently the most advanced method to observe protein–protein interactions at nanometer resolution inside single living cells and in real-time. In the FRET-FLIM approach, the information gained using steady-state FRET between interacting proteins is considerably improved by monitoring changes in the excited-state lifetime of the donor fluorophore where its quenching in the presence of the acceptor is evidence for a direct physical interaction. The combination of confocal laser scanning microscopy with the sensitive advanced technique of time-correlated single photon counting allows the mapping of the spatial distribution of fluorescence lifetimes inside living cells on a pixel-by-pixel basis that is the same as the fluorescence image. Moreover, the use of multiphoton excitation particularly for plant cells provides further advantages such as reduced phototoxicity and photobleaching. In this protocol, we briefly describe the instrumentation and experimental design to study protein interactions within the plant endomembrane system, with a focus on the imaging of plant cells expressing fluorescent proteins and acquisition and analysis of fluorescence lifetime resolved data.

Key words Fluorescence imaging, Microscopy, Proteins, Excited-state lifetime, Multiphoton, TCSPC, FRET, FLIM, Golgi N-glycan processing enzymes

1 Introduction

Knowledge of protein–protein interactions is important to understand the complexity of cellular processes within the organism under study. The development of genetically encoded fluorescent proteins that can be expressed either transiently or stably in living cells, has revolutionized fluorescence-based imaging techniques that are now able to directly visualize protein functions in their natural habitat under physiological conditions. One of these techniques exploits the phenomenon of Förster or fluorescence resonance energy transfer

Marisa S. Otegui (ed.), *Plant Endosomes: Methods and Protocols*, Methods in Molecular Biology, vol. 1209,
DOI 10.1007/978-1-4939-1420-3_6,

(FRET), which was first described by Theodor Förster 50 years ago [1] and relies on the non-radiative energy transfer from an excited fluorescent donor molecule to a different non-excited fluorescent acceptor molecule in its vicinity. FRET only occurs when (1) the donor emission spectrum overlaps sufficiently with the acceptor absorption spectrum, (2) the donor and acceptor dipoles display a mutual molecular orientation, and (3) the donor and acceptor are in close proximity, typically in the range (R) of 2–10 nm. The kinetics of the fluorescence and FRET may be described as:

$$D + A \xrightarrow{h\upsilon} D^* + A \xrightarrow{k_T} D + A^*$$

where D and D* is the non-excited and excited donor, respectively, A and A* is the non-excited and excited acceptor, respectively, $h\upsilon$ is the overall photon energy (h is the Planck's constant and υ is the frequency), and k_T is the rate of energy transfer, which is calculated using Eq. 1:

$$k_T = \left(\frac{1}{\tau_D}\right)\left(\frac{R_0}{R}\right)^6 \quad (1)$$

where τ_D is the donor excited-state lifetime in the absence of the acceptor, R is the distance between D and A, and R_0 is the Förster radius. At the Förster radius, 50 % of the donor molecules will emit fluorescence while the rest will undergo energy transfer. Since the energy transfer process is strongly distance dependent with $1/R^6$, FRET can be used to measure distance and examine molecular interactions on a spatial scale. FRET has been widely utilized in cell biology to detect inter-molecular interactions in living cells by labeling two interaction partners with suitable fluorescent proteins and monitoring changes in the donor and/or acceptor fluorescence. The determination of FRET in the context of microscopy can either be achieved by intensity-based techniques such as fluorescence measurements of the donor (acceptor photobleaching, AP-FRET) or the acceptor (sensitized emission, SE-FRET) or by measurements of the anisotropy or fluorescence lifetime imaging microscopy (FRET-FLIM) [2].

In the FRET-FLIM approach, the information gained using steady-state FRET between interacting proteins is improved by monitoring changes in the excited-state lifetime of the donor. The fluorescence lifetime for a particular environment is fixed for every fluorescent molecule and defines the average time a molecule spends in the excited state before returning to the ground state, which is typically in the range of several nanoseconds (ns) for fluorescent proteins. The decay of donor fluorescence does not depend on its concentration, but is sensitive to events in its local microenvironment, such as FRET between two interacting proteins. Since FRET is a quenching process that depletes the excited state of the donor fluorophore, the donor fluorescence lifetime is shortened by

FRET, which is evidence for a direct physical interaction. The FRET-FLIM approach has successfully been used to study protein–protein interactions in a variety of animal and plant cells at the organelle level [3–12].

In our example, we use genetically encoded fluorescent proteins as FRET probes due to their wide applicability in live cell imaging. When selecting suitable donor and acceptor fluorophores for investigating protein–protein interactions, the conditions necessary for FRET as highlighted above should be considered. Furthermore, the fluorophores need to be monomeric and display a high photostability, high quantum yield and high extinction coefficient (brightness). In the context of FLIM, the donor fluorophore ideally exhibits a mono-exponential fluorescence lifetime decay. Monomeric GFP (mGFP), for example, fulfills these criteria and therefore is well suited as donor for FRET to a monomeric red fluorescent protein (mRFP, DsRed, or mCherry). The color palette of fluorescent proteins, however, is expanding rapidly and with it the selection of suitable FRET pairs with improved photophysical properties is increasing.

We describe here the use of a time-domain FLIM instrument based around a standard confocal laser scanning microscope combined with the highly sensitive advanced technique of time-correlated single photon counting (TCSPC) for the acquisition of fluorescence lifetime resolved data on a pixel-by-pixel basis (Fig. 1). We opted for a two-photon-excitation FLIM setup [4, 7, 8, 11, 12]

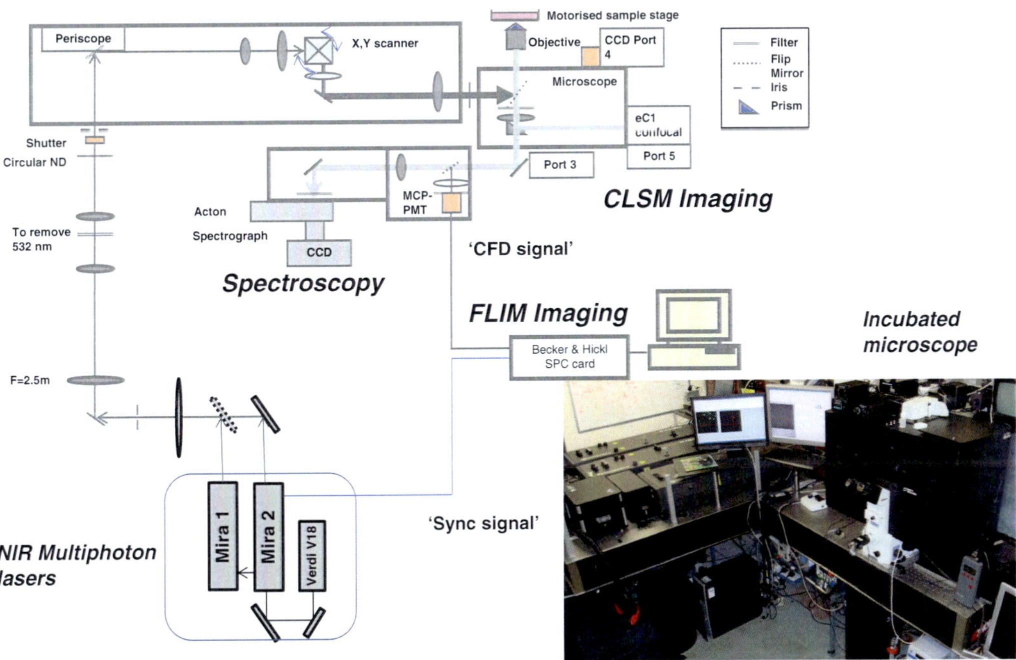

Fig. 1 Schematic of the multiphoton excitation FLIM with 4-channel confocal setup. *CLSM* confocal laser scanning microscopy, *NIR* near infrared, *FLIM* fluorescence lifetime imaging microscopy, *CCD* charged coupled device, *PMT* photomultiplier tube

using near-infrared (NIR) high-repetition rate sub-picosecond lasers, which provides several advantages over the single-photon method, including high-resolution deep tissue imaging, reduced cellular cytotoxicity of the excitation light and reduced photobleaching of the fluorophore. It is also worth noting that time-gated intensified charged coupled devices (ICCD) with pulsed (nanosecond pulse width) light sources may also be used for FLIM in an epifluorescence configuration. The use of frequency-domain lifetime imaging has also been described. However, the spatial resolution and time responses of these systems are poor compared to the TCSPC-FLIM system we describe here. Although our own experiments were conducted on a custom-built FLIM system, the methodology described here can be adapted for other commercial one-photon or two-photon FLIM systems. There are now several manufacturers including Becker & Hickl (http://www.becker-hickl.com/) and PicoQuant (http://www.picoquant.com/) offering complete FLIM systems or as modules for addition to existing confocal and multiphoton microscopes. Whilst the operation of these systems varies, the data interpretation is similar. This report will provide the reader with a basic understanding of the procedures required to investigate protein–protein interactions (independent of their subcellular location) in living plant specimens using the FRET-FLIM method.

As an example we show how the FRET-FLIM method is used to detect the direct interaction between the two Golgi-resident N-glycan processing enzymes *Nicotiana tabacum* β1,2-N-acetylglucosaminyltransferase I (GnTI) and *Arabidopsis thaliana* Golgi α-mannosidase II (GMII) in living tobacco leaf epidermal cells [12].

2 Materials

All the materials used and described here are commercially available. No further purification or treatment was carried out prior to their use. Custom-built instruments are described in detail.

2.1 Equipment

TCSPC-FLIM System

1. Confocal microscope equipped with an argon ion laser to excite GFP-tagged proteins at 488 nm and a helium/neon (HeNe) laser to excite mRFP-tagged proteins at 543 nm (or 561 nm).
2. Inverted microscope coupled to an XY galvanometer scanner (e.g., from GSI Lumonics or Cambridge Technology) or a commercial multiphoton system such as the Nikon A1R or Zeiss multiphoton system.

3. Water immersion objective (60×; numerical aperture (NA) of 1.2).
4. A standard GFP and mRFP filter set, respectively, the latter capable of filtering out chlorophyll emission. For multiphoton FLIM, a short-pass (350–620 nm) emission filter and a short-pass dichroic filter (reflects 700–1,000 nm and transmits 350–660 nm) are used.
5. Confocal image visualization software.
6. Two-photon excitation: laser light at a wavelength of 920 ± 5 nm obtained from a mode-locked titanium sapphire laser (e.g., Mira F900 from Coherent Lasers) producing 180-femtosecond pulses at 76 MHz pumped by a solid-state continuous-wave 532-nm laser (e.g., Verdi V18 from Coherent Lasers).
7. Fast microchannel plate photomultiplier tube (e.g., MCP-PMT R3809U from Hamamatsu) used as detector.
8. Becker & Hickl (BH) SPCM software v.3.2.30 (http://www.becker-hickl.com/software/tcspc/softwaretcspc.htm).
9. BH SPCImage software v.9.3.
10. BH TCSPC personal computer module SPC-830.

2.2 Reagents

1. Crystalline potassium dihydrogen phosphate (KDP) for the calibration of the multiphoton FLIM system.
2. Plant specimen that expresses the GFP-tagged protein of interest (donor).
3. Plant specimen that expresses the mRFP-tagged protein of interest (acceptor) (*see* **Note 1**).
4. Plant specimen that coexpresses the donor and the acceptor tagged fusion proteins in the same subcellular compartment of the same cell, but not necessarily known to interact (*see* **Note 2**).
5. 1 mM latrunculin B (latB) stock solution: 1 mg in 1 ml dimethyl sulfoxide (DMSO). Following reconstitution, divide the stock solution into single-use aliquots and store at −20 °C. Stock solutions are stable up to 2 months at −20 °C. Use at a concentration of 25 μM (in ultrapure water) (*see* **Note 3**).

3 Methods

3.1 Imaging Procedure

Prepare the biological specimens for confocal microscopy as described below. For a protocol on the protein localization of fluorescent protein fusions in plants, refer to [13]. Whatever expression system is used, prepare a donor-alone (control) specimen, an acceptor-alone specimen and one that expresses both the donor- and the acceptor-tagged fusion proteins. Throughout the following sections, we will use as example the transient expression of the GFP-tagged Golgi targeting sequence (cytoplasmic-transmembrane-stem [CTS] region) of GnTI, designated as GnTI-CTS-GFP [12, 14] and henceforth referred to as the donor, and the mRFP-fused CTS region of GMII, designated as GMII-CTS-mRFP [12] and referred to as the acceptor, in leaves of tobacco (*Nicotiana* spp.) using the agroinfiltration technique. If information on the agroinfiltration technique is needed, please refer to [15]. Expression of tagged fluorescent proteins often occurs after 24 h post infiltration. Here, the expression of the protein fusions was analyzed 3 days post infiltration.

1. Verify the subcellular localization and expression levels of the donor and the acceptor separately. Section roughly a 3×3 mm piece of the infiltrated leaf and place it on a glass slide. Add 2–3 drops of ultrapure water and cover with a coverslip (number 1). The ends of the coverslip need to be taped down to reduce movements in the leaf over time during the imaging process. Place the slide with the donor-alone or acceptor-alone specimen on the microscope stage. Use standard confocal microscopy procedures to find the optimal focus, zoom and position in the sample. Find cells with representative donor and acceptor expression levels. Use the argon 488-nm laser line and the green HeNe 543-nm laser line for donor and acceptor excitation, respectively. Use the appropriate optical filter configurations for detecting emission of each of the fluorophores. Use the single-labeled specimens (controls) to assess potential bleed-through by exciting each fluorophore with only one laser at a time, while monitoring the signal in the other emission channel set up for the second fluorophore. Ideally, no signal should come through. Acquire an image of each specimen and record the confocal settings including image resolution (512×512, in our case), brightness, contrast, and zoom settings for future reference (*see* **Notes 4** and **5**).
2. Verify the colocalization of the donor and acceptor in the same cellular compartment. Use the previously determined settings to image the donor and acceptor simultaneously. Check for "real" acceptor emission by scanning only with the 543-nm acceptor excitation wavelength (*see* **Notes 6–8**).
3. Optional Step: Prior to image acquisition, immobilize the subcellular organelle in the biological sample. In samples where the image acquisition requires more than a single scan, the mobility of subcellular organelles such as Golgi bodies needs to be reduced. In our case leaf segments are treated for 45–60 min with the actin-depolymerizing agent latB at a concentration of 25 μM to inhibit Golgi movement [14, 16] (*see* **Note 9**).

Table 1
Nanosecond fluorescence lifetime standards with mono-exponential decay

Standard/medium	λ_{exc} (nm) (MP)	λ_{em} (nm)	τ (ps)	Energy transfer
POPOP/ethanol	360–390 (700)	380–475	1,007 ± 120	n/a
Rhodamine/water	543 (840)	563	1,200 ± 50	n/a
Fluorescein/water	480 (920)	515	4,100 ± 50	n/a
Fluorescein/rhodamine 50:50	920	515	3,100 ± 100	Yes

λ_{exc} excitation wavelength, λ_{em} emission wavelength, τ lifetime, *MP* multiphoton

3.2 FLIM Setup

All FLIM measurements are carried out at 20 °C in a dark room due to the light sensitivity of the PMT detector.

1. Determine the instrument response function (IRF) prior to any FLIM measurement. This will determine whether the data needs to be deconvolved from the characteristics of the instruments. We use KDP to determine the IRF. Focusing the NIR light (920 nm) into the KDP crystal generates nonlinear second harmonic at 460 nm that is detected by the PMT. The IRF in our setup is measured as 25–60 ps. In this case any measured excited-state lifetime of more than 60 ps is free from errors from the instrumentation and does not need a deconvolution step during the data analysis (*see* **Note 10**).
2. Calibrate the multiphoton FLIM system with reference fluorophores of known lifetime. Here, the data from a known standard should help determine inherent effects of poor detector characteristics. Use solutions of fluorophores of known or established excited-state lifetimes. In our laboratory fluorescein, rhodamine, and POPOP (1,4-bis(5-phenyloxazole-2-yl) benzene) are routinely used as standards. Table 1 shows the excitation, emission, and lifetimes of the standard compounds. Place a drop of the solution to be measured on a cover slip and mount it on the microscope stage. The fluorescence excited-state lifetime is obtained both as a single point by keeping the focused laser stationary and as an image by raster scanning the excitation laser to match the conditions of the FLIM experimental samples.

3.3 FLIM Measurement

1. Prepare a slide with the latB-treated donor specimen with a coverslip and place it on the microscope stage. Use standard confocal microscopy procedures to find the optimal focus, zoom, and position in the sample. Identify and image single plant cells for subsequent FLIM measurements. Acquire a reference confocal image of the selected region with the 488-nm donor excitation wavelength (*see* Subheading 3.1, **steps 1** and **3**) (*see* **Note 11**).

2. Use an excitation power level that does not lead to photobleaching of the donor. Specify the number of cycles over which lifetime measurements should be run. You can start the FLIM measurement once the confocal image is recorded to avoid too much movement/drift in the sample (*see* **Note 12**).
3. Collect FLIM data sets from ten cells. It takes us 160–224 s to acquire one data set from a cell expressing only the donor although this is dependent on the expression level of the sample. After run completion save the data set as an ".sdt" file (*see* **Note 13**).
4. Place the slide with the latB-treated specimen expressing both the donor and acceptor on the microscope stage. Find a cell with similar expression levels of the donor and acceptor (or higher in the acceptor channel) to obtain approximately equal levels of interaction partners within the cell. Acquire a reference confocal image of the selected region with the 488-nm donor and the 543-nm acceptor excitation wavelength (*see* Subheading 3.1, **steps 2** and **3**).
5. Repeat **steps 2** and **3** to measure the lifetimes of the donor–acceptor specimen (*see* **Note 14**).

3.4 FLIM Analysis

For a detailed description of the analysis of TCSPC-FLIM data using the BH SPCImage software, please refer to the BH SPCImage software manual (http://www.becker-hickl.com/literature.htm) or [17]. We will describe here the parameters specific for the analysis of Golgi-resident FRET pairs.

1. Start your FLIM analysis with the donor specimen. Import one of the BH FLIM data sets (an ".sdt" file) acquired from cells that express only the donor into the BH SPCImage software.
2. Specify the fitting model (under "Options: Model") and parameters for data analysis (Fig. 2). We used the "Incomplete Multiexponentials" fit model with a laser repetition time of 12.5 ns (*see* **Note 15**).
3. Set the threshold between 15 and 25 to suppress pixels with a poor signal-to-noise ratio.
4. Set the binning factor to two to increase the number of pixels that are summed into each decay trace.
5. Use the software to move the blue crosshair curser in the intensity image onto the center of a single Golgi body to determine the lifetime value of an individual pixel. This hot spot defines the pixel from which the lifetime and decay trace will be shown. The peak value of the decay curve of a single point should be at least 100 for a mono-exponential fit. The Chi-square (χ^2) is

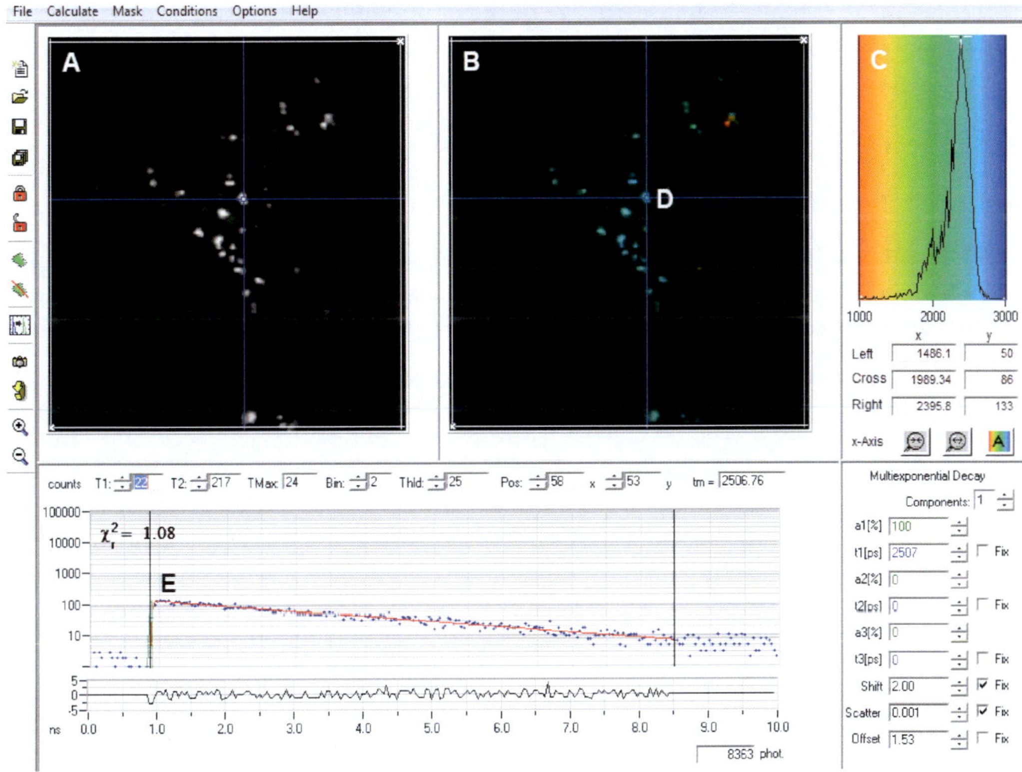

Fig. 2 Analysis of GnTI-CTS-GFP fluorescence lifetime data using the BH SPCImage software. Intensity (**a**) and fluorescence lifetime (**b**) images of a representative tobacco leaf epidermal cell expressing the donor GnTI-CTS-GFP. The donor specimen is used as a control to establish the unquenched lifetime of GFP in the context of the fusion protein. Lifetime values are collected on a single pixel basis from the center of individual Golgi bodies. The distribution curve in the histogram (**c**) depicts the relative occurrence frequency of the lifetimes within the lifetime image (considering all pixels). In the "rainbow" color map here, *red* is low lifetime while *blue* is high lifetime values. The blue cross hair (**d**) defines the pixel from which the current fluorescence decay curve (**e**) is shown

used as an indicator for the best exponential curve fit. Whereas a $\chi^2 = 1.0$ in the structure is the best fit of the data points to the lifetime value generated, values of Chi-square that deviate from 1 indicate poor fitting to the data points when a mono-exponential condition is applied. Since we expect the lifetime value of GFP to be a mono-exponential decay, we discard data under $\chi^2 = 0.8$ and over $\chi^2 = 1.4$ (these indicate multiple decay components) (*see* **Note 16**).

6. Identify a representative Golgi body, fix "Scatter" and "Shift" in the SPCImage software and run fitting on the whole image by calculating the decay matrix. The BH software translates the

intensity image of the donor-alone data set into a false-colored lifetime map in which different colors indicate different lifetimes (*see* **Note 17**).

7. Lock the fit process (sets global fitting characteristics and speeds up the lifetime calculation process). Fit values will not be recalculated when pointing to a new pixel or changing the conditions of the fit.
8. Based on the expected GFP lifetime values to be resolved, set the "Minimum" and "Maximum" values of the color range (used for continuous colors) to 1,000 and 3,000 ps (under "Options: Color").
9. Perform the curve-fitting analysis on the individual Golgi bodies from the color-coded lifetime image according to the guidelines in **step 5**. Record the pixel-by-pixel lifetime values of the donor in Microsoft Excel (*see* **Notes 18** and **19**).
10. Save the analyzed data (an ".img" file in our case).
11. Repeat **steps 1–10** for every data set acquired from cells that only express the donor. Check all analyzed data sets for reproducibility across different cells.
12. Follow the procedures described in **steps 1–11** to analyze the data sets acquired from the cells coexpressing the donor and acceptor.
13. Use the recorded data values to generate a histogram depicting the distribution of lifetime values of all data points within the donor and donor–acceptor specimen.
14. Use the recorded data values determined from the donor and donor–acceptor specimen, respectively, to calculate the mean excited-state lifetime of the donor in the absence (τ_D) and presence of the acceptor (τ_{DA}) in Microsoft Excel. The mean excited-state lifetime of GnTI-CTS-GFP was calculated to be 2,420 ± 70 ps (468 Golgi bodies from 15 cells). This value was reduced to 2,050 ± 90 ps in the presence of GMII-CTS-mRFP (307 Golgi bodies from 13 cells).
15. Use both τ_D and τ_{DA} to calculate the energy transfer efficiency (E), here expressed as percentage, based on Eq. 2:

$$E\% = \left[1 - \left(\frac{\tau_{DA}}{\tau_D}\right)\right] \times 100 \tag{2}$$

Quenching of the mean excited-state lifetime of the donor by 8 % in the presence of the acceptor is indicative of a protein–protein interaction (Fig. 3) (*see* **Notes 20–22**).

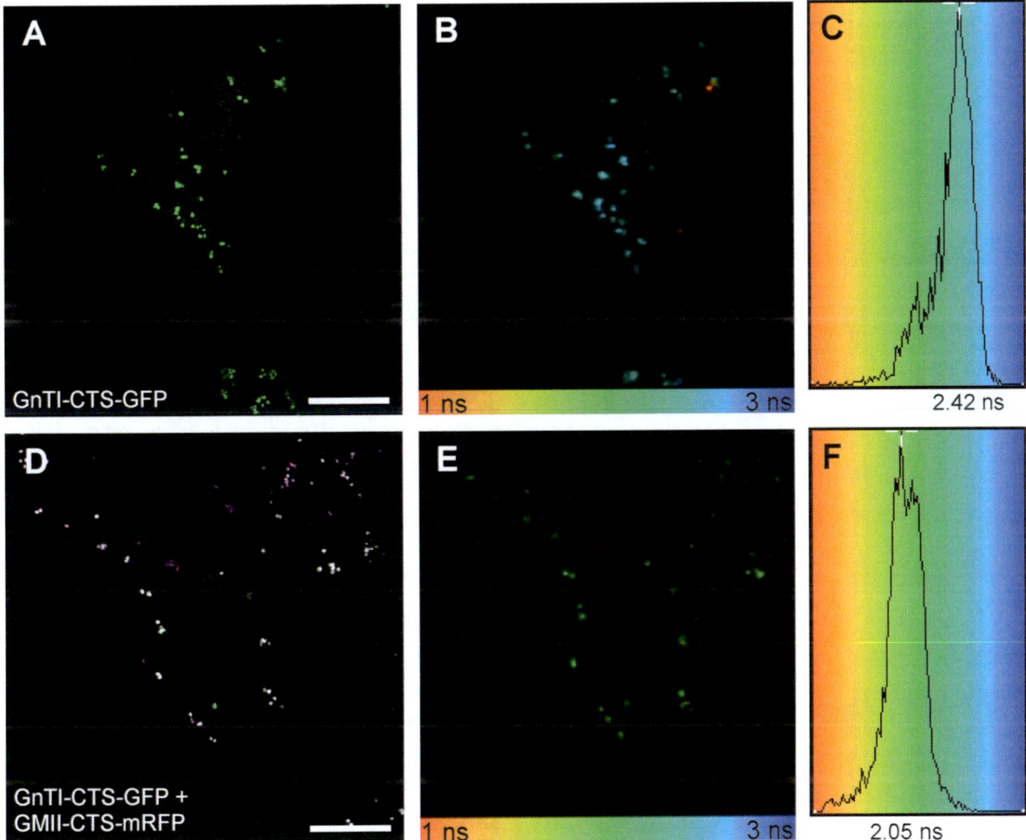

Fig. 3 GnTI-CTS-GFP direct interaction with GMII-CTS-mRFP in living tobacco cells. Confocal (**a**, **d**) and fluorescence lifetime (**b**, **e**) images of GnTI-CTS-GFP alone (**a**–**c**) and in the presence of GMII-CTS-mRFP (**d**-**f**). Colocalization of GnTI-CTS-GFP (in *green*, not shown) and GMII-CTS-mRFP (in *magenta*, not shown) appears in white in the merged confocal image (**d**). The lifetime data points from the entire area in (**b**) and (**e**) are shown in the lifetime histogram in (**c**) and (**f**), respectively. Colors correspond with the different lifetime values following the color code of the lifetime distribution at the *bottom* of images (**b**) and (**e**). The GnTI-CTS-GFP mean excited-state lifetime of 2,420 ± 70 ps (donor alone) is reduced to 2,050 ± 90 ps due to quenching by GMII-CTS-mRFP, thus showing a direct interaction. Bars = 20 μm

4 Notes

1. The acceptor must reside in the same cellular compartment as the donor and assume a favorable dipole–dipole orientation with respect to the donor fluorophore.
2. When working with membrane proteins, first check the membrane topology, either experimentally or using in silico prediction software. Unquenched donor lifetimes in the presence of the acceptor would not allow the discrimination between a non-interaction and an unfavorable dipole–dipole orientation of the fluorophores (i.e., the presence of the donor and acceptor fluorophores on opposing sides of the membrane).

3. LatB is sold as a powder in a glass vial. To dissolve the powder, inject DMSO or ethanol with the help of a needle and syringe directly into the vial by plunging the needle into the septum of the vial. Wear suitable protective clothing (see safety phrase S36) as this chemical is classified as harmful by inhalation, in contact with skin and if swallowed (according to the following risk phrases R20/21/22).
4. Determining the optimal confocal settings for imaging of the donor and acceptor will help to observe any spectral cross talk in the double-labeled donor–acceptor specimen and will also speed up the selection of suitable cells for subsequent FLIM measurements.
5. Make sure the correct coverslip thickness is used or adjusted for the high NA microscope objective if this option is available. This is likely to determine how much fluorescence signal is detected if poorly adjusted.
6. For data reproducibility, it is important to image the donor and donor–acceptor specimen under the same conditions at all times.
7. Although donor levels may exceed that of the acceptor, it is difficult to quantify such differences in confocal microscopy. Therefore ideally a cell should contain sufficient binding partner for the donor. Alternatively, the donor should not be present in excess as this might lead to higher concentrations of free, unbound donor resulting in a mix of quenched (bound donor) and unquenched donor lifetimes (unbound donor). To avoid this problem cells with comparable acceptor/donor ratios or with higher acceptor expression should be selected for fluorescence lifetime measurements to ensure comparability between different interactions. If one of the two proteins' expression levels is considerably higher, try to adjust levels if experimentally possible. In the case of the agroinfiltration technique, the optical density of the infiltrated agrobacterial suspension containing your protein of interest can be increased or decreased accordingly. However, the acceptor expression level may be higher than that of the donor.
8. The colocalization of the donor- and acceptor-tagged fusion proteins is a prerequisite for the application of FRET. Furthermore, it is important to know that the donor fluorophore resides in the same microenvironment in both the donor and donor–acceptor specimen since the fluorescence lifetime decay of the donor is sensitive to changes in its local environment, such as pH or ion concentrations.
9. FLIM analysis involves statistical fitting of fluorescence lifetime decay data and therefore, the number of photons detected needs to be high enough to determine the correct lifetime with accuracy. Consequently, the fluorescent sample must be

as stationary as possible during data acquisition to obtain statistically significant photon counts. Our FLIM technique requires the use of drugs on live plant tissue to inhibit organelle motility, as observed for Golgi bodies. Actin- and microtubule-based movements of organelles can be inhibited by the use of latB and oryzalin, respectively. Fixation of the specimen with formaldehyde-based fixatives is not recommended as such procedures can change the molecular environment of the fluorophore, which might affect its lifetime [18]. In any case, first test the effect of the used chemical drug on your biological sample with regards to changes in the subcellular localization.

10. For example, an IRF that is longer in time than the measured data values will require a deconvolution step in the data analysis to minimize the error and uncertainty in the result. A poor IRF may be a function of some of the following: large radio frequency pick up in the electrical cabling used in the setup, slow response time (>200 ps) of the PMT used in the detection, poor mode-locked laser function for the excitation leading to a spread in the initial rise time or time zero and the zero-crossover of the constant fraction discriminator used. Since the MCP-PMT used here is insensitive beyond 820 nm, it is not possible to use the standard DuPont ludox or light scattering technique of the excitation light [19].

11. We always scan the donor-alone specimen with both the 488-nm donor and the 543-nm acceptor excitation wavelength in order to use the same conditions as used for the donor–acceptor specimen.

12. The number of photons detected for the data analysis needs to be high enough for good exponential decay calculations. In general, a photon count of 500 or more at a count rate of less than 1 MHz (or >100 counts in the peak channel) is sufficient for the current detector counting efficiencies. We therefore recommend running a few more cycles than probably necessary to ensure that data with sufficiently high photon counts will be recorded. The online display allows you to estimate the quality of the data; measurements can be stopped before cycle completion when a good signal-to-noise ratio has been reached.

13. We recommend acquiring data sets from at least 12 different cells to account for the occasional cell with low counts and/or poor χ^2 values (*see* Subheading 3.4) that may affect the final calculation. The time for data acquisition varies dependent on the fluorophore expression levels.

14. The acquisition time for different data sets varies depending on the donor and acceptor expression levels in the analyzed cell. This is due to the fact that the donor- and acceptor-tagged fusion proteins are independently expressed and each cell therefore will have a different donor–acceptor ratio.

15. It is important that the laser repetition rate is considered since the two-photon excitation runs at roughly 80 MHz (~12.5 ns pulse intervals), thus adjusting the decay model in the software.
16. Since the instrument response (IR) in our setup is determined to be less than 60 ps there was no need to deconvolute the IR function from the sample data decay curves. Thus, lifetime differences of larger than 100 ps can be easily resolved.
17. Instead of the whole image only a defined region can be fitted in order to exclude "interfering" lifetimes from chlorophyll-containing plastids that will produce much shorter lifetimes.
18. We used Adobe Photoshop to assign every analyzed Golgi body with a letter in the lifetime image and then we recorded the donor lifetime values in Microsoft Excel with reference to the letters.
19. Data can be extracted from lifetime images either on a single-pixel basis and plotted into distribution histograms [7, 12] or by drawing regions of interest and collecting the average [8].
20. Alternatively or additionally the lifetime difference between τ_D and τ_{DA} can be calculated. A reduction of >200 ps in the excited-state lifetime of the donor represents quenching through protein–protein interaction.
21. In general steady-state FRET, the efficiency of the energy transfer may be determined as Eq. 3:

$$E\% = \left[1 - \left(\frac{I_{DA}}{I_D}\right)\right] \times 100 \tag{3}$$

where I is the steady-state intensities.

22. The TCSPC-FLIM setup used herein is easily able to resolve lifetime differences of larger than 100 ps, which is equivalent to a 4.5 % change in FRET efficiency (assuming an excited-state lifetime of 2,500 ns such as that of GFP). We have previously shown that the reduction of as little as ~200 ps in the excited-state lifetime of the GFP-fused protein represents quenching through protein–protein interaction [4, 7, 8, 12].

Acknowledgements

This work was funded by the Austrian Science Fund (FWF): J2981-B20 (to J.S.), and Oxford Brookes University. Access to the Central Laser Facility, Rutherford Appleton Laboratory, was funded by a Science and Technology Facilities Council Program Access grant. We thank Chris Hawes for carefully reading the manuscript.

References

1. Förster T (1948) Zwischenmolekulare energiewanderung und fluoreszenz. Ann Phys 437: 55–75
2. Becker W (2012) Fluorescence lifetime imaging-techniques and applications. J Microsc 247:119–136
3. Bhat RA, Miklis M, Schmelzer E, Schulze-Lefert P, Panstruga R (2005) Recruitment and interaction dynamics of plant penetration resistance components in a plasma membrane microdomain. Proc Natl Acad Sci U S A 102:3135–3140
4. Stubbs CD, Botchway SW, Slater SJ, Parker AW (2005) The use of time-resolved fluorescence imaging in the study of protein kinase C localisation in cells. BMC Cell Biol 6:22
5. Adjobo-Hermans MJ, Goedhart J, Gadella TW (2006) Plant G protein heterotrimers require dual lipidation motifs of Galpha and Ggamma and do not dissociate upon activation. J Cell Sci 119:5087–5097
6. Aker J, Hesselink R, Engel R, Karlova R, Borst JW, Visser AJ, de Vries SC (2007) In vivo hexamerization and characterization of the Arabidopsis AAA ATPase CDC48A complex using forster resonance energy transfer-fluorescence lifetime imaging microscopy and fluorescence correlation spectroscopy. Plant Physiol 145:339–350
7. Osterrieder A, Carvalho CM, Latijnhouwers M, Johansen JN, Stubbs C, Botchway S, Hawes C (2009) Fluorescence lifetime imaging of interactions between Golgi tethering factors and small GTPases in plants. Traffic 10:1034–1046
8. Sparkes I, Tolley N, Aller I, Svozil J, Osterrieder A, Botchway S, Mueller C, Frigerio L, Hawes C (2010) Five Arabidopsis reticulon isoforms share endoplasmic reticulum location, topology, and membrane-shaping properties. Plant Cell 22:1333–1343
9. Crosby KC, Pietraszewska-Bogiel A, Gadella TW, Winkel BS (2011) Förster resonance energy transfer demonstrates a flavonoid metabolon in living plant cells that displays competitive interactions between enzymes. FEBS Lett 585:2193–2198
10. Berendzen KW, Böhmer M, Wallmeroth N, Peter S, Vesi M, Zhou Y, Tiesler FK, Schleifenbaum F, Harter K (2012) Screening for *in planta* protein–protein interactions combining bimolecular fluorescence complementation with flow cytometry. Plant Methods 8(1):25
11. Yadav RB, Burgos P, Parker AW, Iadevaia V, Proud CG, Allen RA, O'Connell JP, Jeshtadi A, Stubbs CD, Botchway SW (2013) mTOR direct interactions with Rheb-GTPase and raptor: sub-cellular localization using fluorescence lifetime imaging. BMC Cell Biol 14:3
12. Schoberer J, Liebminger E, Botchway SW, Strasser R, Hawes C (2013) Time-resolved fluorescence imaging reveals differential interactions of N-glycan processing enzymes across the Golgi stack *in planta*. Plant Physiol 161: 1737–1754
13. Runions J, Hawes C, Kurup S (2007) Fluorescent protein fusions for protein localization in plants. Methods Mol Biol 390: 239–255
14. Schoberer J, Vavra U, Stadlmann J, Hawes C, Mach L, Steinkellner H, Strasser R (2009) Arginine/lysine residues in the cytoplasmic tail promote ER export of plant glycosylation enzymes. Traffic 10:101–115
15. Sparkes I, Runions J, Kearns A, Hawes C (2006) Rapid, transient expression of fluorescent fusion proteins in tobacco plants and generation of stably transformed plants. Nat Protoc 1:2019–2025
16. Brandizzi F, Snapp E, Roberts A, Lippincott-Schwartz J, Hawes C (2002) Membrane protein transport between the endoplasmic reticulum and the Golgi in tobacco leaves is energy dependent but cytoskeleton independent: evidence from selective photobleaching. Plant Cell 14:1293–1309
17. Sun Y, Day RN, Periasamy A (2011) Investigating protein–protein interactions in living cells using fluorescence lifetime imaging microscopy. Nat Protoc 6:1324–1340
18. Becker W, Su B, Bergmann A (2009) Fast-acquisition multispectral FLIM by parallel TCSPC. Proceedings of the SPIE 7183, multiphoton microscopy in the biomedical sciences IX, 718305
19. Weber G, Teale FWJ (1957) Determination of the absolute quantum yield of fluorescent solutions. Trans Faraday Soc 53:646–655

Chapter 7

Analysis of Rab GTPase–Effector Interactions by Bimolecular Fluorescence Complementation

Emi Ito and Takashi Ueda

Abstract

RAB GTPases interact with specific effector molecules in a spatiotemporally regulated manner to induce various downstream reactions. To clarify the overall picture of RAB GTPase functions, it is important to elucidate the cellular locale where RAB and its effectors interact. Here, we applied a bimolecular fluorescence complementation (BiFC) assay to analyze where RAB GTPase interacted with effectors in endosomal trafficking.

Key words RAB GTPase, Effectors, Bimolecular fluorescent complementation, Transient expression assay, Confocal scanning laser microscopy

1 Introduction

Membrane trafficking is an important system for maintaining the integrity of organelle and cellular functions in eukaryotic cells. Although many of the components involved in membrane trafficking are conserved among eukaryotic lineages, recent studies have highlighted some unique aspects in the organization of membrane trafficking in plants [1, 2]. RAB5 is one of the best-studied RAB GTPases. It is known to localize to early endosomes and regulate a wide variety of endocytotic events in the mammalian system [3–5]. Also in plants, RAB5 acts as an important key regulator in endosomal trafficking. In addition to conventional RAB5, plants possess a family of plant-unique RAB5 homologs, including ARA6/RABF1 in *Arabidopsis thaliana* [6]. Our recent studies demonstrated that ARA6 and conventional RAB5 regulated different membrane trafficking pathways, and ARA6 was required for proper abiotic stress responses [7, 8].

Small GTPases, including RAB5, are known to act as molecular switches by cycling between GTP-bound and GDP-bound states. The GTP-bound state is considered the active form that interacts with effector molecules to evoke downstream reactions. In mammalian cells, various RAB5 effectors have been identified and

Marisa S. Otegui (ed.), *Plant Endosomes: Methods and Protocols*, Methods in Molecular Biology, vol. 1209,
DOI 10.1007/978-1-4939-1420-3_7,

were shown to interact with RAB5 in a spatiotemporally regulated manner. On the other hand, many of the mammalian RAB5 effectors are not conserved in plants; this suggested that, during evolution, plants must have acquired unique RAB5 effectors. Thus, identification and characterization of effector molecules that interact with RAB GTPases have become a key issue in elucidating the molecular basis of RAB GTPase-regulated trafficking pathways in plants.

There are several methods for identifying effector molecules that interact with RAB GTPases, including yeast two hybrid screening and a proteomics approach [9–13]. There are also several methods for clarifying the location of interactions between RAB and its effector molecules; however, experimental challenges and the requirements of a specialized system and specific materials can sometimes hamper the performance of these types of analyses. In some cases, a bimolecular fluorescence complementation (BiFC) assay, combined with transient expression in protoplasts, has been found to be a good choice for investigating protein–protein interactions in vivo. The BiFC assay (also known as the split yellow fluorescent protein [YFP] assay) is a protein-fragment complementation assay, where a fluorescent complex is reconstituted when two nonfluorescent fragments (BiFC probes) are brought together through the interaction between the proteins fused to each probe (Fig. 1) [14–17]. Once the BiFC probes are brought together, the reconstituted fluorescent proteins do not dissociate [15]; thus, the BiFC assay is considered a powerful tool for detecting transient interactions. Here, we describe how to visualize protein–protein interactions in protoplasts of *A. thaliana* cells suspended in culture. The BiFC technique is demonstrated with combinations of ARA7/RABF2b and VPS9a (guanine nucleotide exchange factor for RAB5 in *A. thaliana*) [18], and ARA6 and its effector molecule.

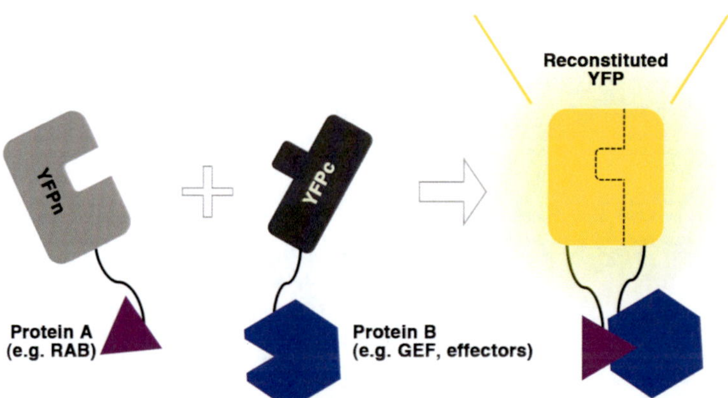

Fig. 1 Bimolecular fluorescence complementation (BiFC). Two proteins are expressed as fusion proteins, each attached to a nonfluorescent fragment of yellow fluorescent protein (YFP; N- and C-terminal halves are indicated as YFPn and YFPc, respectively). When protein A and protein B interact, the YFPn and YFPc fragments associate to reconstitute the fluorescent YFP. Thus, interactions between proteins A and B can be detected by YFP fluorescence in living cells

2 Materials

2.1 Components for the Transient Assay

Prepare the following solutions at room temperature. Use sterilized Milli-Q water and analytical grade reagents.

1. Enzyme solution: 0.4 M mannitol, 5 mM EGTA pH 8.0, 1 % (w/v) cellulase Y-C, 0.05 % (w/v) pectolyase Y-23. Store at −20 °C. Prewarm the solution to 30 °C before use. Typically, 25 ml is required for one experiment.
2. Solution A: 0.4 M mannitol, 70 mM $CaCl_2$, 5 mM MES pH 5.7. Prepare a fresh batch on the day of the experiment. Typically, 50 ml is required for one experiment.
3. MaMg solution: 0.4 M mannitol, 15 mM $MgCl_2$, 5 mM MES pH 5.7. Store at 4 °C.
4. DNA uptake solution: 0.4 M mannitol, 0.1 M $Ca(NO_3)_2$, 40 % (w/v) polyethylene glycol (PEG) 6000. Mix reagents in a conical tube and gently mix by inversion. Prepare 1 ml per sample on the day of the experiment (*see* **Note 1**).
5. Dilution solution: 0.4 M mannitol, 125 mM $CaCl_2$, 5 mM KCl, 5 mM glucose, 1.5 mM MES pH 5.7. Prepare 10 ml per sample on the day of the experiment.
6. B5-vitamin 100× stock solution: Dissolve 2 g myo-inositol, 0.02 g nicotinic acid, 0.02 g pyridoxine hydrochloride, and 0.2 g vitamin B1 (Thiamine) hydrochloride in water, and add water to a final volume of 100 ml. Store the solution in small aliquots at −20 °C.
7. MS mannitol: 0.4 M mannitol, 1× Murasige and Skoog basal salt mixture (catalog number M 5524) (Sigma- Aldrich Corp., St. Louis, MO, USA), 1× B5-vitamin solution, 0.034 % (w/v) KH_2PO_4. Dissolve the reagents in water and adjust to pH 5.7. Autoclave at 120 °C for 15 min. Store at room temperature. Handle with aseptic technique to avoid contamination.
8. Multitest glass slides.

3 Methods

3.1 Plasmid Preparation

Here, we present an example of how to prepare the BiFC vectors. Classical restriction enzyme-based construction is also applicable (*see* **Note 2**).

1. Use an ordinary PCR method to amplify the DNA sequences for the N-terminal (YFPn: 1-154 aa) and C-terminal (YFPc: 155-239 aa) fragments of enhanced YFP (*see* **Note 3**).
2. Subclone the DNA sequences of the YFP fragments into the *Aor51HI* site of the pUGW0 or pUGW2 vector [19] (Fig. 2a) (*see* **Note 4**). Then, insert the cDNA of your chosen protein

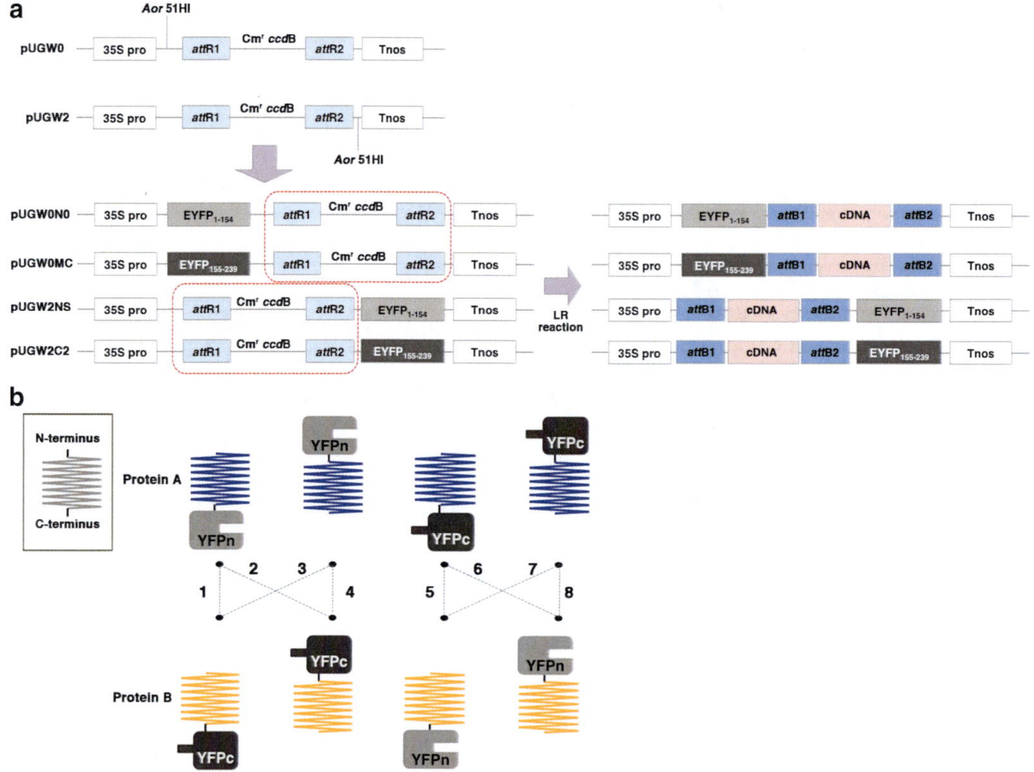

Fig. 2 Designing BiFC plasmid constructs. (**a**) Schematic representation of the BiFC vectors used in our studies. Diagrams represent pUGW0 and pUGW2, modified from Nakagawa et al. [19]. The regions surrounded by *red dotted lines* are replaced by *att*B1-cDNA-*att*B2 after LR reaction. DNA sequences that encode ten amino acids, including those encoded by the *att*B1 or *att*B2 sequences, are inserted as the linker sequence between the enhanced YFP fragment (EYFP) and the protein of interest. (**b**) Combinations of fusion proteins. N- and C-terminal fusions of YFPn and YFPc to two proteins of interest (A and B) will result in eight different pairs to be tested, (indicated by the *dotted lines* between pairs)

into the vector with the Gateway® LR reaction Clonase® Enzyme Mix (catalog number 11791-019) (Invitrogen Life Technologies Corp., Carlsbad, CA, USA) (*see* **Note 5**).

3. Purify the plasmids and adjust to 2 μg DNA/μl. For a typical experiment, 20 μg DNA are required for one assay (*see* **Note 6**).

3.2 Transformation of *Arabidopsis thaliana* Protoplasts Suspended in Culture

Use blunt-ended pipette tips or cut off about 5 mm from the ends of pipette tips when handling protoplasts.

1. Drain the medium of suspension cultures by suction filtration; then, collect about 1 g cells with a spatula and suspend them in 25 ml Enzyme solution (*see* **Note 7**). Incubate in a water bath set at 30 °C for 2 h, under gentle agitation (60 cycles/min).
2. Collect the cells with blunt-ended pipette tips and mount onto Multitest glass slides (Multitest slide, 8 well, 6 mm). Check to make sure the cells are spherical and detached from each other. Incubate further if necessary (*see* **Note 8**).

3. Pass the cells through a nylon mesh (125 μm pore), and then, collect the cells by centrifugation at 250 × *g* for 10 min. Discard the supernatant, resuspend the cells in 25 ml Solution A, and apply gentle pipetting to disperse the cells. Repeat this step to wash the cells. Centrifuge at 200 × *g* to collect cells between washing steps.
4. Resuspend the washed protoplasts with 1 ml MaMg solution. Place the cells on ice.
5. Mix 100 μl protoplast suspension, 50 μg carrier DNA (denatured salmon sperm or calf thymus DNA), and 20 μg each of the plasmids that carry YFPn- and YFPc-fusions (*see* **Note 9**). Then, add 0.4 ml DNA uptake solution. After mixing the solution with gentle pipetting, place the mixture on ice for 20 min.
6. Transfer the protoplast–DNA mixture into 10 ml Dilution Solution and mix by inversion. Centrifuge at 200 × *g* for 10 min to collect the cells.
7. Discard the supernatant and resuspend the cells in 5 ml MS mannitol. Transfer the cells into a petri dish and incubate at 23 °C under gentle agitation (100 cycles/min) in the dark (*see* **Note 10**).

3.3 Confocal Laser Scanning Microscopy

1. Collect 1.5 ml of the cultured cells in tubes and centrifuge the cells at 8,000 × *g* for 5 s. Discard the supernatant.
2. Add 20 μl MS mannitol and gently resuspend the cells (*see* **Note 11**).
3. Mount approximately 3 μl of suspended cells onto a multitest glass slide, and place a glass coverslip of suitable size on top of the cells (*see* **Note 12**)
4. Place the specimen under a confocal laser scanning microscope and observe the subcellular localization of the reconstituted fluorescent probes (Figs. 3 and 4) (*see* **Note 13**).

4 Notes

1. If PEG 6000 does not dissolve, place the container in warm water. Let the solution cool down to room temperature before mixing with the protoplasts.
2. A small vector is preferred (about 3 kbp), because the size of the plasmid affects the transformation efficiency.
3. The enhanced YFP fragments that correspond to 1-154 aa (YFPn) and 155-239 aa (YFPc) possess high complementation efficiency and low background [20]; thus, they are often used as BiFC probes [21–23]. The enhanced YFP fragments that correspond to 1-174 aa and 175-239 aa are also reported to be useful for BiFC [24]. Refer to Dr Tom Kerppolar's Lab

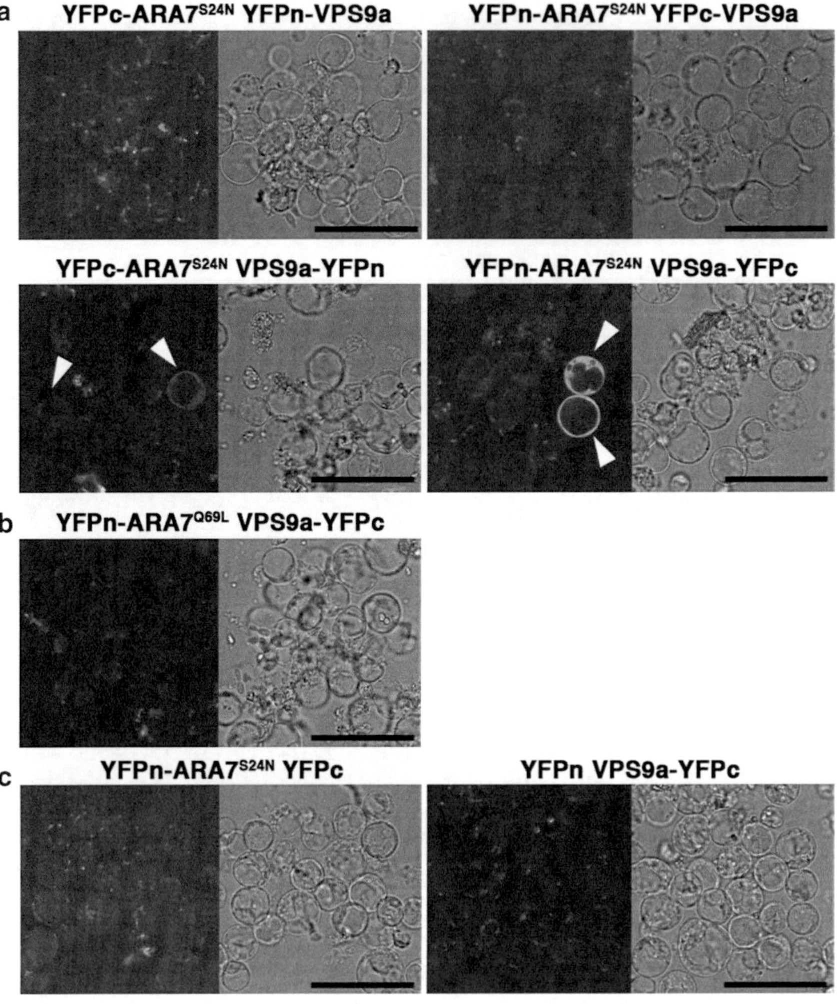

Fig. 3 Visualization of an interaction between ARA7 and VPS9a in the BiFC assay. (**a**) cDNA sequences that encoded a dominant negative mutant of ARA7 (ARA7^{S24N}) and the VPS9a proteins were subcloned into the vectors shown in Fig. 2a and transformed into the protoplasts of *A. thaliana* cells grown in suspension cultures. Four different combinations were tested; YFPc (C-terminal fragment of enhanced YFP) or YFPn (N-terminal fragment of enhanced YFP) were attached to the N-terminus (in front) of ARA7^{S24N}, and YFPc or YFPn are attached to either the N-terminus (in front) or the C-terminus (in back) of VPS9a. Fluorescence from reconstituted YFP is observed in the cytosol (*arrowheads*). Bars = 50 μm. (**b**) Interaction between a constitutively active mutant of ARA7 (ARA7^{Q69L}) and VPS9a. As demonstrated by Goh et al. [18], VPS9a does not interact with active ARA7. Bars = 50 μm. (**c**) YFPn-ARA7 and VPS9a-YFPc do not interact with cYFP or nYFP alone. Bars = 50 μm. ARA7: one of the conventional RAB5s in *A. thaliana*

Web site for more information (http://site-maker.umich.edu/kerppolar.bifc/home).

4. We employ the ten amino acid linker (including amino acids derived from the *att*B1 or *att*B2 sequence). It may be necessary to select an appropriate linker length to increase the flexibility of the BiFC probes.

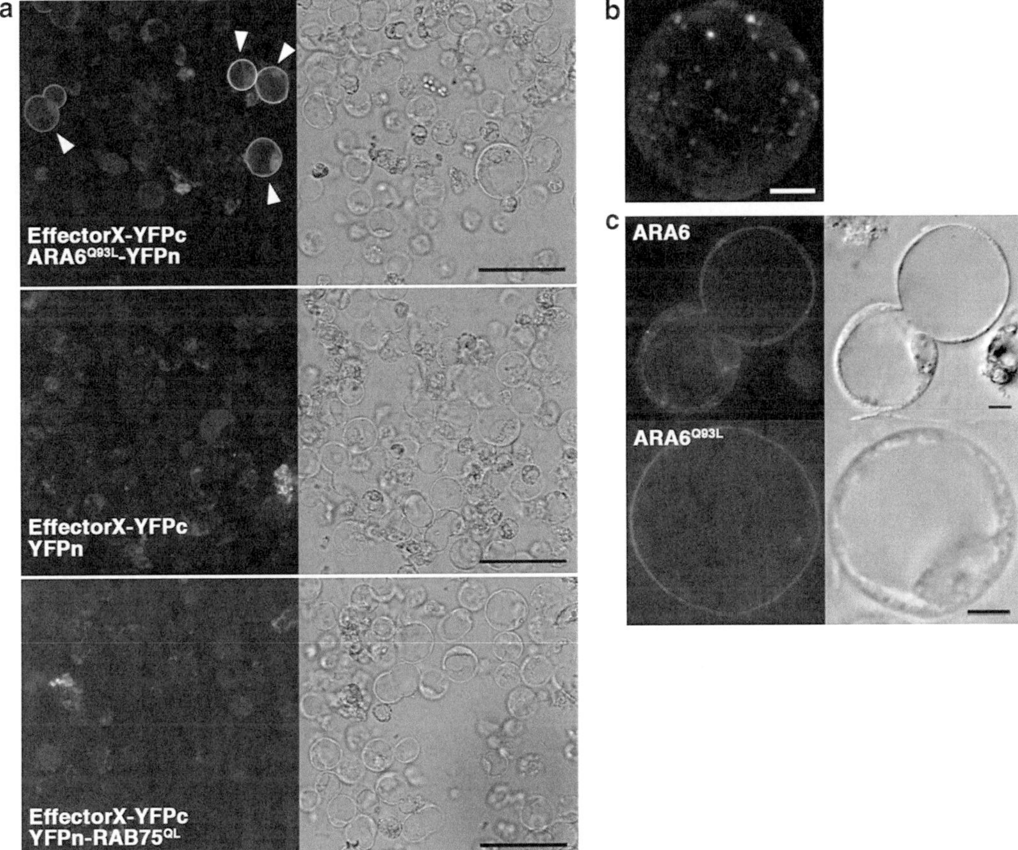

Fig. 4 Visualization of an interaction between ARA6 and its effector in the BiFC assay. (**a**) An effector candidate of ARA6 (EffetorX) was fused to the N-terminus of YFPc (C-terminal fragment of YFP). A constitutively active mutant of ARA6 (ARA6^{Q93L}) was fused to the N-terminus of YFPn (N-terminal fragment of YFP). These constructs were co-transformed into protoplasts of *A. thaliana* cells grown in suspension cultures. (*Upper panels*) Fluorescence from the reconstituted YFP is observed when EffectorX-YFPc is coexpressed with ARA6^{Q93L}-YFPn (*arrowheads*). The YFP signal is not observed when (*middle panels*) EffectorX-YFPc is coexpressed with YFPn alone, or (*lower panels*) when EffectorX-YFPc is coexpressed with a constitutively active mutant of another RAB fused to a YFPn fragment (YFPn-RAB75^{QL}). Bars = 50 μm. (**b**) A maximum intensity projection of confocal images, acquired from a cell that expressed EffectorX-YFPc and ARA6-YFPn. The YFP signal is observed as punctate structures. Bar = 5 μm. (**c**) Different subcellular localizations of reconstructed YFP. (*Upper panels*) The YFP signal is observed at the plasma membrane, tonoplast, and punctate structures, when EffectorX-YFPc is coexpressed with ARA6-YFPn (*left*, fluorescence; *right*, Nomarski). (*Lower panels*) On the other hand, the YFP signal is only evident at the plasma membrane when EffectorX-YFPc is coexpressed with ARA6^{Q93L}-YFPn (*left* fluorescence; *right*, Nomarski). Bars = 5 μm

5. If structural information is lacking for the protein of your interest, it is preferable to prepare all combinations, where the cDNA for probes are placed both C-terminally and N-terminally with respect to the cDNA that encodes the protein of interest (Fig. 2b). It is also recommended that you confirm that the

expressed fusion proteins retain their functions. Most RAB GTPases possess the Cys-motif, which becomes isoprenylated at the C-terminus. This isoprenylation is required for correct membrane targeting. Thus, probes should be fused to the N-terminus of the RAB GTPase to avoid interference with localization of the fusion protein.

6. The expression levels of the fusion proteins are correlated with the amount of plasmid DNA used in the transformation. The amount of DNA should be adjusted to optimize the expression levels of the YFPn- and YFPc-fusion proteins. The insertion of epitope tags (for example, HA- and myc-tags) may facilitate confirming the expression of the fusion proteins, because commercial antibodies are available for immunoblot analysis.
7. Do not collect too many cells, otherwise it takes longer time to digest the cell walls completely.
8. Inadequate cell wall digestion drastically reduces the transformation efficiency; thus, complete digestion must be ensured. It is important to check the cells and adjust the time of incubation in each experiment. Do not incubate the cells for more than 3 h. Prolonged treatment may also reduce the transformation efficiency.
9. BiFC probes occasionally reconstitute a fluorophore by themselves at low frequency; therefore, it is important to estimate the rate of spontaneous reconstitution with a negative control sample. It is preferable to use a non-interactive, mutant protein (for example, dominant negative mutant forms of RAB proteins) as the negative control.
10. Typically, it takes at least 12 h for the protoplasts to express the fluorescent probes.
11. At this time point, the cell wall is starting to regenerate; thus, normal pipette tips can be used here.
12. Glass slide wells are used to prevent the cells from collapsing. Cells may flow when there is too much liquid in a well. Reduce the fluid volume if the cells do not settle.
13. You may need to count the number of cells with the YFP signal and compare it to that in negative control.

Acknowledgement

This work was supported by Grants-in-Aid for Scientific Research from the Ministry of Education, Culture, Sports, Science, and Technology of Japan, and from JST, PRESTO.

References

1. Fujimoto M, Ueda T (2012) Conserved and plant-unique mechanisms regulating plant post-Golgi traffic. Front Plant Sci 3:197
2. Saito C, Ueda T (2009) Chapter 4: functions of RAB and SNARE proteins in plant life. Int Rev Cell Mol Biol 274:183–233
3. Grosshans BL, Ortiz D, Novick P (2006) Rabs and their effectors: achieving specificity in membrane traffic. Proc Natl Acad Sci U S A 103:11821–11827
4. Miaczynska M, Pelkmans L, Zerial M (2004) Not just a sink: endosomes in control of signal transduction. Curr Opin Cell Biol 16: 400–406
5. Zerial M, McBride H (2001) Rab proteins as membrane organizers. Nat Rev Mol Cell Biol 2:107–117
6. Ueda T, Yamaguchi M, Uchimiya H et al (2001) Ara6, a plant-unique novel type Rab GTPase, functions in the endocytic pathway of *Arabidopsis thaliana*. EMBO J 20: 4730–4741
7. Ebine K, Fujimoto M, Okatani Y et al (2011) A membrane trafficking pathway regulated by the plant-specific RAB GTPase ARA6. Nat Cell Biol 13:853–859
8. Ebine K, Miyakawa N, Fujimoto M et al (2012) Endosomal trafficking pathway regulated by ARA6, a RAB5 GTPase unique to plants. Small GTPases 3:23–27
9. Camacho L, Smertenko AP, Pérez-Gómez J et al (2009) *Arabidopsis* Rab-E GTPases exhibit a novel interaction with a plasma-membrane phosphatidylinositol-4-phosphate 5-kinase. J Cell Sci 122:4383–4392
10. Christoforidis S, Miaczynska M, Ashman K et al (1999) Phosphatidylinositol-3-OH kinases are Rab5 effectors. Nat Cell Biol 1: 249–252
11. Horiuchi H, Lippé R, McBride HM et al (1997) A novel Rab5 GDP/GTP exchange factor complexed to Rabaptin-5 links nucleotide exchange to effector recruitment and function. Cell 90:1149–1159
12. Preuss ML, Schmitz AJ, Thole JM et al (2006) A role for the RabA4b effector protein PI-4Kbeta1 in polarized expansion of root hair cells in *Arabidopsis thaliana*. J Cell Biol 172: 991–998
13. Stenmark H, Vitale G, Ullrich O et al (1995) Rabaptin-5 is a direct effector of the small GTPase Rab5 in endocytic membrane fusion. Cell 83:423–432
14. Grinberg AV, Hu CD, Kerppola TK (2004) Visualization of Myc/Max/Mad family dimers and the competition for dimerization in living cells. Mol Cell Biol 24:4294–4308
15. Hu CD, Chinenov Y, Kerppola TK (2002) Visualization of interactions among bZIP and Rel family proteins in living cells using bimolecular fluorescence complementation. Mol Cell 9:789–798
16. Hu CD, Kerppola TK (2003) Simultaneous visualization of multiple protein interactions in living cells using multicolor fluorescence complementation analysis. Nat Biotechnol 21:539–545
17. Weinthal D, Tzfira T (2009) Imaging protein–protein interactions in plant cells by bimolecular fluorescence complementation assay. Trends Plant Sci 14:59–63
18. Goh T, Uchida W, Arakawa S et al (2007) VPS9a, the common activator for two distinct types of Rab5 GTPases, is essential for the development of *Arabidopsis thaliana*. Plant Cell 19:3504–3515
19. Nakagawa T, Kurose T, Hino T et al (2007) Development of series of gateway binary vectors, pGWBs, for realizing efficient construction of fusion genes for plant transformation. J Biosci Bioeng 104:34–41
20. Kerppola TK (2008) Bimolecular fluorescence complementation: visualization of molecular interactions in living cells. Methods Cell Biol 85:431–470
21. Bracha-Drori K, Shichrur K, Katz A et al (2004) Detection of protein–protein interactions in plants using bimolecular fluorescence complementation. Plant J 40:419–427
22. Citovsky V, Lee LY, Vyas S et al (2006) Subcellular localization of interacting proteins by bimolecular fluorescence complementation in *planta*. J Mol Biol 362:1120–1131
23. Walter M, Chaban C, Schütze K et al (2004) Visualization of protein interactions in living plant cells using bimolecular fluorescence complementation. Plant J 40:428–438
24. Hino T, Tanaka Y, Kawamukai M et al (2011) Two Sec13p homologs, AtSec13A and AtSec13B, redundantly contribute to the formation of COPII transport vesicles in Arabidopsis thaliana. Biosci Biotechnol Biochem 75:1848–1852

Chapter 8

In Vivo Imaging of Brassinosteroid Endocytosis in Arabidopsis

Niloufer G. Irani, Simone Di Rubbo, and Eugenia Russinova

Abstract

Increasing evidence shows the involvement of endocytosis in specific signaling outputs in plants. To better understand the interplay between endocytosis and signaling in plant systems, more ligand–receptor pairs need to be identified and characterized. Crucial for the advancement of this research is also the development of imaging techniques that allow the visualization of endosome-associated signaling events at a high spatiotemporal resolution. This requires the establishment of tools to track ligands and their receptors by fluorescence microscopy in living cells. The brassinosteroid (BR) signaling pathway has been among the first systems to be characterized with respect to its connection with endocytic trafficking, owning to the fact that a fluorescent version of BR, Alexa Fluor 647-castasterone (AFCS) has been generated. AFCS and the fluorescently tagged BR receptor, BR INSENSITIVE1 (BRI1) have been used for the specific detection of BRI1-AFCS endocytosis and for the delineation of their endocytic route as being clathrin-mediated. AFCS was successfully applied in functional studies in which pharmacological rerouting of the BRI1-BR complex was shown to have an impact on signaling. Here we provide a method for the visualization of endocytosis of plant receptors in living cells. The method was used to track endocytosis of BRI1-BR complexes in *Arabidopsis* epidermal root meristem cells by using fluorescent BRs. Pulse-chase experiments combined with quantitative confocal microscopy were used to determine the internalization rates of BRs. This method is well suited to measure the internalization of other plant receptors if fluorescent ligands are available.

Key words Brassinosteroids, Endocytosis, Receptor-mediated endocytosis, Confocal microscopy, Live cell fluorescence microscopy, Plant endomembrane systems

1 Introduction

Cells are in constant communication with their environment with the delimiting plasma membrane (PM) forming the perception and interaction platform. Carefully balanced processes of endocytosis and exocytosis at the PM maintain the housekeeping functions of protein and lipid turnover, as well as specialized functions of transport and signaling. In recent years, a wealth of information on the endocytic route of PM proteins, such as the auxin efflux carriers PIN FORMED (PIN), the receptor kinases

Marisa S. Otegui (ed.), *Plant Endosomes: Methods and Protocols*, Methods in Molecular Biology, vol. 1209, DOI 10.1007/978-1-4939-1420-3_8, © Springer Science+Business Media New York 2014

BRASSINOSTEROID INSENSITIVE1 (BRI1) and FLAGELLIN-SENSING2 (FLS2), the tomato ethylene-inducing xylanase (LeEIX2) receptor, nutrient (e.g., boron and iron) uptake transporters and aquaporins, has elucidated cellular trafficking routes, their machinery and regulation in plant cells to be conserved with those of animals [1]. Tools to trace the endocytic route from the "outside" to the "inside" of live plant cells are limited to the lipophilic styryl dye *N*-(3-triethylammoniumpropyl)-4-(*p*-diethylaminophenyl-hexatrienyl) pyridinium dibromide (FM4-64) [2], which upon incorporation into the PM follows an endocytic route into the cell, labeling the *trans*-Golgi network/early endosomes (TGN/EEs) and late endosomes/multivesicular bodies (MVBs) en route to the vacuole, where it labels the tonoplast [3]. Although FM4-64 is an excellent bulk membrane dye to follow endocytosis, tools to specifically label and follow receptors or transporters into cells, and to link the relationship of their signaling or transport activity to their regulated intracellular trafficking, are not available.

The recent development of a synthetic fluorescently labeled brassinosteroid (BR) provides an excellent handle to follow the incorporation of ligand–receptor complexes into live cells, and to elucidate not only its trafficking route to the vacuole but also cellular positional information of the PM as the signaling platform for the receptor [4]. The steroidal hormone BR receptor BRI1 is one of the best studied receptor kinases in plants, which regulates a variety of plant developmental processes through a phospho-relay cascade of receptor/co-receptor and downstream effector phosphorylations [5]. BRI1 undergoes ligand-dependent and -independent endocytosis, constitutively recycling between the PM and TGN/EEs [6, 7], and passes through the MVBs to the vacuoles for degradation [8]. Castasterone (CS) was labeled with Alexa Fluor 647 (AF647) at the C6 keto position to yield Alexa Fluor 647-castasterone (AFCS). This C6 keto group was previously used to generate the biotin-tagged photoaffinity CS [9, 10] and is conducive for labeling according to the crystal structural analysis of the extracellular domain of BRI1 [11, 12]. AFCS is bioactive, although at a much lower level than brassinolide (BL), the most potent BR, and is endocytosed into *Arabidopsis* root cells together with the BRI1-GFP receptor. BRI1-AFCS complexes are present at the PM and transit through EEs (marked by the TGN/EE maker VHAa1-GFP) and MVBs (marked by the MVB marker AtRAB-F2b) to finally accumulate in the vacuole [4]. Genetic and pharmacological perturbations of the trafficking of BRI1-AFCS revealed clathrin- and ARF-GEF-dependent mechanisms of endocytosis. Linking pharmacological and genetic trafficking perturbations to biochemical signaling outputs highlights the PM as being a main signaling platform of BRI1 and endocytosis essential for signal attenuation. The successful application of AFCS in tracking the cell fate of plant hormones in living cells and in positioning BR signaling at the PM emphasizes the need to develop more labeled ligand–receptor couples in order to understand if they undertake similar signaling/trafficking routes or adopt novel mechanisms in plant cells.

Here, we present a method to pulse label and chase the fluorescently labeled BR, i.e., AFCS, and follow its intracellular route into epidermal cells of the *Arabidopsis* root meristem based on the protocol described in [4]. Depending on the pulse times, AFCS can be used to co-label different endomembrane markers or quantitatively measure uptake kinetics. This method can be used to quickly address the rates of BR endocytosis in different mutants and genotypes without the need for further genetic manipulations.

2 Materials

2.1 Solutions

1. Liquid growth medium: Half-strength Murashige and Skoog (1/2 MS) medium with 1 % (w/v) sucrose and 0.05 % 2-(*N*-morpholino) ethanesulfonic acid (MES) monohydrate. For 1 L, 2.2 g MS basal medium salts (Sigma-Aldrich), 10 g sucrose, 0.5 g MES. Adjust the pH to 5.7 with KOH and adjust final volume to 1 L with water. Autoclave to sterilize. Aliquot out liquid medium sterilely into 15-mL tubes for ease of use during treatments and microscopy. Check the pH of liquid medium with pH paper before the start of treatments (*see* **Note 1**).
2. Solid growth medium: 1/2 MS medium prepared as above and supplemented with 0.8 % (w/v) Agar (LabM Limited). Add 500 mL liquid medium to 500 mL DURAN® glass bottles with 4 g Agar. Autoclave to sterilize. The sterilized medium can be stored for several months. Melt medium in a microwave and pour square plates (50 mL per plate) as required.
3. Alexa Fluor 647-Castasterone (AFCS) [4] stock solution: 2–5 mM AFCS in DMSO (Sigma-Aldrich) (batch dependent, *see* **Note 2**). To generate the dose-response curves (*see* Subheading 3.2.2), 200 μL of AFCS ranging from 40 μM to 2.5 μM (by serially diluting the AFCS stock in liquid medium) were used.
4. Brassinazole (BRZ) (Fuji Chemical Industries, Ltd., Toyama, Japan) stock solution: 1 mM BRZ in DMSO (Sigma-Aldrich).
5. Tyrphostin 23 (TyrA23) (Sigma-Aldrich) stock solution: 50 mM TyrA23 in DMSO (Sigma-Aldrich).

2.2 Plant Material

1. *Arabidopsis thaliana* (L.) Heyhn (ecotype Columbia-0, Col-0; referred to as *Arabidopsis*).
2. *bri1-116* [13].

2.3 Equipment and Other Materials

1. Growth chamber.
2. 500 mL DURAN® GL 45 laboratory glass bottles (Schott, Germany).
3. Square plates (Greiner Bio One Labortechnik, Solingen, Germany).
4. Round Petri plates 90 mm (Gosselin).
5. Air-permeable tape 2.5 cm (Micropore™; 3 M, St. Paul, MN, USA).
6. Paper towels.
7. Parafilm "M" (Pechiney Plastic).
8. Microscope slides: 76 mm × 26 mm × 1 mm.
9. Cover slips: 22 × 50 mm, No. 1.5.
10. Confocal microscope with a 635 nm Laser, 10× objective NA 0.4, 60x water immersion objective NA 1.2.
11. NanoDrop ND-1000 UV–Vis Spectrophotometer (Thermo Scientific).

2.4 Analysis Software

1. Software from confocal microscope manufacturer.
2. ImageJ (http://rsbweb.nih.gov/ij/).
3. SigmaPlot (Systat Software Inc.).

3 Methods

3.1 Surface Sterilization and Plant Growth

1. Surface-sterilize not more than 50 μL of seeds in a 1.5 mL microcentrifuge tube. For a quick protocol, add 1 mL of 95 % ethanol, shake, and let it stand for 1 min. Carefully decant ethanol on tissue paper in a sterile flow bench, place the tubes horizontally and let the seeds dry completely (~1 h). The tubes with seeds can be capped and stored sterilely (*see* **Note 3**). With a wetted sterile toothpick, place seeds on the surface of the solid growth medium in square plates. Seal the plates with air-permeable Micropore™ tape.
2. Store the plates with seeds for 2–5 days at 4 °C to synchronize germination.
3. Place the plates upright in a growth chamber at 22 °C with a 16-h/8-h light-dark photoperiod (110 μE/m^2/s photosynthetically active radiation supplied by cool-white fluorescent tungsten tubes; Osram, Munich, Germany). Let the seedlings grow for 4 days.

3.2 Uptake Assay for AFCS

The uptake of AFCS is very rapid: the fluorescent BRs label the PM, followed by the TGN/EEs, the MVBs and finally accumulate in the vacuole. We take advantage of vacuolar accumulation of AFCS

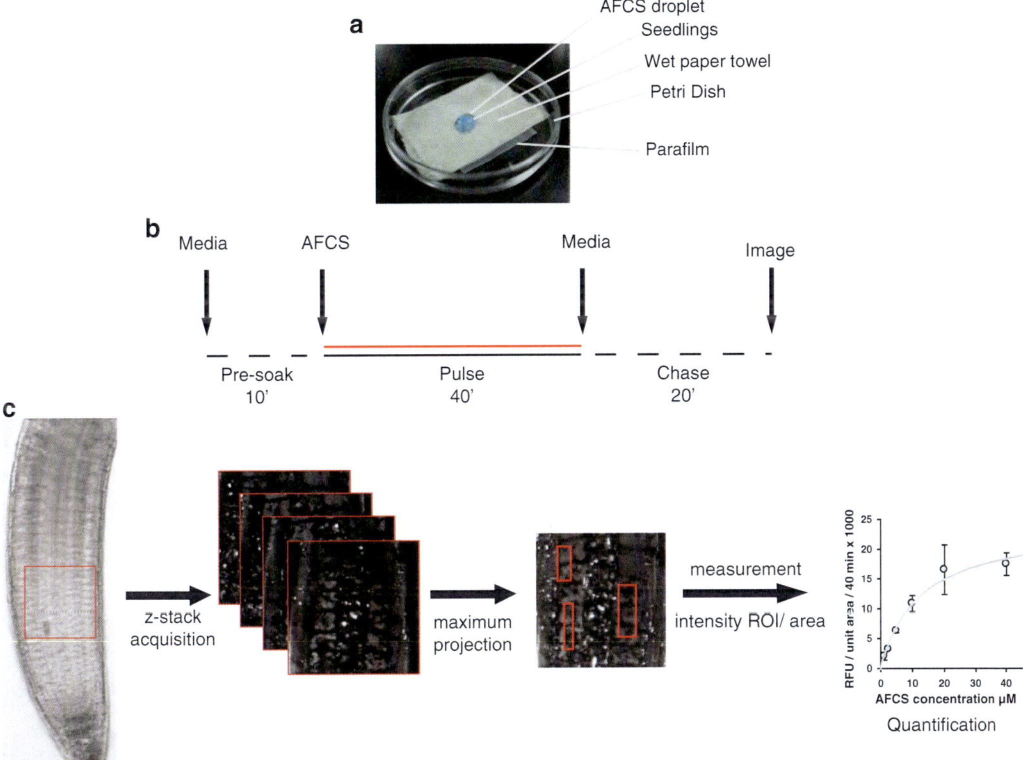

Fig. 1 Uptake and imaging setup for quantification of AFCS internalization. (**a**) Mounted humidified chamber with seedlings pulsed for AFCS uptake. (**b**) Overview of experimental setup for pulse-chase AFCS uptake indicating the timing of each step. (**c**) Schematics of workflow for the imaging and quantification of AFCS accumulation in the vacuole. *Z*-stacks are acquired in the root meristem, 10–15 cells above the QC, as marked by the *red square* (*right*). Maximum projections of four images covering the epidermal layer are used for measuring the average signal intensity/area values of square ROIs. The graph represents an example of a quantitative AFCS dose-response, plotting the signal intensity/area ratio on AFCS concentration. AFCS, Alexa Fluor 647-castasterone. QC, quiescent center. ROI, region of interest

to quantitatively measure the rate of AFCS uptake into the cell [4]. We describe here a basic uptake protocol to follow the endocytic route of AFCS into the *Arabidopsis* root meristem cells (Fig. 1a, b).

3.2.1 Basic AFCS Pulse-Chase Setup

1. Prepare a humidified chamber. Place a stack of 4 paper towels cut to an approximate size (~6 cm × 6 cm) to fit the base of a round 90 mm Petri plate. Soak the towels with Milli-Q water. Place a 5 cm × 5 cm square of Parafilm on the wet paper towels and close the lid of the plate.
2. Place 0.5 mL of liquid growth medium in the center of the Parafilm and pre-soak the *Arabidopsis* seedlings by placing four of them into the drop of medium for 10 min, taking care that all the roots are immersed.

3. Dilute the AFCS stock to 20 μM in 200 μL liquid growth medium in a 0.5 mL microcentrifuge tube (*see* **Note 2**).
4. Remove the medium from the seedlings using a 1-mL Pipetman, making sure that most of the medium, if not all, has been aspirated out.
5. Quickly add the AFCS dilution to the seedlings to start the pulse time. With the Pipetman tip, gently swirl the seedlings in the droplet to ensure that the roots are completely immersed.
6. Replace the lid of the Petri plate and wait for the designated pulse interval.
7. For the chase, aspirate out the AFCS solution and quickly replace with 0.5 mL of liquid growth medium. Repeat this wash step for 6 times ensuring the roots of the seedlings are always immersed in the medium and changing pipette tips between washes to prevent cross contamination of the stock liquid growth medium with residual AFCS (*see* **Note 4**).
8. Either process immediately (chase time ~2 min) for imaging the co-localization with the endomembrane markers [4] or replace the Petri plate lid and wait for longer chase times (~30 min) for kinetic measurements to ensure complete uptake of ligand into the vacuole.

3.2.2 Dose-Response Curves to Measure the Rate of Uptake

1. Set up the *Arabidopsis* seedlings for the basic uptake protocol as described in Subheading 3.2.1.
2. Pulse four 4-day-old seedlings that have been pre-soaked in liquid medium with AFCS as described in Subheading 3.2.1 for 40 min.
3. Wash seedlings thoroughly 6 times with 0.5 mL of liquid growth medium and wait for 20 min. This ensures that AFCS from the PM and endosomal compartments has been internalized into the vacuoles and most, if not all, AFCS has been washed away from the intracellular airspaces, avoiding high backgrounds.
4. Mount the seedlings on a glass slide with liquid medium, cover with a coverslip and image as described in Subheading 3.3.

3.2.3 Uptake Assays in BR Mutants

BR deficient or perception mutants have defects in cell elongation which in turn affects cell shape and vacuolar morphology of the epidermal cells in the root meristem. Vacuoles in these mutant cells will have lower volumes and become more tubular. This change in vacuolar volume/morphology significantly influences the quantification of the AFCS signal, thus leading to false rates of uptake at the PM. This effect of the change of the shape of the vacuoles and cells can be easily normalized across genotypes by growing the seedlings on brassinazole (BRZ) [14], a BR biosynthetic inhibitor, so that cells phenocopy BR deficiency, and the uptake assays can be relatively free from the influence of vacuolar shape as well as inhibition from endogenously synthesized BRs [4].

1. To phenocopy BR deficient mutants, germinate wild-type (Col-0) seeds directly on germination medium with 1 μM BRZ. Assay these seedlings together with the mutants at 7 days.
2. For lines that segregate mutant alleles (e.g., *bri1-116*), germinate heterozygous seeds vertically on half-strength MS with 1 % sucrose for 5 days. Select mutants (cabbage phenotype: stunted hypocotyls, dark green epinastic cotyledons, shorter roots) and transfer to medium with 1 μM BRZ for an additional 2 days.
3. Carry out the basic uptake protocol as described in Subheading 3.2.1 and image (*see* Subheading 3.3).
4. Single time point, single dose uptake values may be performed and AFCS signal intensity quantified to quickly evaluate the relative rates of AFCS uptake across genotypes. Alternatively, uptake kinetics based on dose-response curves can be performed in mutant genotypes following Subheading 3.2.2.

3.2.4 Uptake of AFCS in the Presence of Pharmacological Inhibitors

Pharmacological inhibitors that block trafficking of proteins and cargo at specific sites in the endomembrane system are excellent tools for functional studies of endocytosis [4, 15]. Thus changes in compartment morphology, signaling and trafficking can be very quickly analyzed without the need for genetic manipulations. We use here the example of the endocytosis inhibitor TyrA23, a drug that inhibits clathrin-dependent endocytosis [16]. TyrA23 blocks the entry of AFCS into the cell resulting in no AFCS signal observed in the vacuoles [4].

1. Pretreat 4 seedlings in 0.5 mL of 75 μM TyrA23 or DMSO as a control in liquid growth medium for 30 min in the humidified chamber described in Subheading 3.2.1.
2. Prepare 200 μL of liquid growth medium with 10 μM AFCS and 75 μM TyrA23 or DMSO as the control in 0.5 mL microcentrifuge tubes.
3. Start the pulse by replacing the pretreatment solution with the AFCS solution together with and without the inhibitor TyrA23 for 30 min.
4. Start the chase by washing seedlings 6 times with 500 μL of liquid growth medium containing either 75 μM TyrA23 or DMSO.
5. Mount on microscope slides in TyrA23- or DMSO-containing liquid growth medium and observe immediately as described in Subheading 3.3.

3.3 Image Acquisition

1. Using forceps, carefully place the 4 seedlings on a clean microscope slide with 150–200 μL of liquid growth medium so that the roots point in the same direction. Gently place a glass cover slip avoiding air bubbles and squashing the roots. Wipe out excess medium.
2. Image on a laser scanning confocal microscope using a 10× dry and 60× water immersion objectives.
3. Using the 10× dry objective, mark the position of the seedlings at the root meristem zone of *Arabidopsis* for ease of imaging at 60×.
4. Place a drop of water on the coverslip and switch to the 60× water immersion objective, taking care not to squash the sample.
5. Keep the zone of imaging consistent. Image the epidermal cells of the root meristem zone 10–15 cells above the quiescent center (QC) (Fig. 1c). This can be achieved by adjusting the stage such that the QC of the root apical meristem is at one end of the field of view, rotate the scan field such that the QC is as the bottom of the image. Switch to zoom 3 (or an equivalent zoom to give an *xy* resolution of ~100 nm/pixel).
6. Adjust the Z position so that you start just above the epidermal cells of the meristem (in the lateral root cap) and end below the epidermal cells (cortex). Capture a *Z* stack ~5–6 slices at 2.5 μm apart covering >10 μm of epidermal cells in the *Arabidopsis* root meristem zone (cells in this zone are ~10 μm × 10 μm).
7. Move automatically or manually to the next "marked" seedling and repeat **steps 5** and **6**.
8. For quantitative measurements: confocal settings between all samples should be the same; imaging of all samples must be completed within one hour (~8 min per slide of 4 seedlings). Typically for the quick quantitation of the AFCS emission signal we use an Olympus FluoView 1000 microscope with the following settings: image size 512 × 512, unidirectional scanning, sampling speed at 10 μs/pixel, 30 % 635 nm HeNe laser, either 650 nm long pass (LP) glass filter or set the AOTF emission filters 650–750 nm, pinhole 140 nm, photomultiplier tube (PMT) gain voltage at 740 (keep signal below saturation levels), without averaging.
9. For co-localization experiments: image acquisitions may be adjusted to acquire the best image. Finer stacks at 0.5 μm between stacks may be acquired for 3D reconstruction and co-localization analysis.

3.4 Image Analysis and Quantification

1. Image analysis and signal was quantified using measurement function of the Olympus FluoView software. Alternatively ImageJ may also be used to measure signal intensities.
2. Choose four slices covering the epidermal cell layer, merge them using maximum projection.
3. Mark region of interests (ROIs) in the projection for regions of the stack that uniformly cover entire meristematic cells (sometimes roots are placed obliquely and cells are not flat in 3D). Mask out regions (intracellular spaces, mucilage from lateral root caps) that are heavily stained with the dye to prevent false signals (*see* **Note 5**) (Fig. 1c).
4. Use the measurement tool to extract values for intensity and area from the marked ROIs for each seedling.
5. Export the values into an excel sheet and normalize the signal intensity to the unit area.
6. For dose-response curves, use a statistical software package, such as SigmaPlot, to global fit the data to a regression curve for kinetics.
7. Normalize to wild-type values to calculate the percentage increase or decrease in uptake across genotypes (*see* **Note 6**).

4 Notes

1. Milli-Q water (Millipore) is used for the preparation of all media and solutions.
2. Since the amounts of AFCS synthesized are relatively very small (μg scale) due to the costly starting materials, the dried aliquots are typically dissolved in small amounts of DMSO (~20 μL). A 1/20 to 1/50 dilution is made in 50 μL of liquid medium and the OD_{650nm} is measured on a NanoDrop using 2 μL of the dilution. The AFCS stock concentration is back-calculated using the Beer-Lambert law: $A = \varepsilon\ C\ l$, where the extinction coefficient for AF647 is $\varepsilon_{647} = 239{,}000\ M^{-1}\ cm^{-1}$ and the path length of the NanoDrop is 0.1 cm.
3. If the seeds are heavily contaminated, for surface sterilization use a 50 % (v/v) house-hold bleach to water solution with 0.05 % Tween 20 (freshly made) for 10 min and wash thoroughly 6 times with 1 mL of sterile distilled water. Plate seeds on the germination medium with a pipette using the water from the last wash.
4. For the chase, it is important to wash the seedlings well with liquid growth medium to remove as much unbound AFCS from the extracellular spaces for the acquisition of good images and the quantitation of signals inside the vacuoles.

AF647 (ε_{647} = 239,000 M^{-1} cm^{-1}) is highly fluorescent and nM amounts can be imaged.

5. For convenience of processing, a large number of images for quantitation of the vacuolar AFCS signal in the root meristem cells, rectangular ROIs are drawn over multiple cells including regions of vacuole, cytoplasm and nucleus. This normalizes AFCS intensities in different vacuolar shapes and volumes because multiple cells are chosen. Alternatively, circular ROIs may be drawn in the vacuolar vs cytoplasmic areas to background correct taking care to acquire multiple ROIs for both the vacuolar and cytoplasmic areas.
6. Relative quantities of cellular AFCS uptake can be back calculated from calibration curves generated in a 96-well plate with different dilutions of AF647. AF647-cadaverine can be used instead of AFCS since the castasterone by itself does not contribute significantly to the extinction coefficient of AF647 (ε_{647} = 239,000 M^{-1} cm^{-1}). Briefly, for the calibration of signal intensity, known quantities of AF647-cadaverine ranging from 1 nM to 4 μM were dissolved in 150 μL plant medium and aliquoted into a coverslip bottomed 96-well plate. Wells were imaged with the same settings as described in Subheading 3.3. Fluorescent beads were used to mark the bottom of the plate to acquire image stacks near the coverslip. Fluorescence intensities from ROIs of a maximum projected stack of four slices, 2.5 μm apart, were quantified and normalized to the area. From our experimental setup we estimated the relative amounts of AFCS in the cells to be in the nM range for the indicated time points, which is in the physiological range.

Acknowledgements

We thank A. Bleys for help with manuscript preparation. This work is supported by the Odysseus program of the Research Foundation-Flanders and the BRAVISSIMO Marie-Curie Initial Training Network (predoctoral fellowships to S.D.R).

References

1. Chen X, Irani NG, Friml J (2011) Clathrin-mediated endocytosis: the gateway into plant cells. Curr Opin Plant Biol 14:674–682
2. Vida TA, Emr SD (1995) A new vital stain for visualizing vacuolar membrane dynamics and endocytosis in yeast. J Cell Biol 128:779–792
3. Jelínková A, Malínská K, Simon S et al (2010) Probing plant membranes with FM dyes: tracking, dragging or blocking? Plant J 61: 883–892
4. Irani NG, Di Rubbo S, Mylle E et al (2012) Fluorescent castasterone reveals BRI1 signaling from the plasma membrane. Nat Chem Biol 8:583–589
5. Wang Z-Y, Bai M-Y, Oh E et al (2012) Brassinosteroid signaling network and regulation of photomorphogenesis. Annu Rev Genet 46:701–724
6. Russinova E, Borst J-W, Kwaaitaal M et al (2004) Heterodimerization and endocytosis of

Arabidopsis brassinosteroid receptors BRI1 and AtSERK3 (BAK1). Plant Cell 16:3216–3229

7. Geldner N, Hyman DL, Wang X et al (2007) Endosomal signaling of plant steroid receptor kinase BRI1. Genes Dev 21:1598–1602
8. Viotti C, Bubeck J, Stierhof Y-D et al (2010) Endocytic and secretory traffic in *Arabidopsis* merge in the trans-Golgi network/early endosome, an independent and highly dynamic organelle. Plant Cell 22:1344–1357
9. Caño-Delgado A, Yin Y, Yu C et al (2004) BRL1 and BRL3 are novel brassinosteroid receptors that function in vascular differentiation in *Arabidopsis*. Development 131: 5341–5351
10. Kinoshita T, Caño-Delgado AC, Seto H et al (2005) Binding of brassinosteroids to the extracellular domain of plant receptor kinase BRI1. Nature 433:167–171
11. Hothorn M, Belkhadir Y, Dreux M et al (2011) Structural basis of steroid hormone perception by the receptor kinase BRI1. Nature 474:467–471
12. She J, Han Z, Kim T-W et al (2011) Structural insight into brassinosteroid perception by BRI1. Nature 474:472–476
13. Li J, Chory J (1997) A putative leucine-rich repeat receptor kinase involved in brassinosteroid signal transduction. Cell 90:929–938
14. Asami T, Min YK, Nagata N et al (2000) Characterization of brassinazole, a triazole-type brassinosteroid biosynthesis inhibitor. Plant Physiol 123:93–100
15. Irani NG, Russinova E (2009) Receptor endocytosis and signaling in plants. Curr Opin Plant Biol 12:653–659
16. Banbury AN, Oakley JD, Sessions RB et al (2003) Tyrphostin A23 inhibits internalization of the transferrin receptor by perturbing the interaction between tyrosine motifs and the medium chain subunit of the AP-2 adaptor complex. J Biol Chem 278:12022–12028

Chapter 9

Analysis of Prevacuolar Compartment-Mediated Vacuolar Proteins Transport

Caiji Gao, Yi Cai, Xiaohong Zhuang, and Liwen Jiang

Abstract

Transient expression using protoplasts is a quick and powerful tool for studying protein trafficking and subcellular localization in plant cells. Prevacuolar compartments (PVCs) or multivesicular bodies (MVBs) are intermediate compartments that mediate protein transport between late Golgi or trans-Golgi network (TGN) and vacuole. Both wortmannin treatment and ARA7(Q69L) expression can induce PVC homotypic fusion and PVC enlargement in plant cells. Here, we describe detailed protocols to use transient expression of protoplasts derived from *Arabidopsis* suspension culture cells for studying protein trafficking and localization. Using three GFP-tagged vacuolar cargo proteins and RFP-tagged PVC membrane marker as examples, we illustrate the major tools and methods, including wortmannin treatment, ARA7(Q69L) expression and immunoblot analysis, to analyze PVC-mediated vacuolar protein transport in plant cells.

Key words *Arabidopsis* suspension culture cells, Protoplast, Prevacuolar compartment, Transient expression, Vacuolar transport, Wortmannin treatment, ARA7(Q69L) expression

1 Introduction

Plant vacuole, a membrane-bound organelle crucial for plant growth, has multiple functions including storage of nutrients and metabolites, generation of turgor, protein degradation, and plant defense [1, 2]. It is generally believed that proteins reach vacuoles via a typical ER (endoplasmic reticulum)—Golgi—TGN (trans-Golgi network)—PVC (prevacuolar compartment)—vacuole pathway in the plant endomembrane system. PVCs are intermediate compartments that mediate protein traffic between TGN and vacuoles [3–5]. PVC in the secretory pathway also serves as a late endosomal compartment for protein transport from the plasma membrane (PM) to vacuole in the plant endocytic pathway [4, 6, 7].

By using vacuolar sorting receptor (VSR) proteins and their GFP fusions as markers in immunogold electron microscope (EM) studies, plant PVCs were identified as multivesicular bodies (MVBs) with a typical size of approximately 200 nm in diameter [3].

Marisa S. Otegui (ed.), *Plant Endosomes: Methods and Protocols*, Methods in Molecular Biology, vol. 1209,
DOI 10.1007/978-1-4939-1420-3_9,

In transgenic tobacco BY-2 or *Arabidopsis* cell lines expressing GFP-tagged PVC proteins (e.g., GFP-AtVSR fusion), PVCs were visible as puncta by confocal imaging [3, 8]. Interestingly, treatment of these transgenic cells with wortmannin, an inhibitor of phosphatidylinositol 3-kinases, causes the PVCs to form ring-like structures when observed in a confocal microscope due to wortmannin-induced PVC homotypic fusion [3, 8, 9]. More recently, overexpression of ARA7(Q69L), a GTP-bound, constitutively active Rab5 mutant, has also been shown to induce PVCs to form ring-like structures in plant [10–12]. Therefore, both wortmannin treatment and overexpression of ARA7(Q69L) cause GFP-tagged PVCs to form ring-like structures that are visible under confocal imaging. However, compared to the exogenous wortmannin treatment, ARA7(Q69L) is an endogenous and more specific tool for studying PVC-mediated protein trafficking because its expression does not perturb the endocytic pathway and exerts stronger effect on PVC enlargement [10].

Transient expression using protoplasts derived from suspension culture cells has been a very efficient and reliable system for studying protein subcellular localization and PVC/MVB-mediated vacuolar protein transport in the plant endomembrane system [13–15]. PVC-mediated transport of proteins to vacuole can be quickly evaluated via the transient co-expression of fluorescently labeled cargo proteins and a PVC marker in protoplasts using confocal imaging [11, 15]. However, since PVCs are approximately 200 nm in diameter, they can only be visualized as puncta under confocal imaging, what makes it difficult to study the relative distribution of proteins in the lumen and limiting membrane of PVCs. However, this limitation can be easily solved by treating cells with wortmannin or co-expressing ARA7(Q69L) in a transient expression protoplast system. Indeed, both the use of wortmannin and expression of ARA7(Q69L) have become convenient and reliable methods to distinguish the detailed subcellular distributions and fates of different PVCs-localized proteins in plant cells [11, 15]. Here we explain the use of four fluorescently labeled proteins that trafficked through PVCs to illustrating PVC protein distribution analysis under wortmannin treatment or ARA7(Q69L) expression (Fig. 1): aleurain-GFP (a vacuolar soluble protein consisting of 45 amino acids N-terminal part of aleurain fused to GFP), mRFP-VSR2 (integral membrane protein), EMP12-GFP (a PVC/vacuole-localized integral membrane protein), and LRR84A-GFP (a PM/PVC/vacuole-localized integral membrane protein) [11, 15, 16]. In both wortmannin-treated and ARA7(Q69L)-expressing *Arabidopsis* protoplasts, aleurain-GFP, EMP12-GFP, and LRR84A-GFP were shown to be present in the lumen of the enlarged ring-like PVCs (marked by mRFP-VSR2) (Fig. 2). The lumenal localization of EMP12-GFP and LRR-GFP in the enlarged PVCs

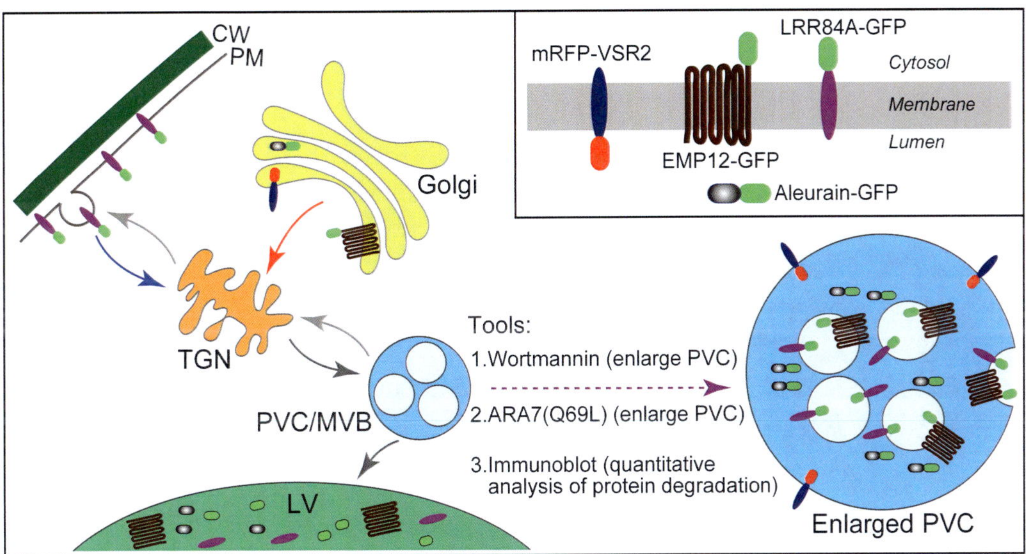

Fig. 1 Tools and methods for studying PVC-mediated vacuolar transport. In this working model, the secretory PVC/MVB also serves as the late endosome for the endocytic pathway from the plasma membrane (PM) that mediated the vacuolar transport of proteins from the endocytic pathway (e.g., LRR84A-GFP) and the secretory pathway (e.g., aleurain-GFP and EMP12-GFP). mRFP-VSR2 is used as a PVC marker. Owing to the small size of PVCs (200 nm) and the limited resolution of confocal microscopy, it is usually very difficult to distinguish the lumenal or membrane localization of proteins in PVCs under confocal imaging. However, in cells subjected to wortmannin treatment or ARA7(Q69L) expression, PVCs are induced to form enlarged ring-like structures that can be easily visualized under confocal imaging. This can be a useful tool to study the distribution of PVC-localized proteins and thus their fate. For example, the functional membrane protein (e.g., mRFP-VSR2) will localize on the limiting membrane of the enlarged PVCs and visualized as ring-like structures, whereas the membrane proteins destined for degradation in the vacuole (e.g., EMP12-GFP and LRR84A-GFP) will be internalized inside the PVC lumen and thus visualized as a central core, which showed similar localization pattern as the soluble vacuolar cargo aleurain-GFP within the enlarged PVCs. The degradation of the internalized proteins (e.g., EMP12-GFP and LRR84A-GFP) can be further verified and quantified via western blotting analysis with anti-GFP antibodies for detection of the cleaved GFP core in the soluble fraction. *CW* cell wall, *LV* lytic vacuole, *MVB* multivesicular body, *PM* plasma membrane, *PVC* prevacuolar compartment, *TGN* trans-Golgi network

indicated the degradation of these two C-terminally GFP-fused membrane proteins. The vacuolar transport of the soluble cargo (aleurain-GFP) and the degradation of these two membrane proteins (EMP12-GFP and LRR84A-GFP) can be further analyzed and quantified by immunoblot analysis with GFP antibodies, showing large amount of cleaved GFP in the soluble fraction (Fig. 3). The following protocol explains the analysis of subcellular localization and distribution of proteins within PVCs and thus, PVC-mediated protein trafficking in plant cells using a combination of transient protoplast expression, wortmannin treatment, ARA7 (Q69L) expression, and immunoblot analysis (Fig. 1).

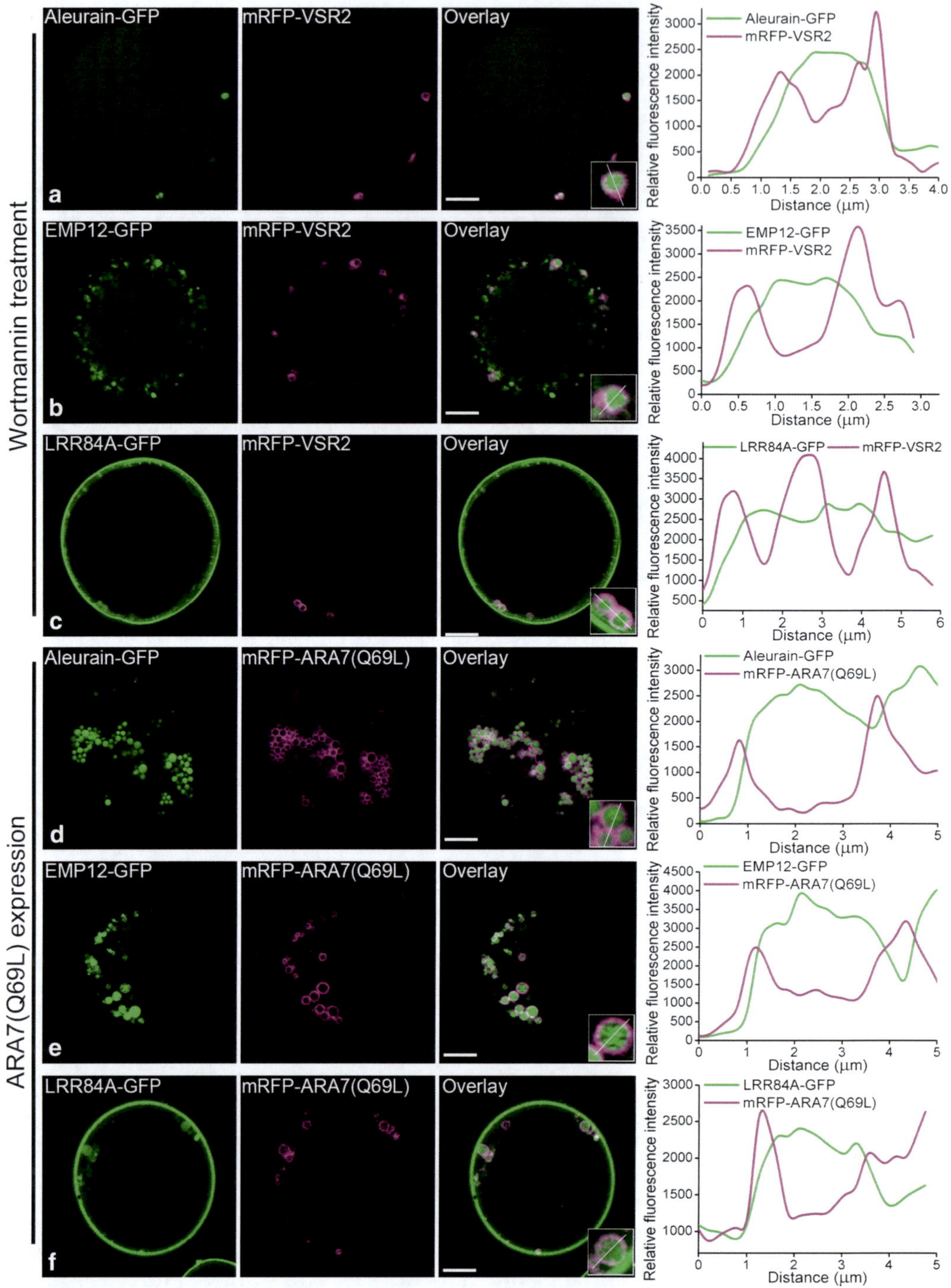

Fig. 2 Cargo proteins destined to vacuolar degradation are trapped in the lumen of the enlarged PVCs. Various GFP fusion proteins destined for the vacuole (aleurain-GFP, EMP12-GFP, and LRR84A-GFP) were co-expressed with the PVC marker mRFP-VSR2 (**a**–**c**) or with the constitutively active Rab5 mutant mRFP-ARA7(Q69L) (**d**–**f**) in *Arabidopsis* protoplasts, followed by confocal imaging at 12–14 h after transfection. Samples in **a**–**c** were subjected to 16.5 μM wortmannin treatment for 1 h prior to confocal imaging. The insets represented 3× enlargement of the selected area for better assessment. The corresponding distributions of GFP and mRFP signals along the line were further analyzed by Olympus FV1000 confocal software (FV10-ASW version 01.07.02.02) and presented on the right panels. Scale bar = 10 μm

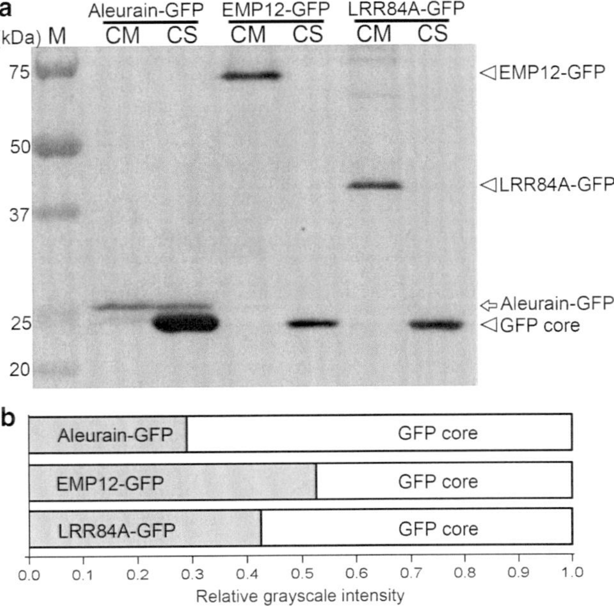

Fig. 3 Immunoblot analysis of protoplasts expressing vacuolar GFP fusion proteins. (**a**) *Arabidopsis* protoplasts expressing three GFP-fused vacuolar cargo proteins were subjected to protein isolation into cell soluble (CS) and cell membrane (CM) fractions, followed by protein separation via SDS-PAGE and western blot analysis using anti-GFP antibodies. *M* Molecular weight marker in kilodaltons. (**b**) The relative gray scale intensity (scale 0–1.0) of the two protein bands representing the full-length GFP fusion and the cleaved GFP core of individual GFP fusion was quantified using ImageJ software, in which the combinational intensity of the full-length GFP fusion and its cleaved GFP core was expressed as 1.0

2 Materials

2.1 Plant Material and Culture Condition

1. *Arabidopsis* MS medium: 4.3 g/l Murashige and Skoog Basal Salt Mixture, 100 mg/l myo-inositol, 0.4 mg/l thiamine hydrochloride, 50 mg/l kinetin, 800 mg/l 1-naphthaleneacetic acid, and 30 g/l sucrose, pH 5.7 (with KOH) (*see* **Note 1**).
2. *Arabidopsis thaliana* suspension cells PSB-D (ecotype Landsberg erecta) are cultured in *Arabidopsis* MS medium in 250 ml flask at 25 °C in light-protected shaker at 130 rpm, and are subcultured once per 5 days by transferring 2.5 ml of old cells into 50 ml of fresh medium (*see* **Note 2**).

2.2 Protoplast Isolation

1. Protoplast culture medium: 4.3 g/l Murashige and Skoog Basal Salt, 0.4 M sucrose (13.7 %), 500 mg/l MES hydrate, 750 mg/l $CaCl_2 \cdot 2H_2O$, 250 mg/l NH_4NO_3, pH 5.7 (with KOH) (*see* **Note 3**).
2. Enzyme solution: 1 % Cellulase "ONOZUKA" RS (Yakult Honsha), 0.05 % Pectinase, 0.2 % Driselase from Basidiomycetes

spp. in protoplast culture medium (adjust pH to 5.7). Filter the enzyme solution through a 0.22 μm syringe filter device into a 50 ml conical tube (*see* **Note 4**).

3. Laminar flow hood.
4. 50 ml conical tube.
5. Centrifuge with swinging bucket rotor for 50 ml conical tubes (for example, Eppendorf 5810R).

2.3 Protoplast Transfection

1. Electroporation buffer: 0.4 M Sucrose (13.7 %), 2.4 g/l HEPES, 6 g/l KCl, 600 mg/l $CaCl_2 \cdot 2H_2O$, pH 7.2 (with KOH) (*see* **Note 3**).
2. Electroporation cuvettes with 4 mm gap.
3. Electroporation system (e.g., Bio-Rad Gene Pulser Xcell System). The machine for electroporation should reach the voltage to 130 V and the capacitance to 1,000 μF.
4. Peristaltic pump.
5. 6-Well cell culture plate.

2.4 Drug Treatment and Confocal Imaging

1. Wortmannin solution: wortmannin dissolved in DMSO at a final concentration of 1.65 mM. Aliquot and store at −20 °C (*see* **Note 5**).
2. Confocal microscope (for example, Olympus Fluoview FV1000).
3. Microscope slides and coverslip.

2.5 Western Blotting Analysis

1. 250 mM NaCl.
2. Protein extraction buffer: 25 mM Tris–HCl pH 7.5, 150 mM NaCl, 1 mM EDTA, 1× protease inhibitor cocktail.
3. Benchtop centrifuge (for example, Effendorf 5417R).
4. SDS–polyacrylamide gel electrophoresis (SDS-PAGE) system.

3 Methods

3.1 Protoplast Isolation

1. Prepare 30 ml enzyme solution in protoplast culture medium and sterilize the enzyme solution by passing through a 0.22 μm syringe filter (*see* **Note 6**).
2. Transfer 50 ml of 5-day-old *Arabidopsis* culture cells into 50 ml conical tubes in a laminar flow hood. Pellet the cells by centrifugation at 100 × *g* for 5 min at room temperature, remove the supernatant.
3. Add the enzyme solution to the cell pellet, carefully resuspend the cells and transfer the mixture to a 150 ml flask. Incubate the flask at 25 °C for 3 h in a light-protected shaker shaking at 130 rpm for cell digestion and protoplast release.

4. After 3 h, check protoplasts under light microscope. If the cell culture is in good condition, usually more than 95 % of the cells will form individual and round shaped protoplasts.
5. Transfer the protoplasts into a 50 ml conical tube and centrifuge the protoplasts at room temperature for 10 min at 100 × *g* using a swinging bucket rotor. The protoplasts in good condition will float to the top after centrifugation (*see* **Note 7**).
6. Connect a Pasteur pipette to a peristaltic pump. Insert the Pasteur pipette tip through the floating protoplasts layer in hood and suck out the underlying solution until the floating protoplasts become close to bottom (*see* **Note 8**).
7. Add 35 ml of electroporation buffer into the protoplasts and mix them gently, centrifuge again at 100 × *g* for 10 min.
8. Remove the electroporation buffer using peristaltic pump and repeat the washing **step** 7 twice.
9. Resuspend the protoplasts gently using electroporation buffer in an appropriate volume to obtain 2–5 × 10^6 protoplasts/ml by counting with a hemacytometer in a light microscope. Generally, this can be achieved by resuspending the protoplasts in two to three volumes of electroporation buffer. For example, 2 ml of healthy floating protoplasts can be diluted with 4–6 ml of electroporation buffer.

3.2 Protoplast Transfection

1. Aliquot 500 μl of the resuspended protoplasts into 4 mm gap electroporation cuvettes using top-cut 1 ml pipette tips (*see* **Note 9**).
2. Mix 40 μg plasmid DNA with electroporation buffer to a final volume of 100 μl (*see* **Note 10**).
3. Mix the plasmid DNA and protoplasts gently by flipping on side, and incubate for 5 min at room temperature.
4. After 5 min, the protoplasts will float to the top. Before electroporation, carefully resuspend the protoplast and DNA mixture by gently tapping the cuvette. Electroporate the protoplasts at 130 V (voltage) and 1,000 μF (capacitance) for one pulse. The pulse time can range from 25 to 50 ms.
5. Incubate the electroporated protoplasts at room temperature for 15 min.
6. Add 1 ml of protoplast culture medium to each cuvette and pour the protoplast suspension into a 6-well plate. Add another 1 ml of protoplast culture medium to wash the cuvette and pour the solution into the above 6-well plate.
7. Incubate the transfected protoplasts at 26 °C incubator for desirable time (*see* **Note 11**).

3.3 Drug Treatment and Confocal Imaging

1. Transfer the protoplasts from the 6-well plate to a 2 ml microcentrifuge tube.
2. For the wortmannin treatment, add 5 μl 1.65 mM wortmannin stock solution to 495 μl of protoplasts to reach the final concentration of the drug at 16.5 μM, mix them gently by inverting the tube for several times. Incubate the mixture at room temperature for 1 h.
3. Usually after 10 min, the protoplasts will float up to the top of buffer in microcentrifuge tube. Transfer 30 μl concentrated protoplasts from the top layer to a microscope slide; carefully cover it with the coverslip for immediate confocal observation (*see* **Note 12**).
4. Fluorescent proteins in the protoplasts can be easily imaged under the 60× water lens or a 100× oil lens. To avoid the movement of protoplasts during confocal observation, the microscope slide and coverslip can be precoated with polylysine.
5. Fluorescence signal distribution can be analyzed using the software provided by the confocal microscope manufacturer or specialized image analysis software. An example of the analysis of confocal results is shown in Fig. 2.

3.4 Western Blot Analysis

1. Transfer 2.5 ml of transfected protoplasts to a 15 ml conical centrifuge tube. Add 7.5 ml of 250 mM NaCl to dilute the protoplasts (*see* **Note 13**).
2. Centrifuge at 100 × *g* for 10 min and collect the protoplasts.
3. To collect the secreted protein in the incubation medium, transfer the supernatant from **step 2** to a new 15 ml conical tube, add 1/5 volume of cold TCA, mix, and incubate on ice for 10 min. Centrifuge at 20,000 × *g* at 4 °C for 15 min. Remove the supernatant and wash the protein pellet using cold acetone twice.
4. To prepare the cell extracts from protoplasts, add 200–500 μl of ice cold protein extraction buffer to the protoplasts pellet from **step 2**. Incubate protoplasts on ice and lysis the protoplasts by passing through a 1 ml syringe with needle (25G × 5/8 in.) for 15 times.
5. Spin the lysis mixture at 100–600 × *g* for 3 min to remove intact cells and large cellular debris.
6. Remove the supernatant to a new microcentrifuge tube and centrifuge at the maximum speed 25,000 × *g* on a benchtop centrifuge at 4 °C for 30 min.
7. The supernatant and pellet can be referred to as soluble and membrane fractions, respectively. Transfer the supernatant to a new microcentrifuge tube. The remaining membrane pellet

can be further dissolved using the protein extraction buffer containing 1 % Triton X-100 or SDS.

8. Proteins can be separated by SDS-PAGE and analyzed by immunoblotting using desirable antibodies.
9. Quantification of the relative gray scale intensity in western blot analysis was done with ImageJ software v1.45 according to the procedure described in http://lukemiller.org/index.php/2010/11/analyzing-gels-and-western-blots-with-image-j/. An example of the result is shown in Fig. 3.

4 Notes

1. *Myo*-inositol and thiamine hydrochloride cannot be autoclaved and should be sterilized by filtration through 0.22 μm syringe filter first and then added to the autoclaved MS medium in a laminar flow hood.
2. The culture condition and subculture schedule of *Arabidopsis* suspension culture cells are critical for the successful protoplast isolation and gene expression. If the cells are maintained in good conditions, 5-day-old culture cells are evenly sized without big aggregations; in addition, 10 ml cell pellet can be obtained from 50 ml of 5-day-old culture cells after centrifugation.
3. Electroporation buffer and protoplast culture medium should be sterilized by filtration through 0.2 μm bottle top filter in a laminar flow hood because of the presence of high concentration of sucrose in the media.
4. In order to reach high efficient of protoplast isolation, the enzyme solution should be prepared fresh.
5. Wortmannin should be aliquoted in small amounts and kept at −20 °C due to short half-life of this drug at room temperature. Each time, pick one vial for the experiment.
6. The volume of enzyme solution is three times more than the volume of pelleted cells. For example, 30 ml of enzyme solution is sufficient for digestion 10 ml of pelleted cells from 50 ml of 5-day-old cultured cells.
7. Poor floating of protoplasts reflects poor protoplasts conditions that will dramatically decrease transformation efficiency. It is important to use the swinging bucket rotor for all washing steps which allow the protoplast layer to float well on top. All centrifugation steps should be carried out without deceleration to prevent the top protoplast layer from being disturbed.

8. Before switching on the pump, use the Pasteur pipette to open a small area on top first to prevent the protoplasts from being absorbed into the Pasteur pipette during the insertion step. Insert the Pasteur pipette upright to the bottom and switch on the pump. Finally, pull up the Pasteur pipette quickly and then switch off the pump.
9. The plant protoplasts are quite fragile, so transfer them gently in order to avoid damaging protoplasts.
10. The quality of the plasmid DNA is crucial for high transfection efficiency. We routinely use lysozyme lysis and phenol–chloroform extraction for maxi-plasmid DNA preparation and purification. This protocol is available upon request. We also find that the plasmid prepared by using maxi-DNA preparation kit (Qiagen) is also good for transient expression in our system.
11. The incubation conditions can vary depending on the experiment purposes. We usually incubate the protoplasts in the incubator at room temperature (22–27 °C). Normally, fluorescent protein signal can be observed after 4–6 h. 12–24 h incubation time is good for confocal imaging or biochemical analysis.
12. A standard electrician's tape (100–150 μm thickness) with a small rectangular square cut out can be placed on the slide; the protoplasts can then be put in the cut out square to avoid squashing between the slide and the coverslip.
13. The protoplasts in protoplast culture medium should be diluted at least threefold with 250 mM NaCl, otherwise the protoplasts cannot sediment to the bottom after centrifugation. Make sure that the bottom conical part of the tube has been mixed thoroughly before centrifugation; otherwise the protoplasts will sediment on top of a small remaining sucrose cushion that remains on the bottom, preventing the formation of a neat pellet.

Acknowledgements

We are grateful to the current and previous members, especially Dr. Yansong Miao of Prof. Jiang's Laboratory and our long-term research collaborator, Dr. Peter Pimpl (University of Tübingen) for their contributions in setting up and improving the protoplasts transient expression system using suspension culture cells. This work was partially supported by grants from the Research Grants Council of Hong Kong (CUHK466610, 466011, 465112, 466313, CUHK2/CRF/11G, HKUST10/CRF/12R, and AoE/M-05/12), NSFC/RGC (N_CUHK406/12), NSFC (31270226), Shenzhen Peacock Project (KQTD201101), and CUHK to L.J.

References

1. Hara-Nishimura I, Hatsugai N (2011) The role of vacuole in plant cell death. Cell Death Differ 18:1298–1304
2. Marty F (1999) Plant vacuoles. Plant Cell 11:587–599
3. Tse YC, Mo B, Hillmer S et al (2004) Identification of multivesicular bodies as prevacuolar compartments in Nicotiana tabacum BY-2 cells. Plant Cell 16:672–693
4. Lam SK, Tse YC, Robinson DG et al (2007) Tracking down the elusive early endosome. Trends Plant Sci 12:497–505
5. Mo B, Tse YC, Jiang L (2006) Plant prevacuolar/endosomal compartments. Int Rev Cytol 253:95–129
6. Robinson DG, Jiang L, Schumacher K (2008) The endosomal system of plants: charting new and familiar territories. Plant Physiol 147:1482–1492
7. Lam SK, Siu CL, Hillmer S et al (2007) Rice SCAMP1 defines clathrin-coated, trans-golgi-located tubular-vesicular structures as an early endosome in tobacco BY-2 cells. Plant Cell 19:296–319
8. Miao Y, Yan PK, Kim H et al (2006) Localization of green fluorescent protein fusions with the seven *Arabidopsis* vacuolar sorting receptors to prevacuolar compartments in tobacco BY-2 cells. Plant Physiol 142: 945–962
9. Wang J, Cai Y, Miao Y et al (2009) Wortmannin induces homotypic fusion of plant prevacuolar compartments. J Exp Bot 60:3075–3083
10. Jia T, Gao C, Cui Y et al (2013) ARA7(Q69L) expression in transgenic *Arabidopsis* cells induces the formation of enlarged multivesicular bodies. J Exp Bot 64:2817–2829
11. Cai Y, Zhuang X, Wang J et al (2012) Vacuolar degradation of two integral plasma membrane proteins, AtLRR84A and OsSCAMP1, is cargo ubiquitination-independent and prevacuolar compartment-mediated in plant cells. Traffic 13:1023–1040
12. Kotzer AM, Brandizzi F, Neumann U et al (2004) AtRabF2b (Ara7) acts on the vacuolar trafficking pathway in tobacco leaf epidermal cells. J Cell Sci 117:6377–6389
13. Denecke J, Aniento F, Frigerio L et al (2012) Secretory pathway research: the more experimental systems the better. Plant Cell 24: 1316–1326
14. Miao Y, Jiang L (2007) Transient expression of fluorescent fusion proteins in protoplasts of suspension cultured cells. Nat Protoc 2: 2348–2353
15. Gao C, Yu CK, Qu S et al (2012) The Golgi-localized *Arabidopsis* endomembrane protein12 contains both endoplasmic reticulum export and Golgi retention signals at its C terminus. Plant Cell 24:2086–2104
16. Miao Y, Li KY, Li HY et al (2008) The vacuolar transport of aleurain-GFP and 2S albumin-GFP fusions is mediated by the same pre-vacuolar compartments in tobacco BY-2 and *Arabidopsis* suspension cultured cells. Plant J 56:824–839

Chapter 10

Evaluation of Defective Endosomal Trafficking to the Vacuole by Monitoring Seed Storage Proteins in *Arabidopsis thaliana*

Tomoo Shimada, Yasuko Koumoto, and Ikuko Hara-Nishimura

Abstract

Vacuolar proteins are synthesized as precursor forms in the endoplasmic reticulum and are sorted to the vacuole. In this chapter, we introduce two easy methods for the evaluation of vacuolar protein transport using *Arabidopsis* seeds. These methods are adequate to detect defects in vacuolar transport mediated by endosomes and other trafficking pathways as well. They include an immunoblot assay that monitors the abnormal accumulation of storage protein precursors, and an immunogold labeling assay that monitors the abnormal secretion of storage proteins. Each method facilitates the rapid identification of defects in the transport of endogenous vacuolar proteins in *Arabidopsis* mutants.

Key words *Arabidopsis thaliana*, Immunoblot, Immunogold labeling, Membrane trafficking, Protein storage vacuole, Seed storage protein

1 Introduction

The vacuole is a unique plant organelle that plays vital roles for survival [1, 2], such as maintenance of turgor pressure, storage of various materials, and programmed cell death. Vacuolar functions depend on the activities of proteins that specifically localize in the vacuole. Generally, vacuolar proteins are synthesized in the endoplasmic reticulum (ER) and sorted to the vacuole via the Golgi apparatus [3–5]. This protein trafficking pathway shows evolutionary conservation with those associated with yeast vacuoles and animal lysosomes. Analyses of protein transport to plant vacuoles have been carried out using leaves, roots, and cultured cells. In some studies, marker proteins that were fused to well-known vacuolar-targeting signals were expressed in transgenic plants to monitor the destination of vacuolar proteins [6]. The use of green fluorescent protein (GFP) has become a very powerful technique to monitor the dynamics of vacuolar protein transport [7, 8].

Marisa S. Otegui (ed.), *Plant Endosomes: Methods and Protocols*, Methods in Molecular Biology, vol. 1209, DOI 10.1007/978-1-4939-1420-3_10,

However, it is still difficult to observe the transport behavior of endogenous vacuolar proteins.

Plant seeds store large amounts of storage proteins for use as a nitrogen source after germination [9]. The majority of storage proteins are specialized types of vacuolar proteins that are sorted to specialized vacuoles, called protein storage vacuoles (PSVs), during seed maturation. After arriving at the PSV, most storage proteins are proteolytically processed by VACUOLAR PROCESSING ENZYME (VPE), which converts the precursor proteins into their mature forms [10, 11]. Consistently, in a pioneering study, an *Arabidopsis thaliana* mutant deficient in VACUOLAR SORTING RECEPTOR 1 (*vsr1*) was shown to accumulate the precursors of vacuolar storage proteins in seeds [12]. Because seed storage proteins are synthesized and transported in bulk, they are suitable endogenous markers that can be labeled to monitor the trafficking of vacuolar proteins. By monitoring the transport of seed storage proteins, several vacuolar trafficking mutants [13–16] have been isolated and characterized in *Arabidopsis*, including those within the *maigo* (*mag*) and *green fluorescent seed* (*gfs*) groups. Here, we describe protocols to monitor the abnormal accumulation of storage protein precursors by immunoblotting and immunogold labeling.

2 Materials

Prepare all solutions using ultrapure water (prepared by purifying deionized water to attain a sensitivity of 18 MΩ/cm) and analytical grade reagents. Prepare and store all reagents at room temperature (unless indicated otherwise). Diligently follow all waste disposal regulations when disposing of waste materials. We do not add sodium azide to reagents, except for antibodies that are stored at 4 °C.

2.1 SDS Polyacrylamide Gel Components

1. 1.5 M Tris–HCl, pH 8.8: Weigh 18.17 g tris-aminomethane (Tris) and add water to a volume of 80 mL. Mix and adjust the pH to 8.8 with HCl. Bring the total volume to 100 mL with the addition of water.
2. 0.5 M Tris–HCl, pH 6.8: 6.06 g Tris in 100 mL of water. Prepare solution as in Subheading 2.1, **item 1**.
3. 40 % (w/v) acrylamide/bis-acrylamide mixed solution (29:1).
4. Ammonium persulfate (APS): 10 % APS solution in water (*see* **Note 1**). Store at 4 °C.
5. *N,N,N′,N′*-Tetramethyl ethylenediamine (TEMED): Store at 4 °C.
6. SDS-PAGE running buffer: 0.025 M Tris–HCl, pH 8.3, 0.192 M glycine, and 0.1 % SDS. Prepare 10× buffer. Weigh 30.3 g Tris and 144 g glycine, and dissolve in 800 mL of water.

Mix and adjust the pH to 8.3 with HCl. Add the detergent (10 g SDS) last, and minimize vigorous mixing to avoid the production of foam. After completely dissolving all chemicals, bring the final volume up to 1 L.

7. 2× SDS sample buffer : Mix 2 mL of 0.5 M Tris–HCl, pH 6.8, 4 mL of 10 % SDS, 1.2 mL of β-mercaptoethanol, 2 mL of glycerol, and 0.8 mL of water. Add a few drops of 1 % bromophenol blue (BPB) to make the solution blue.
8. Molecular weight markers: A prestained molecular weight marker is a convenient guide for cutting the gel. The prestained ladder of low range markers (15–85 kDa) is recommended. Markers that react with peroxidase are convenient.

2.2 Immunoblotting Components

1. PVDF membrane: Immobilon-P (Millipore, IPVH00010).
2. Blotting buffer: 0.048 M Tris, 0.039 M glycine, and 20 % methanol (*see* **Note 2**).
3. Tris-buffered saline (TBS, 10×): 0.2 M Tris–HCl, pH 7.5, and 1.5 M NaCl.
4. TBS containing 0.05 % Tween-20 (TBST).
5. Blocking solution: 5 % skim milk in TBST (*see* **Note 3**).
6. Plastic containers or stainless-steel trays.
7. Western blot detection kit: ECL Western Blotting Detection Reagents (GE Healthcare, RPN2106).
8. Secondary antibody: Horseradish peroxidase-labeled goat anti-rabbit IgG $F(Ab')_2$.
9. A system for the development and fixation of X-ray films, or a lumino-image analyzer.

2.3 Immunoelectron Microscopy Components

1. 10 % paraformaldehyde solution: Weigh 2 g of paraformaldehyde [electron microscopy (EM) grade], transfer into a 50 mL flask, and add 16 mL of water. Place the flask into a 60 °C water bath and add an appropriate volume of 1 M NaOH to completely dissolve the powder. After dissolving, bring it to 20 mL with water, and then filter through a 0.45 μm filter. This solution can be stored for a week or a month at 4 °C.
2. 0.2 M cacodylate buffer, pH 7.4: Mix stock solution A with stock solution B at a ratio of 50:2.7. To prepare stock solution A (0.2 M sodium cacodylate), weigh 42.8 g of sodium cacodylate and bring it to 1 L with water. To prepare stock solution B (0.2 M NaCl), measure 1.68 mL of concentrated sodium chloride and bring it to 100 mL with water.
3. 25 % glutaraldehyde solution (distilled grade): Withdraw the solution from the stock bottle into a plastic syringe with a needle, put a rubber plug on the top of the needle, and store at 4 °C.

4. Fixation buffer: 4 % paraformaldehyde, 1 % glutaraldehyde, 10 % dimethylsulfoxide in 0.05 M cacodylate buffer, pH 7.4 (*see* **Note 4**).
5. Medium grade acrylic resin such as LR white resin.
6. Beam capsules.
7. UV polymerizer.
8. Ultra-microtome and diamond knife.
9. A glass knife with a boat: The top of the knife is surrounded by 9 mm tape. To avoid leakage of water, nail polish is used to seal the surface between the tape and the knife.
10. Adhesive glass microscope slide such as MAS-coated glass slides Matsunami glass Ind. Ltd.
11. A section picker: An eyelash mounted on a toothpick with nail polish.
12. Toluidine blue solution: 1 % toluidine blue in 1 % borax.
13. Copper, preferably nickel, grids with support film. The mesh size is a compromise between support of the section and the viewing area between the mesh.
14. 2 % uranyl acetate solution: Weigh 0.5 g of uranyl acetate and dissolve in 25 mL of water. Leave the solution overnight at 4 °C. Then, transfer the supernatant to a plastic syringe. Attach a needle to the syringe, put a rubber plug on the top of the needle, and store at 4 °C in the dark. RADIOCHEMICAL HAZARD.
15. Lead citrate: Weigh 0.2 g of NaOH, dissolve in 20 mL of degassed water, and filter through a 0.45 μm filter. Weigh 0.2 g of lead citrate and dissolve in the NaOH solution. Again, filter through a 0.45 μm filter. Then, transfer the filtrate into a plastic syringe with a needle, and remove air bubbles (*see* **Note 5**). Put a rubber plug on the top of the needle and store at 4 °C.
16. Phosphate-buffered saline (PBS, 10×): 0.1 M phosphate buffer, pH 7.4, and 8.5 % NaCl. To prepare stock solution A (0.2 M NaH_2PO_4), weigh 15.6 g $NaH_2PO_4 \cdot 2H_2O$ and bring it to 500 mL with water. To prepare stock solution B (0.2 M Na_2HPO_4), weigh 35.8 g $Na_2HPO_4 \cdot 12H_2O$ and bring it to 500 mL with water. Mix 19 mL of stock solution A, 81 mL of stock solution B, and 17 g of NaCl, and bring it to 200 mL with water.
17. Blocking solution: 1 % bovine serum albumin (BSA) and 0.1 % sodium azide (NaN_3). Weigh 0.1 g of BSA and 10 mg of NaN_3, transfer to a graduated cylinder, add 1× PBS to a final volume

of 10 mL, and dissolve completely. Filter the solution using a 0.2 μm filter, and store at 4 °C.

18. Secondary antibody: Immunogold conjugate EM Goat anti-rabbit IgG (15 nm diameter).

3 Methods

Carry out all procedures at room temperature unless otherwise specified.

3.1 15 % SDS-PAGE

1. Prepare a separating gel solution. Mix 2.5 mL of 1.5 M Tris–HCl, pH 8.8, 3.75 mL of 40 % acrylamide solution, 0.1 mL of 10 % SDS, 1.55 mL of water, 2.0 mL of 50 % glycerol, and 100 μL of APS in a 50 mL tube. If necessary, leave the mixture on ice (*see* **Note 6**).
2. Prepare a stacking gel solution. Mix 0.63 mL of 0.5 M Tris–HCl, pH 6.8, 0.63 mL of 40 % acrylamide solution, 0.05 mL of 10 % SDS, 3.6 mL of water, and 50 μL of APS in a 50 mL tube. If necessary, leave the mixture on ice.
3. Cast a mini gel cassette and mark the position 1 cm below the bottom of the gel comb.
4. Add 4 μL of TEMED to the separating gel solution, mix well immediately, and pour it into a mini wide gel cassette. Without a pause, add 5 μL of TEMED to the stacking gel solution, mix well immediately, and gently overlay it on the separating gel. Insert a gel comb immediately without introducing air bubbles.
5. Homogenize ten seeds in 100 μL of 2× sample buffer using microtubes and plastic pestles (*see* **Note 7**). Heat it at 95 °C for 5 min and centrifuge at 20,000 × *g* for 15 min.
6. Apply 10 μL of the supernatant (1 seed equivalent) to a well of the gel. Also apply 5 μL of the molecular weight marker ladder to one lane on each side of the gel (two lanes of molecular weight markers per gel).
7. Perform electrophoresis at 30 mA until the dye front reaches the bottom of the gel.
8. Following electrophoresis, pry the gel plates open with the use of a spatula. The gel remains on one of the glass plates. Remove the wells of gel and measure the size of the gel. Transfer carefully to a container with blotting buffer.
9. Cut a PVDF membrane and a filter paper to the size of the gel. Six filter papers are needed. Immerse the cut PVDF membrane in methanol for 1 min. Transfer the membrane and the filter papers to the same container with the gel.

3.2 Immunoblot Analysis with Antibodies Against Storage Proteins

1. Place three filter papers on the anode of a semidry blotter. Layer the membrane, the gel, and the remaining three filter papers on top without introducing air bubbles (*see* **Note 8**).
2. Place the cathode on the blotter and run at the appropriate current for 1 h (*see* **Note 9**).
3. Transfer the blotted membrane to a container and cover with blocking solution for 1 h on a platform rotary shaker.
4. Wash with TBST once for 15 min and twice for 5 min.
5. Add anti-12S globulin or 2S albumin antibodies [10] to the membrane. The antibodies are diluted to 1:10,000 and 1:5,000, respectively, with TBST.
6. Shake for 1 h.
7. Wash with TBST once for 15 min and twice for 5 min.
8. Add the secondary antibody to the membrane. The antibody is diluted to 1:5,000 with TBST.
9. Wash with TBST once for 15 min and four times for 5 min.
10. Detect the signals of the secondary antibody with the detection kit. Signals are directly detected by the lumino-image analyzer (Fig. 1), or detected as the sensitized spots of an X-ray film that was exposed to the PVDF membrane.

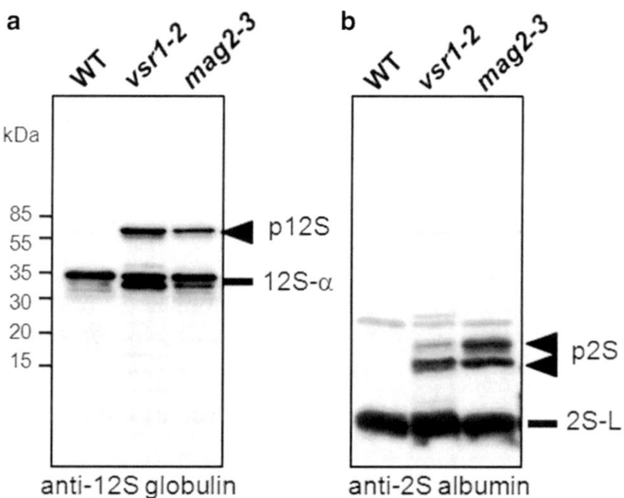

Fig. 1 Immunoblot analysis of *Arabidopsis* seeds. Immunoblot of dry seeds from wild-type (WT), *vsr1-2*, and *mag2-3* plants with anti-12S globulin antibodies (**a**) and anti-2S albumin antibodies (**b**). The *vsr1-2* and *mag2-3* seeds accumulate large amounts of the precursors pro12S globulin (p12S) and pro2S albumin (p2S), whereas WT seeds do not. 12S-α, an acidic chain of 12S globulin; 2S-L, a large chain of 2S albumin. Molecular masses are indicated on the *left* in kDa

3.3 Preparation of the Blocks from Seed Cells

All reagents used for EM processing must be of high purity (EM grade is recommended). It is recommended to section several blocks to determine what is representative of the material.

1. Put a seed in a paraffin film and cut the seed in half using a blade (*see* **Note 10**). Add 4–5 mL of fixation buffer into a vial (approximately 2 cm diameter). Add some drops of fixation buffer to the cut seed, and use tweezers to transfer the two halves into the vials.
2. Put the vials in a desiccator and apply vacuum for 5 min. Stop the pump and open the valve gradually to release the vacuum slowly (*see* **Note 11**). In this step, the fixation buffer penetrates into the samples. Repeat three times (*see* **Note 12**). Leave the vials for 2 h (*see* **Note 13**).
3. Prepare 50 % dimethylformamide in water and store at 4 °C. Prepare 70, 80, and 90 % dimethylformamide in water and store at −20 °C. Approximately 10 mL is needed per sample. Approximately 15 mL of 100 % dimethylformamide is needed per sample (store at −20 °C).
4. Remove the fixation buffer from the vial using a 10 mL syringe with a No.18 needle (*see* **Note 14**). Wash three times with 0.05 M cacodylate buffer (pH 7.4) using the same syringe and needle.
5. Add 4–5 mL of 50 % dimethylformamide solution per sample, and leave for 15 min at 4 °C. Repeat twice. Then, add 4–5 mL of each of the 70, 80, and 90 % dimethylformamide solutions, leave for 15 min at −20 °C, and repeat twice, respectively.
6. Prepare various concentration of LR white in 100 % dimethylformamide. For seeds, prepare 2, 4, 6, 8, 10, 20, 40, 60, and 100 % LR white solutions. Approximately 5 mL of solution is needed per sample, except for 100 % solution, which requires approximately 10 mL. Store all solutions at −20 °C.
7. Add 4–5 mL of 100 % dimethylformamide per sample vial, and leave for 20 min at −20 °C. Repeat three times.
8. Substitute with 5 mL of 2 % LR white solution and leave at −20 °C overnight (*see* **Note 15**).
9. Substitute with 4, 6, 8, 10, 20, 40, 60, and 100 % LR white solutions for 1 h each at −20 °C, respectively (*see* **Note 16**).
10. Substitute with 5 mL 100 % LR white solution again, and leave at −20 °C overnight.
11. Add 1 μL of acceleration agent to 20 mL of 100 % LR white and mix for 15 min with a rotator. Precool this solution for 1 h at −20 °C.
12. Substitute with the LR white containing the accelerator and leave at −20 °C for 24 h. Store the remaining solution at −20 °C.

13. Precool a UV polymerizer at −20 °C.
14. Prepare small pieces of paper labeled with the sample name and block number using a pencil (*see* **Note 17**). The width of the paper is approximately twice the diameter of a beam capsule.
15. Set the labels along the inner walls of the upper part of the beam capsules.
16. Add LR white with accelerator into the beam capsules up to two-thirds of their height.
17. Add 4–5 pieces of samples with the solution (from **step 12**) into the beam capsules using a pipette. The seeds will sink to the bottom of the capsules.
18. Fill the LR white with accelerator to the top of the capsules and close the covers without introducing air bubbles.
19. Set them in the UV polymerizer and polymerize for 24 h at −20 °C.
20. Polymerize for 1 h at 4 °C (*see* **Note 18**).
21. Polymerize for 1 h at room temperature.
22. Pick up the blocks (*see* **Note 19**). Store them in a desiccator.

3.4 Collection of Sections

1. Trim the excess resin from the block face using a glass knife, and from the edge of the block using a single-edged razor blade.
2. Cut a thin section of 750 nm using an ultramicrotome, collect onto water in the boat of a glass knife, and use a section picker to transfer the thin section to a drop of water on a adhesive glass microscope slide.
3. Dry on a hotplate and stain with toluidine blue for 2 min at 70–80 °C until the rim around the drop of the stain is dry. Wash off with water and dry on the hotplate.
4. Check the stained image using an optical microscope and select the block for further processing.
5. Trim the block to a smaller size (*see* **Note 20**).
6. Attach a diamond knife on an ultramicrotome, fill the boat with water, and collect ultrathin sections of 70–80 nm in a ribbon on the surface of the water (*see* **Note 21**).
7. Place the grid in the water beneath the section and raise it at a slight angle so the first section of the ribbon sticks to the edge of the grid. Slowly raise the grid out of the water and the rest of the ribbon will adhere to the grid. Blot the edge of the grid to remove excess water. Do not blot the flat surface of the grid.
8. Place grids in a filter paper-lined petri dish before staining and store in a desiccator.

3.5 Immunoelectron Microscopy

These steps are performed by placing drops of the solutions onto a paraffin film, and then placing the grid onto the drop for the required time. One drop of each solution is required for one grid.

1. Prepare the required number of drops (approximately 50 μL) of blocking solution, and place one drop per grid onto a paraffin film. Place the grid onto the drop for 30 min.
2. Prepare 3 times the required number of drops of PBS. Place the grid onto the drop for 5 min, transfer the grid to the second drop, leave for 5 min, and then transfer the grid to the third drop for 5 min.
3. Dilute anti-12S globulin or 2S albumin antisera with PBS (*see* **Note 22**).
4. Prepare the required number of drops of the primary antibody, transfer the grid to the drop, and leave overnight at 4 °C.
5. Prepare 6 times the required number of drops of PBS. Place the grid onto the drop for 5 min, repeat 6 times.
6. Prepare the required number of drops of the secondary antibody, transfer the grid to the drop, and leave for 30 min.
7. Prepare 6 times the required number of drops of PBS. Place the grid onto the drop for 5 min, repeat 6 times.
8. Prepare 6 times the required number of drops of water. Place the grid onto the drop for 5 min, repeat 6 times.
9. Prepare the required number of drops of uranyl acetate using appropriate radioactive handling procedures (*see* **Note 23**). Place the grid onto the drop for 3 min.
10. Transfer the grid onto drops of water to stop the staining.
11. Pick up the grid with forceps. Wash the grid by streaming water from a washing bottle for 10–15 s.
12. Transfer the grid onto the drops of water until the next step.
13. Prepare the required number of drops of lead citrate (*see* **Note 24**). Place the grid onto the drop for 3 min (*see* **Note 25**).
14. Transfer the grid onto the drops of water to stop the staining.
15. Pick up the grid with forceps. Wash the grid by streaming water from a washing bottle for 10–15 s.
16. Blot the edge of the grid to remove excess water. Place grids in a filter paper-lined petri dish before drying, and store in a desiccator.
17. After drying, the sections are examined with a transmission electron microscope (JEOL, model JEM-1015B) at 100 kV (Fig. 2).
18. Put the sections in a case for EM grids and store in a desiccator.

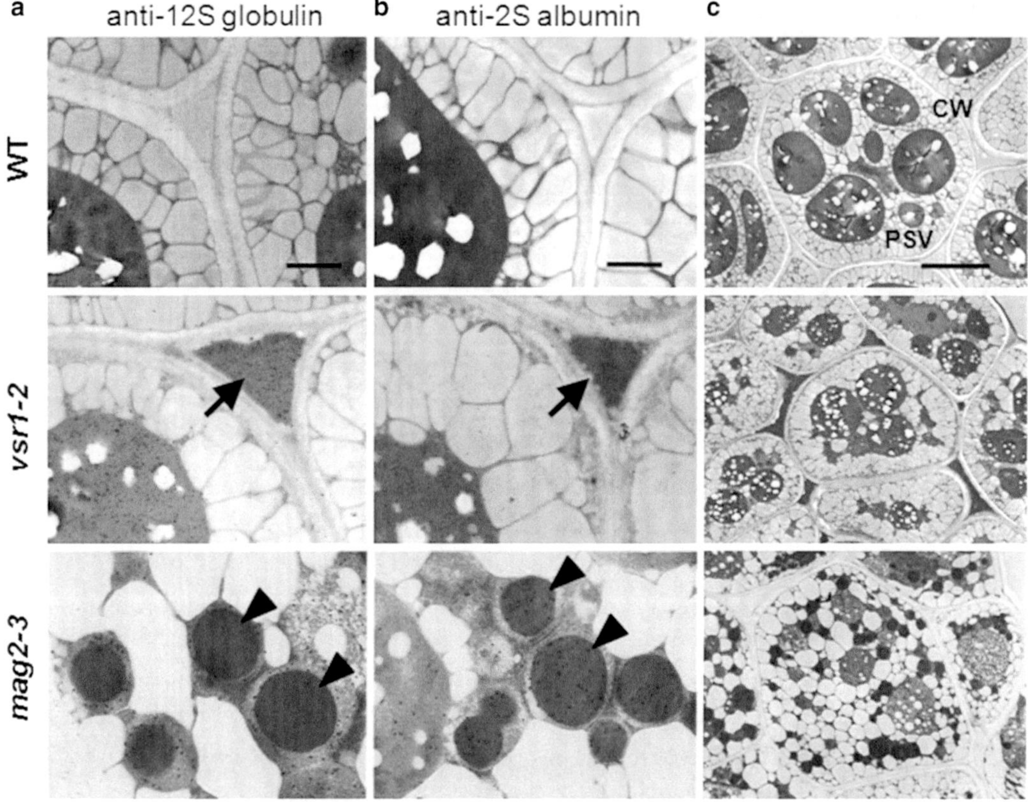

Fig. 2 Immunogold analysis of *Arabidopsis* seeds. (**a**, **b**) Immunoelectron micrographs of wild-type (WT), *vsr1-2*, and *mag2-3* seeds with anti-12S globulin antibodies (**a**) and anti-2S albumin antibodies (**b**). The *vsr1-2* mutant abnormally missorts the storage proteins by secreting them from cells (*arrows*), whereas the *mag2-3* mutant develops abnormal structures composed of storage proteins (*arrowheads*). (**c**) Electron micrographs of WT, *vsr1-2*, and *mag2-3* seed cells. *PSV* protein storage vacuole, *CW* cell wall. Scale bars: (**a**, **b**) 10 μm; (**c**) 5 μm

4 Notes

1. It is best to prepare fresh solution every month.
2. SDS is not included, because it reduces the binding of 2S albumin to the membrane.
3. It is best to prepare this solution fresh.
4. It is best to prepare this solution fresh for each experiment. If reactivity of the antibody is weak, decrease the glutaraldehyde concentration.
5. Lead nitrate reacts with carbon dioxide to precipitate as lead carbonate.
6. Polymerization starts with the addition of TEMED. Because the speed of polymerization depends on the temperature, you may cool the mixture to slow polymerization.

7. Alternatively, an electrical metal pestle can be used.
8. Proteins denatured by SDS are negatively charged and move from the cathode to the anode. We use a semidry blotter equipped with the anode in the body and the cathode as the cover.
9. A value of 1.5 times of the size of gel (cm^2) is used as the value of the appropriate current (mA). For example, 75 mA is used for the gel of 50 cm^2.
10. Approximately 20 seeds are needed to prepare 4–5 blocks.
11. Under vacuum, bubbles will be released from the samples.
12. Vacuum treatment must be continued longer if the penetration of the fixation buffer is not adequate. When completely penetrated, no bubbles emerge from the sample.
13. Leaving for more than 2 h causes hyper-fixation and produces worse results.
14. Do not drain the buffer completely. The seeds should always be covered with buffer.
15. Samples in 100 % dimethylformamide may also be left overnight.
16. Because resin does not penetrate seeds easily, it is best to increase resin concentration gradually, especially during the first steps up to 10 % resin.
17. Pencils are recommended (ink can be partially dissolved by the resins).
18. It is best to change the position of capsules to prevent non-homogeneous polymerization.
19. First, remove the cover. Then, cut the side of the capsules using a box cutter and peel them off, or pick up the blocks using tweezers. The bottom of the blocks is the important side (take caution not to touch).
20. For ultrathin sectioning, 0.5 mm^2 is recommended.
21. When a continuous section ribbon cannot be made, ensure that the sides of the block are not jagged; recut with a fresh razor blade if needed.
22. It is best to test various dilutions of antibodies (e.g., 1:50, 1:200, and 1:1,000).
23. The uranyl acetate is directly supplied from the syringe. The first 0.1 mL must be discarded in an appropriate radioactive waste container.
24. The lead citrate is directly supplied from the syringe. The first 0.1 mL must be discarded.
25. If the presence of small (approximately 10 nm) electron-dense material is observed throughout the tissue, this may be from the lead citrate stain. Do not stain for a longer time.

Acknowledgement

We thank Maki Kondo and Hideyuki Takahashi for technical assistance with the EM protocol. This work was supported by Specially Promoted Research of Grant-in-Aid for Scientific Research to I.H-.N. (22000014).

References

1. Neuhaus JM, Martinoia E (2011) Plant vacuoles, eLS. Wiley, Chichester, UK. doi:10.1002/9780470015902.a0001675.pub2
2. Shitan N, Yazaki K (2013) New insights into the transport mechanisms in plant vacuoles. Int Rev Cell Mol Biol 305:383–433
3. Pedrazzini E, Komarova NY, Rentsch D, Vitale A (2013) Traffic routes and signals for the tonoplast. Traffic 14:622–628
4. Rojas-Pierce M (2013) Targeting of tonoplast proteins to the vacuole. Plant Sci 211:132–136
5. Xiang L, Etxeberria E, Van den Ende W (2013) Vacuolar protein sorting mechanisms in plants. FEBS J 280:979–993
6. Pimpl P, Hanton SL, Taylor JP, Pinto-daSilva LL, Denecke J (2003) The GTPase ARF1p controls the sequence-specific vacuolar sorting route to the lytic vacuole. Plant Cell 15: 1242–1256
7. Hunter PR, Craddock CP, Di Benedetto S, Roberts LM, Frigerio L (2007) Fluorescent reporter proteins for the tonoplast and the vacuolar lumen identify a single vacuolar compartment in *Arabidopsis* cells. Plant Physiol 145: 1371–1382
8. Wolfenstetter S, Wirsching P, Dotzauer D, Schneider S, Sauer N (2012) Routes to the tonoplast: the sorting of tonoplast transporters in *Arabidopsis* mesophyll protoplasts. Plant Cell 24:215–232
9. Muntz K (1998) Deposition of storage proteins. Plant Mol Biol 38:77–99
10. Shimada T, Yamada K, Kataoka M, Nakaune S, Koumoto Y, Kuroyanagi M et al (2003) Vacuolar processing enzymes are essential for proper processing of seed storage proteins in *Arabidopsis thaliana*. J Biol Chem 278:32292–32299
11. Gruis D, Schulze J, Jung R (2004) Storage protein accumulation in the absence of the vacuolar processing enzyme family of cysteine proteases. Plant Cell 16:270–290
12. Shimada T, Fuji K, Tamura K, Kondo M, Nishimura M, Hara-Nishimura I (2003) Vacuolar sorting receptor for seed storage proteins in *Arabidopsis thaliana*. Proc Natl Acad Sci U S A 100:16095–16100
13. Li L, Shimada T, Takahashi H, Ueda H, Fukao Y, Kondo M et al (2006) MAIGO2 is involved in exit of seed storage proteins from the endoplasmic reticulum in *Arabidopsis thaliana*. Plant Cell 18:3535–3547
14. Shimada T, Koumoto Y, Li L, Yamazaki M, Kondo M, Nishimura M et al (2006) AtVPS29, a putative component of a retromer complex, is required for the efficient sorting of seed storage proteins. Plant Cell Physiol 47: 1187–1194
15. Fuji K, Shimada T, Takahashi H, Tamura K, Koumoto Y, Utsumi S et al (2007) *Arabidopsis* vacuolar sorting mutants (green fluorescent seed) can be identified efficiently by secretion of vacuole-targeted green fluorescent protein in their seeds. Plant Cell 19:597–609
16. Takagi J, Renna L, Takahashi H, Koumoto Y, Tamura K, Stefano G et al (2013) MAIGO5 functions in protein export from Golgi-associated endoplasmic reticulum exit sites in *Arabidopsis*. Plant Cell 25(11):4658–4675

Chapter 11

Trans-species Complementation Analysis to Study Function Conservation of Plant Endosomal Sorting Complex Required for Transport (ESCRT) Proteins

Francisca C. Reyes

Abstract

ESCRT (Endosomal Sorting Complex Required for Transport) proteins are required for the sorting of biosynthetic and endocytic proteins at multivesicular bodies (MVBs). Here, I describe an assay to evaluate conservation of endosomal sorting functions of plant ESCRT proteins by trans-species complementation analysis. The assay is based on the imaging of a fluorescent biosynthetic MVB cargo, the carboxypeptidase S (CPS) fused to the green fluorescent protein GFP, in yeast ESCRT mutants expressing putative plant orthologues of the missing yeast ESCRT components.

Key words Endosomal sorting, Yeast, ESCRT mutants, MVB cargo, Confocal microscopy

1 Introduction

The components of the Endosomal Sorting Complexes Required for Transport (ESCRT) and their functions in cargo sorting and multivesicular bodies (MVB) formation were first described in yeast (*Saccharomyces cerevisiae*) [1, 2]. Genes coding for the subunits of the ESCRT complexes 0, I, II, and III [2–6] have been functionally characterized in yeast through the analysis of a vast collection of mutants identified as *vacuolar protein sorting* (*vps*) mutants, since they fail to sort carboxypeptidase Y (CPY) or a CPY-invertase to the vacuole [7, 8]. These mutants can be used to study function conservation of putative ESCRT components from other organisms.

I present here an imaging assay to test for function conservation of plant ESCRT proteins by trans-species complementation. The assay consists of imaging a fluorescent MVB cargo, GFP-CPS [9], to determine its localization in control yeast strains as well as in characterized yeast ESCRT mutants [9]. In control cells, GFP-CPS is sorted into intraluminal vesicles of MVBs and delivered into

Marisa S. Otegui (ed.), *Plant Endosomes: Methods and Protocols*, Methods in Molecular Biology, vol. 1209, DOI 10.1007/978-1-4939-1420-3_11, © Springer Science+Business Media New York 2014

the vacuolar lumen. In contrast, ESCRT mutants fail to sort cargo into MVB vesicles and GFP-CPS is mis-sorted to the vacuolar membrane. Thus, GFP signal in the vacuolar lumen indicates normal MVB sorting, whereas GFP labeling of the vacuolar membrane indicates MVB sorting defects.

By using this imaging assay, it is possible to quantify the number of cells with normal and abnormal GFP-CPS sorting and estimate the degree of rescue when putative plant ESCRT proteins are introduced into yeast mutant strains defective for ESCRT function.

As an example, I describe the assay using yeast mutants for two ESCRT components that are involved in the late membrane invagination and intraluminal vesicle release during MVB sorting: the AAA ATPase Vps4p and its positive regulator Vta1p, and the putative *Arabidopsis thaliana Vta1* orthologue, *LIP5* [10].

2 Materials

2.1 Yeast Strains and Plasmids

1. Yeast strains: *Saccharomyces cerevisiae* BY4742 (*MATα, his3Δ1 leu2Δ0 lys2Δ0 ura 3Δ0*), *Δvta1/ΔG418*, and *Δvps4* [9]. Other mutant strains can be used depending of the putative plant ESCRT protein to be tested.
2. Plasmids: URA plasmids p416GPD for the expression of HIS-tagged plant ESCRT proteins such as Arabidopsis LIP5 (putative *Vta1* orthologue) [10]; p416GPD-His-Vta1; and LEU plasmid pRS425-GFP-CPS [9].

2.2 Media and Solutions

All culture media and solutions should be prepared with ultrapure water, sterilized by autoclaving, and stored at room temperature until use (unless otherwise specified).

1. Amino acids stock solution: Leucine 7.2 mg/ml; Lysine 3.6 mg/ml; Histidine 2.4 mg/ml; Tryptophan 4.8 mg/ml; Uracil 3.6 mg/ml. Filter-sterilized all solutions.
2. YPD medium: 1 % yeast extract (w/v), 2 % peptone (w/v), 2.2 % glucose (w/v), 3.7 % KCl (w/v), and 40 μg/ml adenine.
3. Minimal medium: 0.67 % yeast nitrogen base (w/v), 2.2 % glucose (w/v), 0.5 % NH_4SO_4 (w/v), 3.7 % KCl, 40 μg/ml adenine (w/v) (filter-sterilized solution), and 0.825 % amino acid solution (v/v).
4. Minimal medium selective plates: Minimal medium as the one described above supplemented with 1.5 % agar (w/v) and poured on plates. Once autoclaved, adenine and amino acid solution (without uracil and/or leucine, depending on the transformation selection conditions), should be added.

5. 10× TE (Tris–EDTA) pH 7.5: 100 mM Tris–HCl, 10 mM EDTA.
6. 1 M lithium acetate (LiAc) stock solution (10×): 10.2 g of LiAc dissolved in 100 ml of 1× TE Buffer.
7. 50 % (v/v) PEG 4000 stock solution: 25 ml of PEG 4000 dissolved in water to complete 50 ml of solution.

The following solutions need to be prepared right before use and must be sterilized by filtration.

8. Denatured salmon sperm DNA: prepare a 10 mg/ml stock solution, boil it for 10 min and place it immediately on ice.
9. 100 mM LiAc in TE: 1.5 ml of 10× TE buffer, pH 7.5, 1.5 ml 10× LiAc solution, and 12 ml sterile water.
10. PEG–LiAc solution: 1 ml of 10× TE buffer, pH 7.5; 1 ml 10× LiAc, and 8 ml 50 % PEG 4000 stock solution.

2.3 Equipment for Microscopy Imaging

1. Dissecting microscope with epifluorescence attachment and filter sets for detection of GFP.
2. Confocal microscope for GFP detection and a 63× oil-immersion NA 1.4 objective.

3 Methods

3.1 Yeast Transformation for Expression of GFP-CPS

1. Grow control and mutant strain (*Δvta1* and *Δvps4*) in YPD-agar plates for 48 h at 30 °C.
2. Inoculate 15 ml of YPD liquid medium with a single yeast colony coming from the YPD-agar plate. Grow overnight at 30 °C in a shaker, until the culture reaches an OD_{600} of 1.
3. Transfer the 15 ml to a flask with 300 ml of YPD. Grow until an OD of 0.2–0.3 and centrifuge at 1,000 × *g* for 10 min.
4. Wash twice the pellet by resuspending in 15 ml of sterile water followed by centrifugation.
5. After the second wash, centrifuge at 1,000 × *g* for 10 min.
6. Resuspend in 1.5 ml of freshly prepared 100 mM LiAc in TE solution. The cells are competent for transformation.
7. In a 1.5 ml plastic tube add 0.1 μg of pRS425-GFP-CPS plasmid, 100 μg of denatured salmon sperm, and 100 μl of competent yeast from **step 6** and mix by inverting the tube.
8. Add 600 μl of freshly prepared PEG–LiAc solution and vortex.
9. Incubate in a shaker at 30 °C for 30 min.
10. Add 70 μl of 100 % DMSO and mix by inverting the tube. Put on ice for 10 min.

11. Incubate at 42 °C for 15 min and immediately put on ice. Keep on ice for 5 min.
12. Centrifuge for 5 s at max speed in a tabletop centrifuge and remove the supernatant.
13. Resuspend pellet in 1 ml of minimal medium and incubate overnight at 30 °C in a shaker.
14. Centrifuge as in **step 12** and plate on a minimal medium plate lacking LEU for selection. After 2–3 days at 30 °C, transformant colonies should be visible.

3.2 Transformation of Δ vta1 and Δ vps4 Strains with Yeast VTA1 and Arabidopsis LIP5

1. Grow *Δvta1* and *Δvps4* (*see* **Note 1**) cells expressing GFP-CPS in YPD(−LEU)-agar plates for 48 h at 30 °C.
2. Inoculate 15 ml of YPD (−LEU) liquid medium with a single yeast colony coming from the YPD (−LEU)-agar plate. Grow overnight at 30 °C in a shaker, until the culture reaches an OD_{600} of 1.
3. Transfer the 15 ml to a flask with 300 ml of YPD(−LEU). Grow until an OD of 0.2–0.3 and centrifuge at 1,000 × *g* for 10 min.
4. Wash twice the pellet by resuspending in 15 ml of sterile water followed by centrifugation.
5. After the second wash, centrifuge at 1,000 × *g* for 10 min and resuspend in 1.5 ml of freshly prepared 100 mM LiAc in TE solution.
6. Transform both mutant strains with 0.1 μg of plasmids p416GPD-HIS-LIP5, p416GPD-His-Vta1, and empty p416GPD as described in Subheading 3.1, **step 7–13**.
7. Centrifuge for 5 s at max speed in a tabletop centrifuge, remove the supernatant, and plate cells on a minimal medium plate lacking LEU and URA for selection. After 2–3 days at 30 °C, transformant colonies should be visible.

3.3 Selection and Screening of Transformant Yeast Colonies

1. Detect GFP-positive colonies using a dissecting microscope with epifluorescence attachment and a filter set for GFP detection.
2. Select 3–5 GFP-positives colonies per transformation event and grow them in 4 ml of minimal medium with appropriate amino acids for selection and at 30 °C.
3. After 16–18 h, image cells in a confocal microscope by placing 20 μl of the culture on a glass slide or imaging chamber (*see* **Note 2**).
4. Obtain *Z*-stack images of at least 200 cells of each strain (*see* **Note 3**).

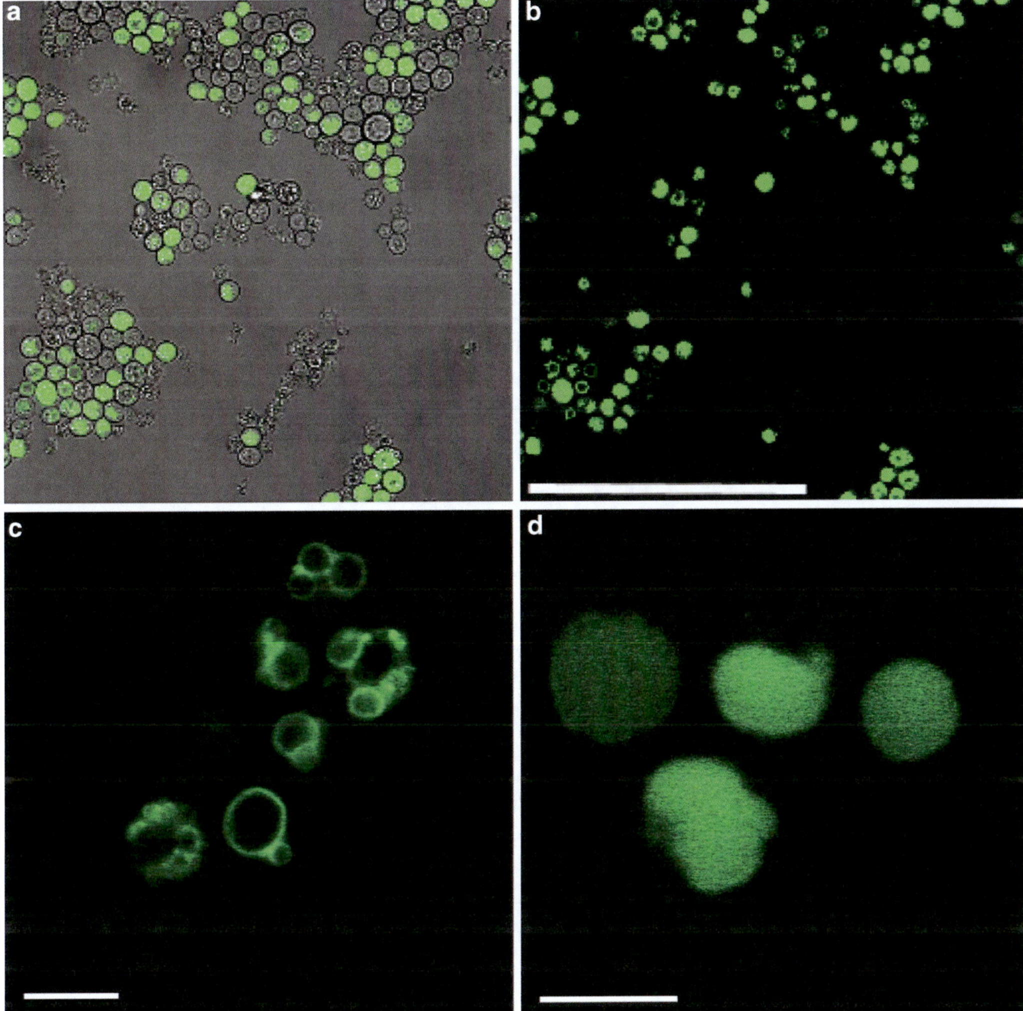

Fig. 1 Trans-species complementation of yeast ESCRT mutants using the GFP-CPS reporter. (**a**, **b**): bright field and fluorescence images of yeast transformed with the GFP-CPS marker. All yeast cells were positive for GFP signal (**a**, **b**). Only at higher magnification it is possible to discern between GFP-CPS mis-sorting to the vacuolar membrane in a *Δvta1* mutant cells (**c**) and proper sorting to the vacuolar lumen in a control cell (**d**)

5. Score cells as a normal GFP-CPS sorting (GFP signal in vacuolar lumen) or abnormal GFP-CPS sorting typical of ESCRT mutants (GFP signal on vacuolar membrane), depending on the localization of the GFP (*see* **Note 4**, Fig. 1). If LIP5 is able to restore normal GFP-CPS sorting in the *Δvta1* but not *Δvps4* background, it can be concluded that Arabidopsis LIP5 and yeast Vta1 have conserved ESCRT functions.

4 Notes

1. The Δvps4 strain is used as control to test whether any change in the sorting of GFP-CPS caused by the expression of the plant ESCRT protein depends on Vps4p function.
2. To prevent cell movement and drying during confocal observation, samples can be covered with a thin layer of 0.7 % agar in water
3. At low magnification and depending on the focal plane, the GFP signal coming from the vacuolar membrane can be mistaken by signal coming from inside the vacuole. To avoid this, collection of *Z*-stacks are recommended.
4. Statistical analysis is strongly recommended in order to determine significance in the changes in GFP-CPS localization.

References

1. Raymond CK, Howald-Stevenson I, Vater CA, Stevens TH (1992) Morphological classification of the yeast vacuolar protein sorting mutants: evidence for a prevacuolar compartment in class E vps mutants. Mol Biol Cell 3:1389–1402
2. Katsumata T, Hasegawa A, Fujiwara T, Komatsu T, Notomi M, Abe H, Natsume M, Kawaide H (2008) Arabidopsis CYP85A2 catalyzes lactonization reactions in the biosynthesis of 2-deoxy-7-oxalactone brassinosteroids. Biosci Biotechnol Biochem 72: 2110–2117
3. Babst M, Katzmann DJ, Estepa-Sabal EJ, Meerloo T, Emr SD (2002) ESCRT-III: an endosome-associated heterooligomeric protein complex required for MVB sorting. Dev Cell 3:271–282
4. Babst M, Katzmann DJ, Snyder WB, Wendland B, Emr SD (2002) Endosome-associated complex, ESCRT-II, recruits transport machinery for protein sorting at the multivesicular body. Dev Cell 3:283–289
5. Bilodeau PS, Winistorfer SC, Kearney WR, Robertson AD, Piper RC (2003) Vps27-Hse1 and ESCRT-I complexes cooperate to increase efficiency of sorting ubiquitinated proteins at the endosome. J Cell Biol 163:237–243
6. Katzmann DJ, Stefan CJ, Babst M, Emr SD (2003) Vps27 recruits ESCRT machinery to endosomes during MVB sorting. J Cell Biol 162:413–423
7. Bankaitis VA, Johnson LM, Emr SD (1986) Isolation of yeast mutants defective in protein targeting to the vacuole. Proc Natl Acad Sci U S A 83:9075–9079
8. Rothman JH, Stevens TH (1986) Protein sorting in yeast: mutants defective in vacuole biogenesis mislocalize vacuolar proteins into the late secretory pathway. Cell 47:1041–1051
9. Azmi I, Davies B, Dimaano C, Payne J, Eckert D, Babst M, Katzmann DJ (2006) Recycling of ESCRTs by the AAA-ATPase Vps4 is regulated by a conserved VSL region in Vta1. J Cell Biol 172:705–717
10. Haas TJ, Sliwinski MK, Martínez DE, Preuss M, Ebine K, Ueda T, Nielsen E, Odorizzi G, Otegui MS (2007) The *Arabidopsis* AAA ATPase SKD1 is involved in multivesicular endosome function and interacts with its positive regulator LYST-INTERACTING PROTEIN5. Plant Cell 19:1295–1312

Chapter 12

A Re-elicitation Assay to Correlate flg22-Signaling Competency with Ligand-Induced Endocytic Degradation of the FLS2 Receptor

Michelle E. Leslie and Antje Heese

Abstract

In the model plant *Arabidopsis*, the best studied Pattern-triggered immunity (PTI) system is perception of the bacterial pathogen-associated molecular pattern (PAMP) flagellin, or its active peptide-derivative flg22, by the plasma membrane-localized receptor FLAGELLIN SENSING 2 (FLS2). Flg22 perception initiates an array of immune responses including the fast and transient production of reactive oxygen species (ROS). In addition, FLS2 undergoes ligand-induced endocytosis and subsequent degradation within 60 min of flg22-treatment.

Luminol-based assays are routinely used to measure extracellular ROS production within minutes after flg22 treatment. Many mutants in flg22-response pathways display defects in flg22-induced ROS production. Here, we describe a luminol-based ROS Re-elicitation Assay that can be utilized to quantitatively assess flg22-signaling competency of FLS2 at times during which FLS2 is internalized, trafficked through endosomal compartments, and degraded in response to flg22. This assay may also be employed to correlate FLS2 signaling competency with receptor accumulation in vesicular trafficking mutants that either affect FLS2 endocytosis or replenishment of FLS2 through the secretory pathway. In addition, this assay can be extended to studies of other PAMP (ligand)–receptor pairs.

Key words Signaling competency, Ligand-induced endocytosis, Pattern recognition receptor (PRR), FLAGELLIN SENSING2 (FLS2), Pathogen-associated molecular pattern (PAMP), Flagellin, flg22, Desensitization, Reactive oxygen species (ROS), Luminol-based assay

1 Introduction

Plants are continuously exposed to a broad range of microbial pathogens including bacteria and fungi. As a first line of defense, plants use pattern recognition receptors (PRRs) to perceive extracellular pathogen-associated molecular patterns (PAMPs), also referred to as microbial-associated molecular patterns [1, 2]. The PRR FLAGELLIN SENSING2 (FLS2) undergoes ligand-induced endocytosis from the plasma membrane upon recognition of the bacterial PAMP flagellin, or the derived 22-amino-acid peptide flg22 [3, 4],

Marisa S. Otegui (ed.), *Plant Endosomes: Methods and Protocols*, Methods in Molecular Biology, vol. 1209, DOI 10.1007/978-1-4939-1420-3_12,

which is followed by degradation of FLS2 within 1 h post-elicitation [4–6]. Consistent with ligand-induced removal of the receptor from the site of perception (the plasma membrane), the plant can no longer perceive the flg22 peptide [6]. Flg22-induced endocytosis and degradation of FLS2 may serve important roles in attenuating immune responses, removing activated receptors from the plasma membrane, and/or desensitizing cells to the flg22 stimulus. In what seems to be a mechanism to prepare plants for additional rounds of flg22 perception, mRNA and protein levels of FLS2 are upregulated following the initial flg22-perception and FLS2 endocytic degradation, leading to resensitization of cells to flg22 within 3 h [6].

Activation of the FLS2 receptor by flg22 triggers a wide range of immune responses aimed at restricting pathogen growth. Early immune responses include the production of reactive oxygen species (ROS) and activation of a mitogen-activated protein kinases (MAPK) cascade [2, 7]. Taking advantage of the rapid and transient nature of these early flg22 responses, we have developed assays to correlate FLS2 signaling competency with flg22-induced endocytic degradation [6]. Here, we describe in detail our ROS Re-elicitation Assay for quantitatively measuring whether cells are able to sense flg22 at different times after an initial flg22 treatment. This protocol is based on earlier described luminol-based methods for measuring PAMP-induced ROS production, in which extracellular release of ROS facilitates a chemiluminescent reaction that can be detected using a luminometer [8, 9]. We have extended these methods to include additional, subsequent rounds of flg22-elicitation ("re-elicitation") of the same tissue during periods at which FLS2 undergoes ligand-induced endocytic degradation [4–6] and at which FLS2 protein levels are replenished for resensitization to flg22 (3+ hours) [6].

Vesicular trafficking proteins are emerging as key regulators of plant responses to pathogens [10, 11]. However, only few vesicular trafficking proteins that facilitate FLS2 endocytosis and further trafficking for degradation are known [3, 5, 12]. The described ROS Re-elicitation Assay may be used for identifying proteins, including novel vesicular trafficking proteins, that are involved in the desensitization of cells to flg22, possibly through regulating ligand-induced endocytosis and degradation of FLS2. In addition, by measuring ROS production at later time points (3+ hours), vesicular trafficking proteins that are required for replenishment of FLS2 at the plasma membrane may be identified.

2 Materials

1. *Arabidopsis* tissue obtained from well-expanded rosette leaves (*see* **Notes 1** and **2**).
2. White Costar assay plates (96 well, no lid, flat bottom, non-treated, non-sterile, white polystyrene; Corning/Fisher Scientific, USA; catalog number 3912).

3. 96-well, transparent plate lid.
4. Sterile, deionized water (room temperature; neutral pH).
5. Metal plant puncher with a 5-mm diameter, with sharpener (*see* **Note 3**).
6. Metal rod (i.e., handle of a bacteria plate spreader).
7. Fine forceps.
8. Glass plate for cutting tissue (*see* **Note 4**).
9. Razor blades.
10. Low-binding tubes (0.5 and 1.5 mL volumes) (*see* **Note 5**).
11. Filter tips.
12. 15 mL conical tubes.
13. 25 mL plastic reservoirs (optional).
14. Aluminum foil.
15. Flg22 peptide (*see* **Note 6**): lyophilized peptide is dissolved in DMSO to a concentration of 10 mM, adjusted to 100 % purity (*see* **Note 7**). Make serial dilutions of peptide in DMSO for 1,000× stocks (e.g., 100 μM 1000× stock will be used at a working concentration of 100 nM in ROS assays). To avoid contamination, use filter tips for pipetting of all peptide stock solutions. Make and store all the peptide stocks in low-binding tubes as 10–30 μL aliquots at −20 °C (*see* **Notes 8** and **9**).
16. Horseradish Peroxidase (HRP; essentially salt free, lyophilized powder; Sigma Chemical Company, St. Louis, MO, USA; catalog number P6782; Store at 4 °C prior to resuspension): for 500× HRP stock solution, dissolve 10 mg/mL in sterile deionized water. Store HRP solution in 10–30 μL aliquots at −20°C (*see* **Note 8**).
17. 500× Luminol solution: 17 mg luminol (≥97 % purity-HPLC; Sigma Chemical Company, St. Louis, MO, USA; catalog number A8511; stored at room temperature) dissolved in 1 ml of 200 mM KOH (*see* **Note 10**). Luminol is light-sensitive and should be protected from light by wrapping tube in aluminum foil. Make 500× luminol stock solution just prior to running the first assay, and keep solution in dark at 4 °C in between experiments. Discard daily.
18. *Elicitation Solution*: 1× luminol, 1× HRP, and 1× peptide in sterile, deionized water (*see* **Note 11** for information on peptide concentration). For example, for 10 mL elicitation solution, you will add 20 μL luminol (from 500× stock), 20 μL HRP (from 500× stock), and 10 μL peptide (from 1,000× stock, i.e., for 100 nM final concentration of flg22, add 10 μl of 100 μM flg22 stock). Invert tube to mix several times. Keep solution in the dark at room temperature (*see* **Note 12**).

19. 96-well microplate Luminometer Plate Reader with very high sensitivity, such as Promega GloMax® 96 Microplate Luminometer (*see* **Note 13**).
20. Computer that meets technical specifications and is equipped with necessary software for luminometer.
21. Graphing software of choice.

3 Methods

3.1 Leaf Tissue Preparation

Samples are prepared from fully expanded *Arabidopsis* rosette leaves (*see* Fig. 1a; **Note 1**). More reproducible results with less variability will be obtained by using at least two plants per genotype, as well as ensuring that you have at least 12 leaf samples (n-value ≥ 12) for each genotype–treatment combination in your experimental setup (*see* Fig. 2).

1. Pipette 150 μL sterile, deionized, room temperature water into each well of a white 96-well Costar assay plate. If available, use a multichannel pipette.
2. Using a well-sharpened, 5 mm-diameter plant tissue puncher, punch out 2–3 leaf disks on either side of the central vein of fully expanded rosette leaves (*see* Fig. 1b). If a leaf disk remains in the puncher, gently slide a metal rod through the puncher to push the leaf disk out the bottom. Do not allow leaf disks to accumulate in the puncher as this can lead to additional tissue damage and desiccation. Oftentimes, the leaf disks remain attached to the plant by way of the vein architecture. Carefully tease out leaf disks one at a time from the plant using forceps

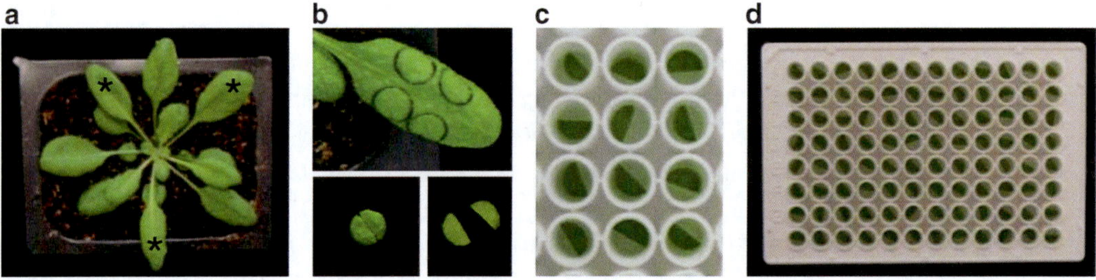

Fig. 1 Preparation of *Arabidopsis* leaf tissue for ROS Re-elicitation Assay. (**a**) Five-week-old wild type (Col-0) plant. The three well-expanded rosette leaves of similar developmental stage that we routinely use for ROS assays are marked with *asterisks*. (**b**) Using a 5-mm leaf punch, leaf disks are punched out of the medial region of the leaf, avoiding the central vein as well as the leaf margin (*outer edge*). Each leaf disk is placed on a glass surface and cut in half using a sharp razor blade. (**c**) Close-up of several wells containing leaf disk halves. One leaf disk half is carefully placed in each well of the 96-well Costar assay plate, floating on the surface of the water. The leaf tissue is orientated such that the abaxial (*lighter green*) underside of the leaf is touching the water. (**d**) A completed 96-well Costar Assay plate containing tissue samples

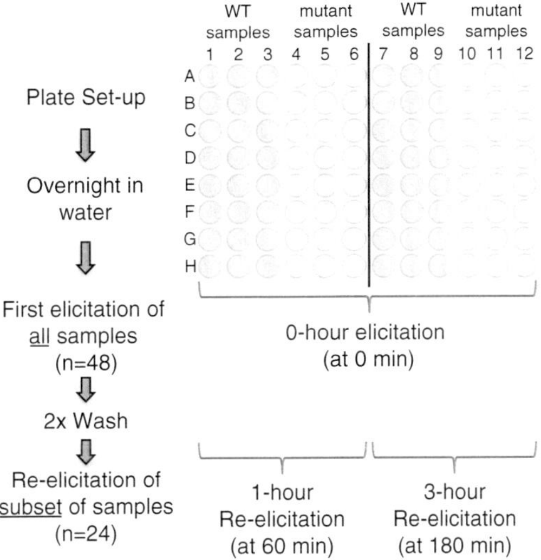

Fig. 2 ROS Re-elicitation Assay setup for wild-type and single mutant analysis. A 96-well Costar Assay plate is depicted containing leaf disk samples of wild type (WT; *gray*) and mutant (*white*) placed in the indicated water-containing wells. For this particular experimental setup, you will collect data for each genotype from independent samples, specifically $n = 48$ for 0-h elicitation and $n = 24$ for either 1 or 3-h Re-elicitations. After a 16–20 h recovery in a growth chamber with continuous light, all samples are treated with *Elicitation Solution* at 0 min (0-h elicitation). After the luminometer run of the 0 h-elicitation has been completed, all the samples are washed 2× with water. Half of the plate is re-elicited at 60 min (1-h re-elicitation) after the start of the first elicitation at 0 h. The tissue samples in the other half of the plate are allowed to rest in water until re-elicitation at 180 min (3-h re-elicitation) after start of the first elicitation at 0 h. This experimental setup will allow for comparison between ROS bursts during the initial elicitation (0-h), the desensitization period (1-h), as well as the resensitization phase (3-h)

to carefully grasp the edge of the leaf disk only. Take care not to rip the leaf disks or visibly damage the tissue. Discard any ripped or damaged leaf disks.

3. Working with tissue from one leaf at a time, place the leaf disks on the glass plate adaxial (darker green) side up. Avoid desiccation of leaf tissue by working quickly or by punching out only 1–2 leaf disks at a time.
4. Cut each leaf disk in half (*see* Fig. 1b; **Note 14**). We have found the following method to cause the least amount of damage to the tissue, leaving an exposed margin of viable cells. Obtain a new razor blade, flipping or replacing whenever the blade begins to pull on the tissue resulting in ripped and damaged tissue. We often use 4 razor blades to fill a 96-well plate. Use one finger to gently hold the leaf disk in place, touching the edge of the leaf disk only and not applying excessive pressure.

Place one edge of the razor blade on the glass plate, directly at the center of a leaf disk. In a singular motion (keeping the razor blade in contact with the glass plate), slide the blade through the center of the leaf disk. Discard ripped and severely damaged leaf disks.

5. Transfer leaf disks to appropriate wells of the 96-well Costar assay plate, such that they float adaxial (darker green) side up (*see* Fig. 1c, d). It is best to take advantage of the concavity of the leaf to scoop up the leaf disks one at a time using forceps. If necessary, you may also use forceps to grasp the curved edge of the leaf disk, but this method often causes additional tissue damage. If the leaf disk flips when placed into the well, gently use your forceps to scoop the leaf disk over. Again, discard and replace any severely damaged leaf disks during this step.
6. Once all leaf disk halves have been transferred, you may gently tap the plate on your bench top to ensure that each sample is freely floating on the surface of the water and not adhered to the side of the well. Cover the plate with a transparent cover, such as a lid from a 96-well tissue culture plate, to prevent water evaporation during the resting period. You may need to break off a corner of the lid in order to fit the Costar assay plate. Lids may be rinsed and reused.
7. Place the Costar assay plate into a 22 °C growth chamber with continuous light. Allow the tissue to rest for 16–20 h.

3.2 Luminol-Based Reactive Oxygen Species Re-elicitation Assay

The luminol-based ROS re-elicitation assay should be performed 16–20 h following the cutting and placement of leaf tissue in the 96-well Costar assay plate to allow for robust, consistent, and reproducible ROS production (*see* **Note 15**).

1. Prepare fresh 500× luminol solution.
2. Turn on GloMax Luminometer and select a protocol (*see* GloMax Manual and **Note 16**).
3. Prepare fresh *Elicitation Solution* for 0-h elicitation.
4. Obtain the 96-well Costar assay plate with the leaf disk samples from growth chamber. Following **step 3** without delay, remove water from each sample well of the 96-well plate. Tilt the plate as you go in order to remove as much as water as possible. To avoid additional ROS-producing stress, it is important to not touch (thereby damage) the leaf disks with the pipette tip and prevent dehydration by working quickly (*see* **Note 17**).
5. Working in a dimly lit room, quickly add 100 μL *Elicitation Solution* to each well of your 96-well plate containing leaf samples (*see* **Note 18**).
6. Ensure that all the leaf samples are exposed to the elicitation solution. Using a pipette tip, gently move samples down so that they are touching solution if necessary (*see* **Note 19**).

7. Immediately place the plate into luminometer and start run. Peak ROS production often occurs after 10–12 min of exposure to the flg22 elicitation buffer. Timing of peak ROS production is dependent upon the flg22 concentration used (the higher the flg22 concentration, the earlier the ROS peak) (*see* **Note 20**).
8. Allow the luminometer to acquire luminescence measurements for 30–40 min.
9. Remove the plate from the luminometer at the end of the run.
10. Save data to Microsoft Excel (or appropriate software program for your plate reader).
11. As for **step 4**, remove solution from each leaf disk sample in the 96-well Costar assay plate.
12. Wash all the samples with 150 μL sterile, deionized water. Repeat wash once. Keep the leaf disks in water until time of re-elicitation (*see* Fig. 2). Avoid dehydration of tissue by ensuring that all the leaf samples are exposed to the water.
13. About 10 min prior to the time of re-elicitation, prepare fresh Elicitation Solution for 1-h elicitation. Volume of buffer required will depend upon your experimental setup (*see* Fig. 2). Keep the solution in the dark at room temperature.
14. As described in **steps 4–6**, carefully remove solution from each leaf disk sample in the assay plate. Working in a dimly lit room, add 100 μL *Elicitation Solution* to each well with a "1-hour elicitation sample" (*see* Fig. 2). Ensure that all the leaf samples are exposed to the elicitation solution.
15. Place the plate immediately into the luminometer, start run and data collection as described in **steps 7–10**.
16. In between the 1- and 3-h elicitations (*see* Fig. 2), store the plate containing leaf tissue floating on distilled water in 22 °C growth chamber with continuous light. About 10 min prior to the start of the 3-h elicitation, prepare fresh *Elicitation Solution*. Treat 3-h tissue samples as described for the 0- and 1-h elicitations (**steps 4–6**). Obtain luminescence readings as in **steps 7–10**.
17. If using genetically modified plant material, you should dispose of the plate and its contents according to your institutional policies at the end of the experiment. None of the components of the elicitation solution require special disposal methods. Do not reuse Costar assay plates due to the possibility of peptide contamination from one experiment to the next and pathogen growth in plate wells that will interfere with ROS production.
18. Data obtained from the luminometer will include a reading of the ROS-induced luminescence (relative light units, RLUs) for each sample over the 0, 1, and 3-h time courses (*see* **Notes 20** and **21** for discussion of expected results).

Time points are typically obtained every 2 min depending upon the number of samples and the total "plate run time" of the luminometer (*see* **Note 16**). Inspect data for obvious outlying RLU values (*see* **Note 22**). Using a graphing software of your choice, plot data as the change in relative light (ROS production) units over time. For example: x = time (minutes) versus y = RLU (*see* Fig. 3a). The data can also be

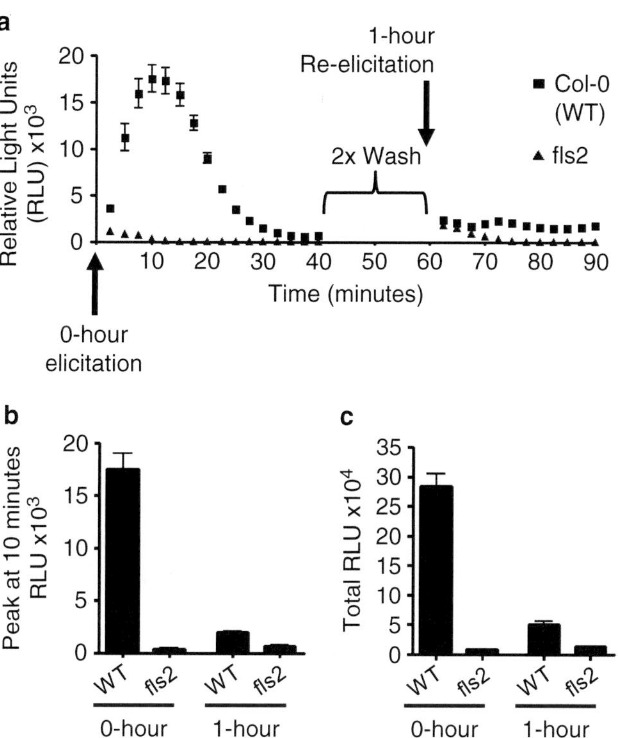

Fig. 3 Graphical representations of data obtained from the ROS Re-elicitation Assay. Leaf disks of wild-type (Col-0) and *fls2* mutants that do not express functional FLS2 receptor were treated with 100 nM flg22 at 0 min (0-h elicitation), washed twice with water, and then re-elicited at 60 min (1-h Re-elicitation) with 100 nM flg22. (**a**) Time-course of ROS production. All data points are graphed as times post-elicitation (minutes) (x-axis) versus relative light units (RLU) (y-axis). The initial ROS production is observed between 5 and 25 min post-elicitation in wild type tissue, peaking at 10–12 min. Consistent with ligand-induced endocytosis and degradation of the FLS2 receptor at 60 min after the initial round of elicitation (0 h-elicitation), wild-type cells are unable to re-elicit ROS production burst when re-elicited at 60 min with flg22 (1-h Re-elicitation). *fls2* plants lacking the FLS2 receptor are unable to perceive flg22, and thus do not produce any ROS after either elicitation. (**b**) Peak ROS production. Peak RLU values at 10 min post-elicitation are graphed for both the 0- and the 1-h elicitations, showing ROS production at 10 min and 70 min, respectively. (**c**) Total ROS production. RLU values are obtained from an integration of the 0- and 1-h elicitation curves for each n value (sample) shown in **a**

analyzed as a bar graph of average RLU at the time of peak elicitation (typically 10–12 min post-elicitation) (*see* Fig. 3b). Finally, you may choose to graph average total ROS production for the entire run of each elicitation by obtaining an integration of the area under the curve for each sample (*see* Fig. 3c). We use a Student's *t*-test for pair-wise comparisons between average RLU values obtained from experimental groups on the same Costar assay plate (*see* **Notes 23** and **24**).

19. *Variations*: The described methods and data analysis can be extended to studying additional receptor–ligand signaling pathways that produce a ROS response, as well as the effect of vesicular trafficking inhibitors upon ligand de-sensitization and recovery of signaling (*see* **Notes 25** and **26**). Results obtained with the ROS Re-elicitation Assay may lead to further analysis of FLS2 signaling competency using established biochemical and cell biological assays (*see* **Notes 27** and **28**).

4 Notes

1. For obtaining consistent results in the described ROS Re-elicitation Assay, particular care should be taken in ensuring healthy plant growth. In our laboratory, we germinate *Arabidopsis* seeds on half-strength MS plates with 1 % sucrose under continuous light for 7 days at 22 °C. Seedlings are transplanted into soil (Sunshine Mix or similar) and grown under short-day conditions (8-h light) for approximately 4–5 additional weeks at 22 °C. Well-spaced plants will produce well-expanded leaves, ideal for obtaining tissue samples. We grow only one plant per 7 (w)×9 (l)×5.5 (h)-mm pot (*see* Fig. 1a). To avoid water stress, water potted plants only when soil becomes dry to the touch, taking care to drain excess water after a period of 1–4 h.
2. When choosing rosette leaves for tissue samples, we typically use three leaves per plant of similar developmental stage (*see* Fig. 1a). Avoid any leaf with previous damage, chlorosis, or visible cell death. In addition, do not take plants that show any symptoms of any apparent pathogen infection. Due to variability between plants, it is best to have at least two plants per genotype, evenly spread out between treatments on the plate.
3. For smaller leaves, one may use a smaller metal punch (i.e., a diameter of 3–4 mm). However, cutting the leaf disks in half and transferring to the Costar assay plate may prove more difficult.
4. It is important to cut plant tissue on a hard, nonporous surface such as a glass or plastic plate to reduce excess tissue damage.

For ease of cutting, the minimum glass surface size that we use is approximately 7 × 10 cm, with larger surfaces being easier to work with.

5. Flg22 peptides are inherently "sticky" [13]. Using low-binding tubes for dilutions and storage of peptide stocks prevents peptide loss.
6. The active 22-amino-acid flg22 peptide used for eliciting an immune response in *Arabidopsis* leaf tissue is: QRLSTGSRINS AKDDAAGLQIA [14]. As an inactive control, we use a 22-amino-acid peptide derived from *Agrobacterium tumefaciens* flagellin: ARVSSGLRVGDASDNAAYWSIA [14].
7. We order our flg22 peptides from Genscript (Piscataway, NJ, USA) at a guaranteed purity of ≥90 %, and take the actual purity into account when making our initial 10 mM peptide stock solutions. For example, if our peptide is 95 % pure, we will resuspend 5 % more peptide than we would for resuspending a peptide with a theoretical purity of 100 %. Due to the potential risk for peptide contamination within the lab or at the time of peptide synthesis, a recent study highlights the importance of testing peptides for purity and expected cellular responses [13].
8. Peptide and HRP stock solutions are best stored in small, single use aliquots. For example, we routinely use 10 μL of our 1,000× stock peptides in each experiment; therefore, we freeze peptide stock solutions as 11–12 μL aliquots to avoid multiple freeze-thaw cycles (*see* **Note 9**).
9. Avoid leaving a peptide stock at room temperature for more than a few minutes. Also, repeated freeze-thaw cycles may cause peptide degradation. We never refreeze a peptide stock more than once. If you have a large amount of remaining peptide stock after an experiment, you may want to divide the remaining peptide into smaller aliquots for refreezing and subsequent reuse.
10. Simple method for preparation of luminol solution: Measure 1–3 mg of luminol powder directly into light-protected microcentrifuge tube such as an aluminum-foil covered tube. Add appropriate volume of 200 mM KOH directly to tube (100 μL for every 1.7 mg of luminol). Resuspend luminol by rigorous vortexing, checking periodically to see that all luminol is resuspended. You may need to use a pipette tip to physically break up clumps of luminol, and then vortex again. Make fresh daily and store at 4 °C between uses within the same day. In our hands, luminol resuspended in KOH produces a more reproducible and higher ROS production compared to luminol resuspended in DMSO stored at −20 °C. In addition, luminol is not soluble in water. Be sure to make enough luminol

to finish all steps of the Re-elicitation Assay since making a new luminol stock in the middle of an assay may introduce variability.

11. Prior to setting up a ROS Re-elicitation Assay, you may want to test a dilution series of your flg22 peptide (i.e., 1, 10, 100, and 1,000 nM) in a ROS single-elicitation assay, comparing your mutant(s) of interest to the wild type. This will allow you to test the peptide for ability to produce a ROS burst in a dose-dependent manner. For correlating FLS2 signaling competency with ligand-induced endocytic degradation and new protein synthesis of FLS2, we use 0.1 or 1 μM flg22, concentrations commonly used for physiology assays, and which we have found to induce FLS2 degradation within 1 h of elicitation [6]. In general, we avoid using higher flg22 concentrations such as 10 μM.
12. Since abiotic stresses, such as cold, can cause rapid and transient ROS production similar to a PAMP response, it is necessary to ensure that all materials used for PAMP elicitation are brought to room temperature prior to use.
13. We use a Promega GloMax® 96 Microplate Luminometer (Madison, WI, USA) which has a high enough sensitivity (3×10^{-21} mol luciferase) to detect PAMP-induced ROS production in plants. We do not utilize an injector for solution exchanges since this may cause unnecessary wounding.
14. While an uncut leaf disk will produce a ROS burst upon elicitation with flg22, the razor-cut edge increases the cellular surface area that is exposed to the elicitation solution. We have found this to be a key step for obtaining reproducible responses with less variability within an experiment as well as when performing experiments over an extended amount of time, such as the Re-elicitation Assay.
15. Since the initial wounding of the tissue during cutting causes a ROS burst, the resting period of 16–20 h is necessary for recovery and "re-setting" to the response baseline.
16. For a 96-well Costar plate assay using the Promega GloMax Luminometer, we use the following 40-min protocol:
    (a) Delay before first run: 0 s.
    (b) Integration Time: 0.5 s.
    (c) Number of runs: 16 (approximately 2.5 min each).
    (d) Delay between runs: 0 s.
17. All solution exchanges should be added in the direction in which the luminometer reads the plate. For example with the Promega GloMax Luminometer, all solution exchanges should be completed in a left to right and top to bottom orientation, as this is the direction that the luminometer obtains data.

18. Because ROS responses are fast and transient, it is important to add all elicitation solutions swiftly to the samples to obtain consistent and meaningful results with relatively low variability. We recommend that you use an automated single-channel pipette or a multi-channel pipette (manual or automated) for quick addition of elicitation solution. Avoid pipetting solution directly onto leaf disk, which could cause nonspecific wounding/ stress response. If using a multichannel pipette, you will need to pour elicitation solution into a clean and dry plastic reservoir.
19. When adding elicitation solution, some leaf disk samples may flip such that the adaxial (darker green) side of leaf is now exposed to the solution. Using your pipette tip, you can try to gently flip the leaf disk over. If the disk resists a gentle flipping, it is better to leave the sample as is than to risk damaging the tissue. You may want to make a note of the well to possibly eliminate data of these samples from subsequent data analysis.
20. Based on our experience, flg22 concentrations of at least 100 nM in the *0-h Elicitation Solution* are necessary to result in consistent and significant endocytic degradation of FLS2. Consistent with ligand-induced endocytic degradation of FLS2, cells rapidly become desensitized to further flg22 stimulation (*see* Fig. 3). Within 1 h of exposure to the *0-h Elicitation Solution*, the majority of FLS2 receptors have undergone ligand-induced endocytosis and degradation [4–6]. In this case, you would not expect additional flg22-induced ROS production at the 1-h time point (*see* Fig. 3). A comparison between the wild type and your mutant of interest may uncover differences in flg22-induced desensitization of the cells.
21. Flg22-induced replenishment of FLS2 through new protein synthesis occurs over extended periods of time [6]. Signaling recovery can be tested by re-elicitation of samples with flg22 at various time points (e.g., 3, 9, 16, 24 h), comparing the wild type to your mutant of interest. Your mutant of interest may affect the recovery of resensitization following desensitization of FLS2 to flg22.
22. Visual inspection of your data points may reveal extreme outliers that you choose to remove from the data set. Oftentimes, these samples were damaged during the washes or were not properly exposed to elicitation/wash solutions, which may be confirmed by inspecting your plate at the end of the luminometer runs (*see* **Note 19**).
23. When analyzing our data, we use a statistical cutoff of p-value ≤ 0.05 obtained with a Student's t-test. We consider differences observed between experimental groups to be consistent and reproducible when we obtain statistically significant results from at least three independent experiments.

24. For data analysis, you can only directly compare experimental results obtained from the same Costar assay plate because absolute RLU values obtained from one experiment to the next will vary (i.e., peak RLU of 20,000 versus 30,000). Therefore, it will be necessary to compare trends between experiments (i.e., from different plates) rather than absolute values.
25. While FLS2 is the best-studied PAMP receptor and one of few plant receptors known to undergo ligand-induced endocytosis, this ROS Re-elicitation Assay can be extended to studying ligand-induced desensitization of additional signaling pathways [1, 2] and correlating signaling competency to ligand-induced endocytosis of receptors. In addition, comparing re-elicitation of samples with different PAMPs that are recognized by different PRRs will provide specificity of observed ROS response (or lack thereof) [6].
26. To gain a better understanding of the vesicular trafficking pathway(s) affected by any particular protein of interest, chemical inhibitors can be utilized in the ROS assay [6]. Both Wortmannin and Tyrphostin A23 were previously shown to negatively affect flg22-induced endocytosis and degradation of FLS2 when applied to tissue for a 1-h pretreatment, just prior to elicitation with flg22 [3, 6].
27. Results obtained with the described ROS Re-elicitation Assay can be complemented by determining whether or not cells remain signaling competent for different flg22-responses, such as the ability to re-phosphorylate MAPK proteins. Flg22 treatment of *Arabidospis* leaf tissue causes a fast and transient phosphorylation of the MAPKs MPK3 and MPK6. Peak phosphorylation of these MAPKs can be detected 10 min post elicitation using an antibody that specifically detects activated, phosphorylated MAPKs [15]. FLS2-flg22 signaling competency can therefore be tested at the same time points used for the ROS Re-elicitation Assay using a standard MAPK assay [6].
28. To correlate the effect of a ROS defect upon flg22-induced endocytosis of FLS2, any observed differences in the ROS Re-elicitation Assay could be followed up with an analysis of FLS2 protein abundance and/or FLS2 protein localization [3–6].

Acknowledgements

This work was supported by grants received from the National Science Foundation (NSF-IOS-1147032) and University of Missouri Research Board to A.H. We thank Daniel Salamango for contributions to assay development and John Smith for critical reading of this chapter.

References

1. Monaghan J, Zipfel C (2012) Plant pattern recognition receptor complexes at the plasma membrane. Curr Opin Plant Biol 15:349–357
2. Nicaise V, Roux M, Zipfel C (2009) Recent advances in PAMP-triggered immunity against bacteria: pattern recognition receptors watch over and raise alarm. Plant Physiol 150:1638–1647
3. Beck M, Zhou J, Faulkner C et al (2012) Spatio-temporal cellular dynamics of the Arabidopsis flagellin receptor reveal activation status-dependent endosomal sorting. Plant Cell 24:4205–4219
4. Robatzek S, Chinchilla D, Boller T (2006) Ligand-induced endocytosis of the pattern recognition receptor FLS2 in Arabidopsis. Genes Dev 20:537–542
5. Lu D, Lin W, Gao X et al (2011) Direct ubiquitination of pattern recognition receptor FLS2 attenuates plant innate immunity. Science 332:1439–1442
6. Smith JM, Salamango DJ, Leslie ME et al (2014) Sensitivity to flg22 is modulated by ligand-induced degradation and de novo synthesis of the endogenous flagellin-receptor FLS2. Plant Physiol 164:440–454
7. Tena G, Boudsocq M, Sheen J (2011) Protein kinase signaling networks in plant innate immunity. Curr Opin Plant Biol 14:519–529
8. Gómez-Gómez L, Felix G, Boller T (1999) A single locus determines sensitivity to bacterial flagellin in Arabidopsis thaliana. Plant J 18: 277–284
9. Keppler LD, Baker CJ, Atkinson MM (1989) Activated oxygen production during a bacteria-induced hypersensitive reaction in tobacco suspension cells. Phytopathology 79:974–978
10. Beck M, Heard W, Mbengue M et al (2012) The INs and OUTs of pattern recognition receptors at the cell surface. Curr Opin Plant Biol 15:367–374
11. Frei dit Frey N, Robatzek S (2009) Trafficking vesicles: pro or contra pathogens? Curr Opin Plant Biol 12:437–443
12. Choi SW, Tamaki T, Ebine K et al (2013) RABA members act in distinct steps of subcellular trafficking of the Flagellin Sensing2 receptor. Plant Cell 25:1174–1187
13. Mueller K, Chinchilla D, Albert M et al (2012) Contamination risks in work with synthetic peptides: flg22 as an example of a pirate in commercial peptide preparations. Plant Cell 24:3193–3197
14. Felix G, Duran JD, Volko S et al (1999) Plants have a sensitive perception system for the most conserved domain of bacterial flagellin. Plant J 18:265–276
15. Heese A, Hann DR, Gimenez-Ibanez S et al (2007) The receptor-like kinase SERK3/BAK1 is a central regulator of innate immunity in plants. Proc Natl Acad Sci U S A 104: 12217–12222

Chapter 13

Preparation of Enriched Plant Clathrin-Coated Vesicles by Differential and Density Gradient Centrifugation

Gregory D. Reynolds, Ben August, and Sebastian Y. Bednarek

Abstract

Methods for the subcellular fractionation and enrichment of specific intracellular compartments are essential tools in the analysis of compartment composition and function. In vitro characterization of isolated cell organelles and other endomembrane intermediates, including exploration of the compartment protein ensemble, offers strong clues of in vivo function identity. Here, we describe methodology for the isolation of clathrin-coated vesicles from *Arabidopsis thaliana* suspension-cultured cells on the basis of differential and density centrifugation.

Key words Clathrin, Fractionation, Negative stain electron microscopy, Plant, Vesicles

1 Introduction

Clathrin-coated vesicles (CCVs) are membranous organelles that function in the biosynthetic and endocytic trafficking of soluble and membrane proteins in all eukaryotes. Clathrin is a coat protein composed of three heavy chains (CHC) and three light chains (CLC) assembled into a three-legged structure called a clathrin triskelion. With the aid of distinct adaptor protein (AP) complexes, including AP1 and AP2, which selectively package protein and lipid cargos at the *trans*-Golgi network (TGN) and the plasma membrane, respectively, CCVs provide a route in which cellular materials are internalized by endocytosis and are transported within the late secretory pathway. The AP complexes along with numerous other accessory proteins also help direct the clathrin triskelia to form curved polygonal lattices around budding membrane invaginations, which eventually form CCVs.

In plant and mammalian cells, clathrin-mediated endocytosis is important for the uptake of nutrients and signaling receptors, intracellular communication, and signal transduction. In addition, clathrin-mediated trafficking is critical for the formation of the cell plate during cytokinesis and for the delivery of membrane and cell

Marisa S. Otegui (ed.), *Plant Endosomes: Methods and Protocols*, Methods in Molecular Biology, vol. 1209,
DOI 10.1007/978-1-4939-1420-3_13, © Springer Science+Business Media New York 2014

wall materials necessary for cell expansion in plants (for review *see* ref. 1). Our understanding of the mechanism of clathrin-mediated trafficking in plants is underdeveloped relative to animal and yeast. One approach to gaining a better understanding of the formation and trafficking of plant CCVs and their cargo is through the analysis of enriched preparations of isolated CCVs.

Methodology for the isolation of CCVs from mammalian microsomal membrane fractions by differential centrifugation was first described by Pearse [2] and subsequently improved upon by the use of a density gradient comprised of Ficoll and D_2O [3]. Here we describe a protocol for the isolation of CCVs from Arabidopsis suspension-cultured cells. This procedure was adapted from previously established protocols for the isolation of plant CCVs [4, 5]. In brief, the protocol describes enrichment of CCVs by differential centrifugation, velocity sedimentation, and equilibrium density gradient centrifugation from Arabidopsis suspension-cultured cell microsomal membrane extracts (*see* Fig. 1). In addition, the use of negative stain electron microscopy for the assessment of the purity and enrichment of CCV subcellular fractions is described.

2 Materials

Prepare all solutions using purified, deionized water (resistivity ≥18 at 25 °C), and reagents of analytical grade or higher (unless otherwise noted). Follow all appropriate disposal requirements with special attention directed to discarding the (mildly radioactive) D_2O and (highly toxic) OsO_4 solutions used in this experiment.

2.1 Isolation Buffers, Reagents, and Solutions (See Note 1)

1. Ethylenediaminetetraacetic acid (EDTA) Stock: To 80 mL of water in a clean beaker, add 18.61 g of $Na_2{\cdot}EDTA{\cdot}2H_2O$ (m. mass 372.20 g/mol). While stirring, add 1.8 g solid NaOH. Upon NaOH dissolution, pH to 8.0 with 10 N NaOH. When EDTA has completely dissolved, adjust final volume to 100 mL with water. Store in a plastic container at room temperature.
2. Ethylene glycol tetraacetic acid (EGTA) Stock: To 80 mL of water in a clean beaker, add 19.02 g of EGTA (m.mass 380.35 g/mol). While stirring, add 3.5 g solid NaOH. Upon NaOH dissolution, adjust pH to 8.0 with 10 N NaOH. When EGTA has completely dissolved, adjust final volume to 100 mL with water. Store in a plastic container at room temperature.
3. Magnesium Chloride ($MgCl_2$) Stock: To 80 mL of water, add 9.52 g of anhydrous $MgCl_2$ (m.mass 95.211 g/mol). Upon dissolution, adjust volume to 100 mL with water. Store at room temperature.

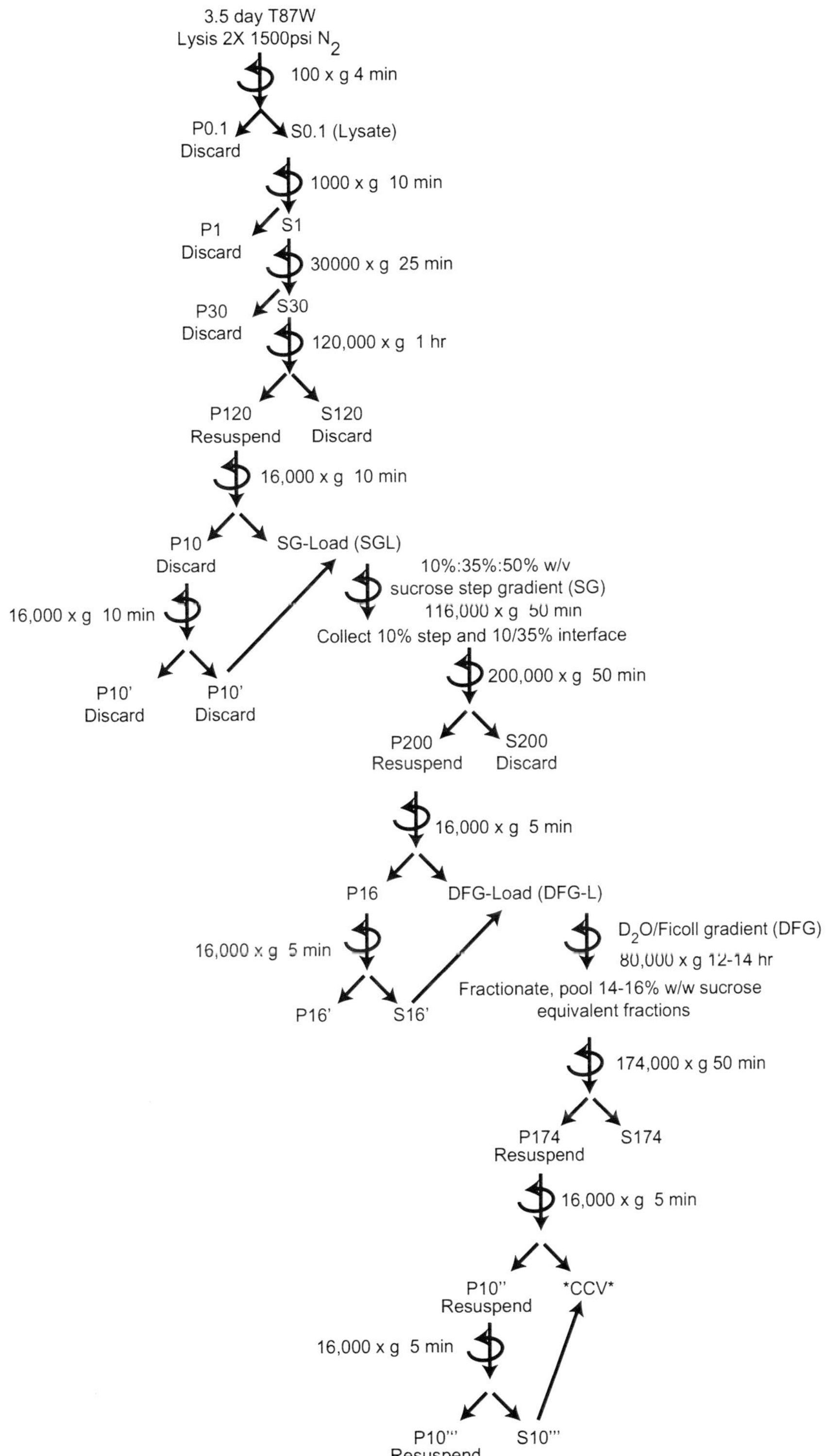

Fig. 1 Flowchart summarizing fractionation and centrifugation steps involved in the preparation of the final enriched CCV fraction

4. 1× Clathrin Isolation Buffer (CIB) (*see* **Note 2**): 100 mM 2-(*N*-morpholino)ethanesulfonic acid (MES), 0.5 mM $MgCl_2$, 3 mM EDTA, 1 mM EGTA, pH 6.4. To 750 mL water in a 2 L beaker, add 19.523 g of MES hydrate (m.mass 195.24 g/mol), 500 μL of 1 M $MgCl_2$ stock, 6 mL of EDTA stock pH 8.0, and 2 mL of EGTA stock pH 8.0. Adjust volume to 900 mL and adjust pH to 6.4 using 5 N KOH. Dilute to 1 L final volume with water. Filter sterilize (pore size ≤0.2um). Store at 4 °C.
5. 1,000× Protease Inhibitor Cocktail Dimethyl Sulfoxide (DMSO)-soluble (PIC-D) (*see* **Note 3**): 500 mM Phenylmethylsulfonyl fluoride (PMSF), 5 mg/mL pepstatin A, 1 mg/mL chymostatin. To 1 mL DMSO in a sterile 15 mL screw-top conical tube, add 176 mg PMSF, 10 mg pepstatin A, and 2 mg chymostatin. Dilute to 2 mL with DMSO and mix thoroughly. Distribute into 500 μL aliquots. Store at −80 °C [6].
6. 1,000× Protease Inhibitor Cocktail Water-soluble (PIC-W). 1 M *p*-Aminobenzamidine, 1 M e-Aminocaproic acid, 5 mg/mL aprotinin, 1 mg/mL leupeptin. To 1 mL water in a sterile 15 mL conical tube, add 416 mg *p*-Aminobenzamidine, 262 mg *ε*-Aminocaproic acid, 10 mg aprotinin, and 2 mg leupeptin. Dilute to 2 mL with water and mix thoroughly. Distribute into 500 μL aliquots. Store at −80 °C [6].
7. 100× E-64: 1 mg/mL epoxysuccinyl-L-Leucylamido (4-guanidino) butane (E-64). To 8 mL 1× CIB (see above), add 10 mg E-64. Dilute to 10 mL final volume with 1× CIB. Distribute into 1 mL aliquots. Store at −80 °C.
8. 1,000× DTT: 1 M dithiothreitol (DTT). To 1 mL water, add 308.5 mg DTT, dissolve and adjust to a final volume of 2 mL with water. Store in 500 μL aliquots at −80 °C.
9. 1× CIB* solution: 1× CIB containing 1× PIC-W, PIC-D, DTT, and E-64.
10. 50, 35, and 10 % w/v Sucrose Density Gradient solutions: To three individual, immaculately clean 100 mL graduated cylinders, add 50, 35, or 10 g of molecular biology grade sucrose (for 50 %, 35 %, and 10 % w/v solutions, respectively). Dilute to 90 mL with 1× CIB, seal the cylinder with parafilm, and mix by inversion until complete dissolution. Dilute to 100 mL total volume, seal, and mix thoroughly by inversion. Filter sterilize (pore size ≤0.2 μm). Store at 4 °C (*see* **Note 4**).
11. 90 %(w/v) D_2O/30 % Ficoll Density Gradient Solution: Into each of two clean 50 mL conical flasks, add: 22.5 g D_2O, 7.5 g Ficoll, 0.488 g MES hydrate (m.mass 195.24 g/mol), 12.5 μl of 1 M $MgCl_2$ stock, 150 μL of 500 mM EDTA stock pH 8.0, and 50 μL of 500 mM EGTA stock pH 8.0. Incubate at 60 °C while mixing (e.g., using a rotary mixer), for 1 h. Combine the two flasks and adjust pH to 6.4 using 10 N NaOH. Store at 4 °C or for long-term storage at −20 °C (*see* **Note 5**).

12. 9 %(w/v) D_2O/2 % Ficoll Density Gradient Solution: Into a clean 50 mL conical flask, add: 4.5 g D_2O, 1.0 g Ficoll (GE Healthcare), 0.976 g MES hydrate (mol mass 195.24 g/mol), 25 μL of 1 M $MgCl_2$ stock, 300 μL of 500 mM EDTA stock pH 8.0, and 100 μL of 500 mM EGTA stock pH 8.0. Incubate at 60 °C while mixing (e.g., using a rotary mixer), for 1 h. Adjust pH to 6.4 using 10 N NaOH and bring final volume to 50 mL with water. Store at 4 °C or for long-term storage at −20 °C.

2.2 Centrifugation Equipment (See Note 6)

1. Low speed/floor model centrifuge (e.g., *Beckman J2-21*) and accompanying rotor (e.g., *Beckman JA-20*) capable of accommodating 200 mL of sample at 30,000 × *g* with appropriate sample holders (e.g., *Beckman 29 × 104 mm polypropylene thick-wall tube ×8*).
2. Low speed/tabletop centrifuge and accompanying rotor (e.g., *Jouan CR 3 22 with 50 mL screw-top conical tube compatible inserts*) capable of accommodating 500 mL of sample at 10,000 × *g* with appropriate sample holders *(50 mL conical screw-top tubes ×10)*.
3. High speed ultracentrifuge (e.g., *Beckman L-60*) and accompanying rotors (e.g., Beckman *50.2Ti, SW40Ti*) capable of accommodating 200 mL of sample at 200,000 × *g* and three 10 mL gradients at 120,000 × *g* with appropriate tubes (e.g., *Beckman 26.3 mL polycarbonate capped bottles ×8, Beckman Ultraclear 14 × 95 mm tubes ×6*).
4. High speed tabletop ultracentrifuge (e.g., *Beckman TL-100*) and accompanying rotor (e.g., *Beckman TLA 100.3*) capable of accommodating 20 mL of sample at 174,000 × *g* with appropriate sample holders (*Beckman Ultraclear 13 × 51 mm tubes ×2*).
5. Microcentrifuge (e.g., *Eppendorf 5415 D*) with appropriate sample holders (*1.5 mL microfuge tubes with O-ring sealed screw-top*).

2.3 Other Purification Equipment

1. 2× 45 mL Parr Cell Disruption Vessels (Parr Instrument Company, Moline, IL) with accompanying gas regulator and pressurized N_2 storage tank.
2. 15, 5, 2 mL Dounce tissue homogenizers.
3. Linear Gradient Pouring Apparatus (e.g., *Labconco AUTODENSIFLOW Density Gradient Fractionator*).
4. Fractionator.

2.4 Negative Stain Materials

1. Pioloform-coated 200 mesh nickel grids.
2. 4 % OsO_4 in water, 2 mL ampoule.
3. Nano-W® methylamine tungstate (Nanoprobes, Yaphank, NY).
4. Transmission electron microscope.
5. Nonmagnetic steel forceps.

6. Filter paper, 10 cm diameter, cut into eight wedges.
7. ImageJ software (http://rsb.info.nih.gov/ij/, NIH) with "Cell Counter" plug-in (http://rsbweb.nih.gov/ij/plugins/cell-counter.html, Kurt De Vos).

3 Methods

Keep all solutions and carry out all procedures on ice unless otherwise noted. Prechill cell disruption vessel, centrifuge rotors, and centrifuges to 4 °C before steps in which they are utilized. Allow premade sucrose and D_2O/Ficoll solutions to equilibrate to room temperature (20 °C) prior to forming density gradients.

3.1 Cell Collection and Lysis

1. Collect 500 mL of 3.5-day-old *Arabidopsis thaliana* T87 [7] suspension culture (*see* **Note 5**) and distribute into 50 mL conical vials or other appropriate centrifugation container. Collect cells by centrifugation for 4 min, at 100 × *g*, 4 °C. Decant and discard supernatant (*see* **Note 7**).
2. Resuspend and combine all pellets to a total volume of 150 mL with 1× CIB in a 250 mL beaker. Distribute into 50 mL conical or other appropriate containers and collect cells by centrifugation: 4 min at 100 × *g*, 4 °C.
3. Measure total cell volume and decant supernatant. Resuspend all pellets to a total volume of 100 mL with 1× CIB* and collect cells via centrifugation: 4 min at 100 × *g*, 4 °C.
4. Resuspend and pool both pellets to 70 mL total with 1× CIB* (*see* **Note 8**). Load 35 mL of the suspension into the disruption vessel(s) prechilled on ice; rinse conical with 5–10 mL of 1× CIB* and add to bomb(s). Ensure a complete seal (including the exit valve). Pressurize with N_2 to 1,500 psi and hold on ice for 20 min.
5. With constant pressure from the N_2 tank (an open line into the disruption vessel), slowly discharge the disruption vessel(s) via the exit valve into 50 mL conical flasks on ice. CAUTION: This procedure is extremely sensitive and great care must be taken to avoid sample loss (*see* **Note 9**). Analyze approximately 15 μL of the lysate by bright field microscopy to assess and verify cell breakage efficiency (*see* **Note 10**).
6. Centrifugate the lysate for 4 min at 100 × *g*, 4 °C. Collect and combine all 100 × *g* lysate supernatants in a 250 mL beaker on ice (lysate pool). Resuspend cell debris with 25 mL of 1× CIB*; spin 4 min at 100 × *g*, 4 °C. Add supernatant to the lysate pool.
7. Repeat **steps 5–7** with the cell debris pellets. Thus, cells are subjected to two passes through the disruption vessel(s). Total volume of lysate should be approximately 200 mL. Save 500 μL labeled as "Lysate" for future analysis.

3.2 Differential and Density Gradient Centrifugation

1. Apportion lysate into 6 × 50 mL conical tubes (or other appropriate containers) and centrifuge 10 min at 1,000 × *g*, 4 °C.
2. Transfer supernatant to Beckman JA-20 rotor-compatible tubes, or equivalent, and centrifuge 25 min at 30,000 × *g*, 4 °C.
3. Save 500 μL of supernatant labeled "S_{30}" for subsequent analysis. Transfer supernatant to 6 Beckman 50.2Ti rotor-compatible tubes, or equivalent, and centrifuge 1 h at 120,000 × *g*, 4 °C.
4. During the 1 h 120,000 × *g* centrifugation step (**step 3**, above), prepare the sucrose step gradients as follows: add 10 μL of 1,000× PIC-D, PIC-W, and DTT, and 100 μL of 100× E64 stocks to 9.87 mL of the 10 % (w/v)sucrose in 1× CIB solution; add 16 μL of PIC-D, PIC-W, and DTT, and 160 μL of E64 stocks to 15.8 mL of the 35 % (w/v) sucrose in 1× CIB solution; add 4 μL of PIC-D, PIC-W, and DTT, and 40 μL of E64 to 3.95 mL of the 50 % (w/v)sucrose in 1× CIB solution. In a Beckman 14 × 95 mm Ultra-Clear centrifuge tube, layer 1 mL of the 50 % (w/v) sucrose solution + protease inhibitors, 5 mL of 35 % (w/v), and 3 mL of 10 % (w/v) (*see* **Note 11**). Make three gradients and chill on ice once completed.
5. Save 500 μL of supernatant from the 120,000 × *g* spin, labeled "S_{120}" for subsequent analysis. Discard remaining supernatant. Resuspend 120,000 × *g* pellets into 5 mL total volume of 1× CIB* (*see* **Note 10**). Homogenize in a 15 mL dounce homogenizer; save 200 μL of suspension labeled "P_{120}" for subsequent analysis (*see* **Note 12**). Clear homogenate by centrifugation: 10 min at 16,000 × *g* and 4 °C.
6. Transfer supernatant to a 15 mL conical tube and determine volume. Calculate 11.3 mL minus the volume of the supernatant (XmL). Resuspend the loose pellet with XmL 1× CIB*; homogenize. Clear once more by centrifugation 10 min at 16,000 × *g*, 4 °C.
7. Pool supernatant with that from **step 6** in a 15 mL conical. Mix well by pipetting. Save 200 μL of supernatant labeled "Sucrose Gradient Load (SGL)" for subsequent analysis. Total volume should be as close to 11.1 mL as possible without exceeding this volume.
8. Load 3.7 mL of the SGL onto each of the three sucrose step gradients prepared earlier. When programming the centrifuge, select slow deceleration, if available. Load gradients into the Beckman SW40Ti rotor, or equivalent, and centrifuge 50 min at 116,000 × *g*, 4 °C (*see* **Note 13**).
9. Inspect gradients and compare to Fig. 2. Draw off and discard the top (0 % (w/v) sucrose) layer from the gradient (Fig. 2, A).
10. Collect the 10 % (w/v) layer (Fig. 2, B) and the 10/35 % interface (Fig. 2, C), the latter which is easily visualized as a

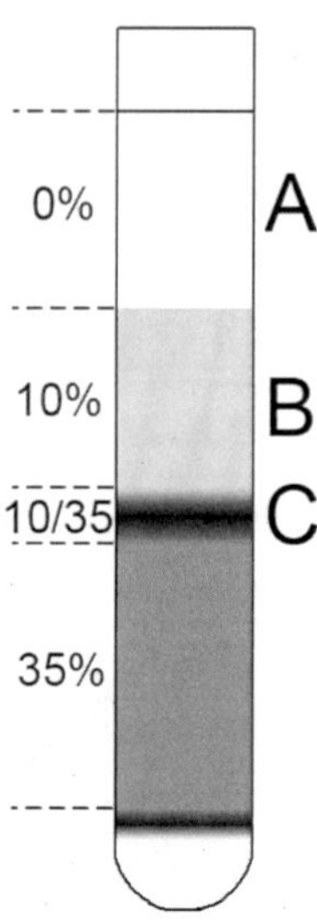

Fig. 2 Schematic of sucrose step gradient after 116,000 × *g* spin (*see* **step 8**). Layer *A* (0 % layer) is discarded whereas layers *B* and *C* (10 % layer and 10 %/35 % interface) are collected as described

thick green/yellow pad, from two of the gradients into each of two 26.3 mL capacity, 50Ti/70Ti rotor-compatible tubes. Evenly distribute the 10 % fraction and 10/35 % (w/v) interface of the remaining gradient equally between the two tubes (i.e., half of a 10 % fraction and half of a 10/35 % (w/v) interface into each tube (*see* **Note 14**).

11. Dilute the contents of each tube with 1× CIB* to ≤5 % (w/w) sucrose as measured by a refractometer (*see* **Note 15**). Centrifuge 50 min at 200,000 × *g*, 4 °C.
12. During the 200,000 × *g* spin, prepare the D_2O/Ficoll gradients (DFGs) as follows: assemble the gradient making rig composed of a GE Lifesciences SG-15 Hoefer gradient former, or equivalent, in tandem with a peristaltic pump and a deposition apparatus of choice, such as the (discontinued) Labconco Auto Densiflow Density Gradient Fractionator (*see* **Note 16**). Prepare the gradients in 14 × 95 mm Beckman Ultra-Clear™ centrifuge tubes.
13. In each of two clean 15 mL conical tubes, add protease inhibitors (11 μL of 1,000× PIC-D, PIC-W, DTT and 110 μL of 100× E-64) to 11 mL of each D_2O/Ficoll solution. Follow the gradient making procedure specific to the device in use. For the apparatus described above (SG15 + Auto Densiflow), ensure the connector and delivery stopcocks are closed and load 5 mL of the 9 %/2 % (w/v)D_2O/Ficoll (light) solution + protease inhibitors in the reservoir (back) chamber. Briefly open the stopcock to fill the channel between the back

and front chambers with the light solution and transfer any excess in the mixing chamber to the reservoir. Add 5 mL of the 90 %/30 % (w/v) (heavy) solution + protease inhibitors to the mixing (front) chamber.

14. Add an appropriately sized stir bar to the mixing chamber and place the apparatus on a magnetic stir plate. Adjust stir bar speed to ensure thorough mixing in the main chamber while avoiding bubble formation. First open the delivery stopcock and then simultaneously open the reservoir stopcock and start the peristaltic pump at a low speed (≤2 mL/min). Starting at the bottom of the gradient tube, raise the output point such that solution being delivered by the pump is deposited at the rising surface of the solution. After the gradient is complete, keep on ice.
15. Decant and discard 200,000 × *g* supernatant; resuspend pellets into 2 mL total volume with 1× CIB* and homogenize in 5 mL homogenizer (*see* **Note 17**). Distribute homogenate into two 1.5 mL microfuge tubes and centrifuge 5 min at 16,000 × *g*, 4 °C.
16. Transfer supernatant to a virgin 15 mL conical tube labeled "Deuterium/Ficoll Gradient Load (DFGL)." Resuspend the large, diffuse pellets by adding 500 μL 1× CIB* (250 μL to each pellet) and homogenize. Distribute homogenate into the same two 1.5 mL microfuge tubes and spin 5 min at 16,000 × *g*, 4 °C.
17. Transfer and pool supernatants in the "DFGL" conical. Discard pellets and microfuge tubes. Adjust volume of DFGL to exactly 2.7 mL with 1× CIB*. Save 200 μL of DFGL pool for subsequent analysis.
18. Load 2.5 mL of DFGL on the D_2O/Ficoll gradient. Centrifuge overnight (12–14 h) at 80,000 × *g*, 4 °C. When programming the centrifuge, select slow deceleration, if available. Flash-freeze aliquots from previous steps with liquid nitrogen (i.e., 500 μL "lysate" sample). Store at −80 °C.
19. Compare the centrifuged D_2O/Ficoll gradient to Fig. 3; the CCVs are present in the diffuse band indicated by the asterisk. Using a fractionation method of choice, collect the entire gradient into 0.75 mL fractions. Record the density of each fraction (sucrose % w/w equivalent) as measured by a refractometer.
20. CCVs are present in fractions between 14 and 16 % (w/w) sucrose equivalent (refractive index $\eta_D^{20} = 1.33514–1.33542$). Collect and pool fractions in this range from both gradients, and dilute to 8.5 mL with 1× CIB*, or as necessary to achieve a density equivalent to ≤5 % sucrose w/w as measured by a refractometer (*see* **Note 18**). Centrifuge 1 h at 174,000 × *g*, 4 °C in the TLA100.3 rotor or equivalent.

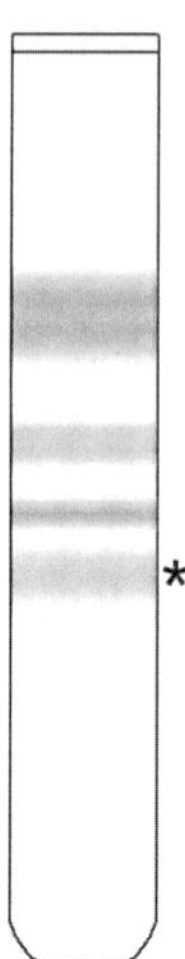

Fig. 3 Appearance of linear D_2O/Ficoll gradient after 80,000 × *g* spin (*see* **step 17**). Fractions containing CCVs (corresponding to ~14 % sucrose (w/w) equivalent) are visualized as the band indicated by the *asterisk*

21. Resuspend pellet(s) in 1× CIB* to 250 μL (*see* **Note 19**). In a microfuge tube, centrifuge resuspension 5 min at 16,000 × *g*, 4 °C.
22. Transfer supernatant to a fresh microfuge tube labeled "CCV." Resuspend pellet by adding 100 μL 1× CIB*. Centrifuge resuspension 5 min at 16,000 × *g*, 4 °C.
23. Pool supernatants. This is the final purified CCV fraction. Save 10 μL for TEM analysis. Further aliquot if desired (*see* **Note 20**). Flash-freeze and store at −80 °C.

3.3 Negative Stain TEM Analysis (See Note 21)

1. Immobilize a pioloform-coated 200-mesh nickel grid by clasping with nonmagnetic steel forceps and securing using a hinged paper binding clip (or other such method), *see* **Note 22**.
2. In an appropriate fume hood, mix one part CCV fraction and three parts 4 % (w/v) OsO_4 solution (i.e., 1 μL and 3 μL) on a bit of parafilm and mix thoroughly by pipetting (the solution should darken).
3. Gently deposit 1 μL of the CCV mixture from **step 2** on the face of the immobilized grid. Using a wedge of Whatman paper, wick away excess fluid (*see* **Note 23**). Allow to dry, approximately 2 min.
4. Place a droplet of the Nano-W solution on the parafilm piece from **step 2**. Take 1 μL from this droplet and gently pipette on the (now dry) mixture-coated grid. Wick away excess as before and allow to cure, approximately 5 min.

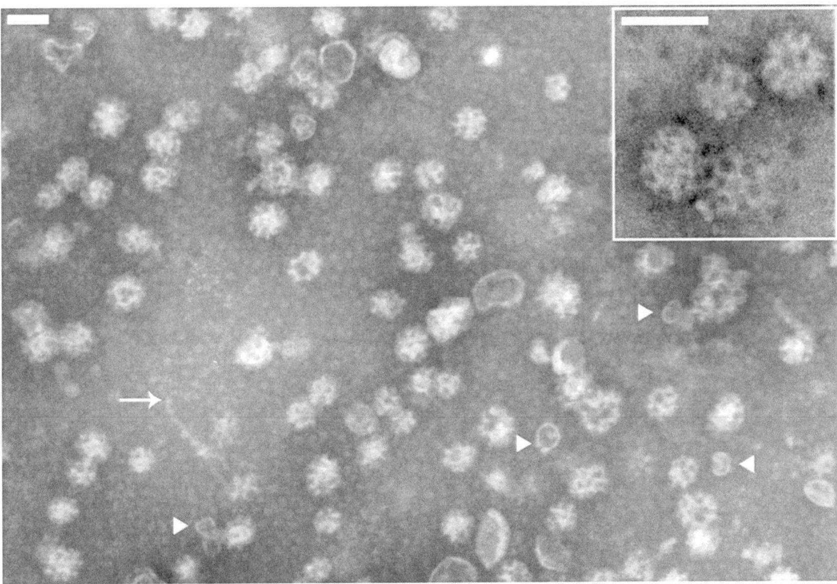

Fig. 4 Negative TEM micrograph of the final CCV fraction, prepared as described. The geometric arrangement characteristic of assembled clathrin triskelia is apparent on coated vesicles. A small proportion of uncoated vesicles (*arrowheads*) and negligible fibrous contamination (*arrow*) are also present. *Inset*: increased magnification of coated vesicles showing striking clathrin lattice organization. Scale bars 100 nm

5. Image the grid using a transmission electron microscope of choice (*see* **Note 24**). Depending on success of the wicking procedures in **steps 3** and **4**, one will observe a nonuniform distribution of both biological material and negative stain. If the enrichment is successful, the striking geometry of the clathrin triskelia (appearing as miniature soccer balls or buckyballs) should be immediately apparent on coated vesicles (Fig. 4).
6. The number of images collected should reflect the desired statistical significance. 10–20 fields are often sufficient to determine a reasonably accurate mean.
7. Open the first image with ImageJ and initialize the Cell Counter plug-in.
8. Select a counter type from the Cell Counter GUI and begin manually designating each coated vesicle by clicking on it. Once all CCVs have been labeled, select a second counter type from the Cell Counter GUI and begin to designate all apparent uncoated vesicles (*see* **Note 25**).
9. Repeat for each image and compile data (manually) to determine a mean coated/uncoated ratio with accompanying standard deviation (*see* **Note 26**).

4 Notes

1. To process the amount of cells described in this experiment, approximately 1.5 L of 1× CIB, 1 mL each of PIC-D, PIC-W, and DTT, and 8 mL of E-64 are required. Therefore, it is advisable to prepare 2 L of 1× CIB, 2 mL of PIC-D, PIC-W, and DTT, and 10 mL of E-64.
2. Early attempts to purify CCVs from plant tissue relied on a lengthy RNAse treatment (1 h at 30 °C) to address extensive ribosome contamination of the final CCV fraction [4]. Due to potential proteolytic side effects, as well as CCV uncoating, this step is omitted in this protocol. Rather, as suggested by Demmer [8], the presence of 3 mM EDTA in the isolation buffer serves both as a general inhibitor of metal-dependent proteases, as well as to promote ribosome instability by chelating structural magnesium ions which serves to reduce their levels in the final CCV fraction.
3. During the experiment, PIC-D must be kept out at room temperature (e.g., on the bench) as DMSO solidifies on ice.
4. 50 % w/v is approaching the saturation of sucrose in water. Avoid the temptation to add additional CIB to speed the dissolution process, lest one find himself with 102 mL of a 49 % sucrose solution. Confirm the sucrose concentrations using a refractometer. *Note*: refractive indices of 10 %, 35 %, and 50 % w/v sucrose are $\eta_D^{20} = 1.3473$, 1.3827, and 1.4039, respectively, at 20 °C
5. With the components listed, the 90 %/30 % D_2O/Ficoll (w/v) solution will be very nearly 50 mL. Thus it is essential to use a concentrated NaOH solution to adjust the pH of the stock solution to pH 6.4 to minimize volume changes. Take care to use lower concentrations in series (i.e., 10, 5, 1 N) as the pH approaches 6.4 to avoid overshooting the target.
6. Although we have specifically defined the centrifugation equipment and conditions employed in our lab for the preparation of CCVs the general centrifugation guidelines provided (e.g., volumes, and centrifugation times/g force) are compatible with other centrifuges, rotors, and tubes .
7. Arabidopsis suspension-cultured cells (T87) were maintained in Murashige and Skoog media supplemented 0.2 mg/L 2,4-D (MS0.2) with shaking at 140 rpm, 22 °C, under constant light. Four days before the fractionation date, dilute a saturated 1-week-old T87W culture 1:10 into fresh MS0.2. The dilution can be modified to accommodate different growth rates between cell lines or media compositions. In the end, aim for 35–45 mL collected cell volume from 500 mL of 3.5-day-old cultures.

8. We have written this protocol under the assumption that the experimenter possesses a pressure-based cell disruption vessel(s) with a minimum capacity ≥90 mL total. Of course, this protocol can be adjusted to meet higher or lower capacities.
9. The use of protective eyewear when discharging the disruption vessel is highly advised. Care must be taken such that the rapid release of nitrogen from the exit valve immediately following complete drainage of the sample does not discharge directly into the sample, as this will result in significant splashing and loss of sample. We find the most success by bleeding approximately 40 mL of sample (our pressure vessel has a 45 mL capacity) into one conical tube and collecting the remainder in another empty tube.
10. After the first pass of the intact cells through the cell disruption vessel, we typically observe ≥60 % of all cells to be lysed as determined by visual estimation. After the second pass, this value approaches ≥90 %.
11. Layering the gradients takes a steady hand (and steadier thumb if using an adjustable micropipettor). With each successive layer, avoid disturbing that which lies below with slow but consistent addition. It might be a good idea to prepare a practice gradient with the 10/35/50 % (w/v) solutions lacking expensive protease inhibitors.
12. Starting with a small volume, roughly 500 μL, resuspend 3–4 pellets and transfer to the homogenizer; repeating for all tubes will yield a volume of approximately 2–3 mL. Subsequently, wash tubes 3–4 times with roughly 750 μL of 1× CIB* until 5 mL total volume is achieved. The S_{120} and P_{120} fractions can be used to determine the presence of soluble and membrane-bound fractions of a given protein.
13. Gently handle the loaded gradients during balancing and when placing the rotor in the ultracentrifuge. The Ultra-ClearTM tubes are very full (to prevent tube collapse) and easily spill with sufficient perturbation.
14. When collecting the 10/35 % (w/v) sucrose gradient interface, avoid taking excess amounts of the 35 % (w/v) layer which is depleted of CCVs and is rich in contaminating denser membranes.
15. If the experimenter is using Beckman polycarbonate, 26.3 mL, capped bottles, the appropriate dilution can be achieved by filling each tube to the base of the neck. Make sure to check the refractive index to confirm sucrose %; failure to dilute sufficiently (<6 %(w/v) sucrose) will impede membrane pelleting in the next step.

16. Regardless of deposition methodology (either automated or by hand using a needle), the key is that the solution being delivered by the peristaltic pump is deposited at the rising surface of the gradient (as described in **step 14**).
17. As with other resuspension steps, it is prudent to begin with approximately half the desired final resuspension volume. In this case, we recommend that pellets be resuspended with 1 mL of buffer and that the residual material that adheres to the centrifuge tube walls be collected by washing the tubes with two additional volumes of 500 μL buffer.
18. In our experience, two 0.75 mL fractions fall in this range. Thus, the pooled two fractions would be diluted to ~4.3 mL.
19. The membrane pellet is a significant contributor of volume in this resuspension step. Accordingly, resuspending in 250 μL as described might actually result in 300 μL final volume. Take this into consideration if there is a desired final volume.
20. The enrichment of CCV in the final fraction relative to contaminating uncoated membranes is determined by the following two methods: (1) immunoblotting for known CCV-associated proteins (e.g., clathrin light chain) and other organelle markers (e.g., chloroplast, ER) to determine enrichment or depletion, respectively, between collected aliquots from various steps in the procedure (i.e., "lysate," "S_{30},"), and (2) examining a sample from the final CCV fraction via negative stain transmission electron microscopy (as described in this protocol).
21. This portion of the procedure involves working with OsO_4, an extremely toxic compound. Accordingly, perform **steps 1–4** in an appropriate fume hood and carefully follow all appropriate safety measures.
22. The benefits of this arrangement are twofold: (1) the grid is secure and stable for the following steps and (2) holding the grid instead of lying it on a surface improves airflow and thus drying times.
23. Avoid touching the surface of the grid with the paper wedge. Instead, very gently touch edges of the fluid droplet to "coax" the mixture to cover all or most of the grid.
24. To determine the relative enrichment of coated vs. uncoated vesicles, we suggest capturing images at 25,000× magnification, wherein coated and uncoated vesicles can be clearly differentiated, yet a large number of vesicles are in the field of view.
25. In our experience, the TEM fields contain materials other than coated and uncoated vesicles, including long fibers and apparently fragmented membranes of small size (10–20 nm). It is up to the user to determine what constitutes an uncoated vesicle versus membrane debris. One metric might be diameter,

achieved by labeling membranes similar to the size of CCVs (50–150 nm) as "uncoated vesicles" and excluding all others.

26. Undoubtedly, this ratio will fluctuate between CCV preparations due to a number of factors (differences in biological sample, minor alterations in experiment procedure, etc.). Over many preps ($n = 11$), we observe 76 ± 10 % of all vesicles to be coated.

Acknowledgments

This work was supported by funding provided by the National Science Foundation (No. 1121998) and a Vilas Associate Award (University of Wisconsin, Madison, Graduate School) to S.Y.B. The authors would also like to thank Nou Vang for technical assistance in the preparation and quantitative analysis of CCVs.

References

1. McMichael CM, Bednarek SY (2013) Cytoskeletal and membrane dynamics during higher plant cytokinesis. New Phytol 197: 1039–1057
2. Pearse BMF (1975) Coated vesicles from pig brain: purification and biochemical characterisation. J Mol Biol 97:93–98
3. Pearse BMF (1982) Human placental coated vesicles carry ferritin, transferrin and immunoglobulin G. Proc Natl Acad Sci U S A 79: 451–455
4. Depta H, Robinson DG (1986) The isolation and enrichment of coated vesicles from suspension-cultured carrot cells. Protoplasma 130:162–170
5. Harley SM, Beevers L (1989) Isolation and partial characterization of clathrin coated vesicles from pea (Pisum sativum L.) cotyledons. Protoplasma 150:103–109
6. Aris JP, Blobel G (1991) Isolation of yeast nuclei. Methods Enzymol 194:735–749
7. Axelos M, Curie C, Mazzolini L et al (1992) A protocol for transient gene expression in Arabidopsis thaliana protoplasts isolated from cell suspension cultures. Plant Physiol Biochem 30:123–128
8. Demmer A, Holstein SHE, Hinz G et al (1993) Improved coated vesicle isolation allows better characterization of clathrin polypeptides. J Exp Bot 258:23–33

Chapter 14

Proteomics of Endosomal Compartments from Plants Case Study: Isolation of *Trans*-Golgi Network Vesicles

Eunsook Park and Georgia Drakakaki

Abstract

A detailed understanding of endomembrane processes and their biological roles is vital for a complete picture of plant growth and development; however their highly dynamic nature has complicated comprehensive and rigorous studies so far. Recent pioneering efforts have demonstrated that isolation of vesicles in their native state, paired with a quantitative identification of their cargo, offers a viable and practicable approach for the dissection of endomembrane trafficking pathways. The protocol presented in this chapter describes in detail the isolation of the SYP61 *trans*-Golgi network vesicles from *Arabidopsis*. With minor alterations, in a few key parameters, it can be adopted to yield a universal procedure for the broad spectrum of plant vesicles.

Key words Vesicle isolation, Immunoisolation, *Trans*-Golgi network (TGN), SYP61, Endomembrane trafficking, Proteomics

1 Introduction

Plant growth, development, and environmental responses are highly dependent on the endomembrane system [1]. Vesicle trafficking facilitates the movement of proteins, membrane lipids, and the transport and deposition of cell wall components to the plasma membrane (PM) and the apoplast [2]. During the last two decades cell biological and genetic studies have improved our knowledge of plant endomembrane trafficking. However, genetic redundancy and lethality are major challenges for studies aiming at its comprehensive understanding [3]. The co-development of confocal microscopy and recombinant fluorescent proteins (in particular GFP and its plethora of derivatives) has provided a toolset for the visualization and quantitative measurement of endosomal trafficking, leading to a great appreciation of its highly dynamic nature [4, 5]. The generation of multiple, specific fluorescent marker proteins,

Marisa S. Otegui (ed.), *Plant Endosomes: Methods and Protocols*, Methods in Molecular Biology, vol. 1209, DOI 10.1007/978-1-4939-1420-3_14, © Springer Science+Business Media New York 2014

each identifying specific membrane compartments, has the potential to establish subcellular localization maps [6].

Recently, a large set of small molecules altering vesicle trafficking in *Arabidopsis* was identified, by employing the emerging technique of chemical genomics. Chemical genomics, using specific pharmacological inhibitors rather than mutations to perturb the function of proteins, is a promising approach for dissecting intersecting secretory pathways. This large-scale approach provides a platform for a systems view of trafficking in plant cells [7].

The direct cargo analysis of individual subcellular compartments requires their isolation and subsequent proteomic analysis. Several organelles have been isolated by sucrose gradient ultracentrifugation and their proteomes have been identified; those include nuclei, chloroplasts [8, 9], vacuole [10], mitochondria [11], peroxisomes [12] and *cis*-Golgi membranes [13]. More elaborate Golgi separations have been performed using free flow electrophoresis (FFE) and localization of organelle proteins by isotope tagging (LOPIT) [14–16]. However, vesicle isolation has been very challenging due to the similarity of physicochemical properties between different vesicle populations. Further complicating factors in the dissection of the highly dynamic vesicular network are the multidirectional movements of vesicles and the convolution of resident proteins and transient cargo. Recently, the isolation of TGN compartments and identification of their proteome was demonstrated [17]. Since TGN is a major intersection in post-Golgi trafficking, its comparison with the Golgi proteome [15, 16] can offer major insights into protein distribution within the secretory pathway [18].

This protocol describes in detail the separation of vesicles by immunopurification optimized for CFP-SYP61 (Fig. 1). Adopting this approach, a variety of compartments can be isolated, allowing the identification of proteins, lipids, and various other compounds, providing vital information about vesicle trafficking and the whole plant endomembrane system.

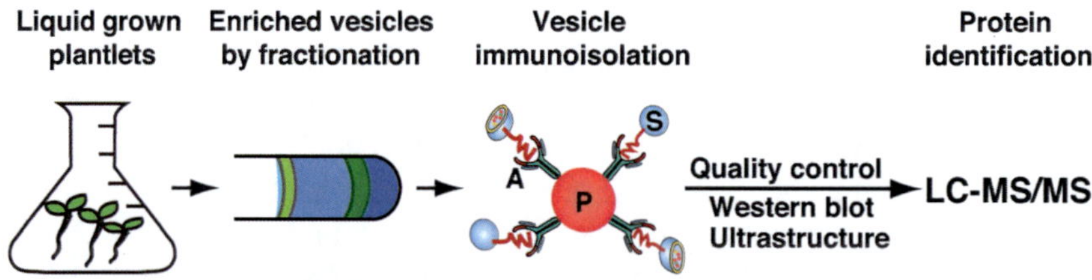

Fig. 1 Schematic representation of vesicle isolation and identification of their proteome. Plant extracts derived from liquid grown platelets are subjected to sucrose fractionation. The enriched vesicles or compartments are isolated from cell lysate with the aid of an antibody against a target protein. Following isolation, Western blot with antibodies against subcellular markers is performed to test the quality of the purified vesicles and ultrastructure of isolated vesicles is observed by transmission electron microscopy. Finally LC-MS/MS analysis is used to identify the proteome. *P* Protein-A agarose bead, *A* antibody, *S* SYP61 vesicles

2 Materials

2.1 Plant Material

1. 50–100 transgenic *Arabidopsis* seeds expressing CFP-SYP61 [17] (*see* **Notes 1** and **2**).
2. Seed sterilization solution: 75 % Ethanol, 0.1 % Triton X-100, 24.9 % autoclaved deionized water.
3. Liquid Murashige and Skoog (MS) medium: 1× MS medium, 1 % sucrose. Dissolve one pouch of MS minimal media (4.26 g) and 10 g of sucrose in 1,000 mL of deionized water. Aliquot 200 mL of media in a 500 mL Erlenmeyer flask and autoclave (*see* **Notes 3** and **4**).
4. Flask shaker in temperature and photoperiod controlled environment, set to a short day light cycle (8 h) at 22 °C.

2.2 Vesicle Fractionation Components

For vesicle preparation, maintain all buffers and rotors at 4 °C.

1. Vesicle immunoprecipitation extraction buffer (VIB): 50 mM HEPES, pH7.5, 0.45 M sucrose, 5 mM $MgCl_2$, 1 mM DTT, 0.5 % PVP, protease inhibitors.
2. Sucrose gradient solutions: 38 % (w/v) sucrose (1.1 M), 33 % (w/v) sucrose (0.96 M), and 8 % (w/v) sucrose (0.23 M) in 50 mM HEPES pH 7.5 (*see* **Note 5**).
3. Mortar and pestle, razor blade, Miracloth, small funnel.
4. Refrigerated benchtop centrifuge.
5. 50 mL conical centrifuge tube.
6. Ultracentrifuge (e.g., Optima L-90 K Beckman Coulter, or equivalent) with rotors SW28 and 70Ti.
7. Centrifuge tubes for SW28 rotor (thickwall, polyallomer, 32 mL tubes).
8. Centrifuge tubes for 70Ti (polycarbonate aluminum bottle with cap assembly).
9. Disposable serological 5, 10, and 25 mL pipettes.
10. Pasteur pipettes.

2.3 Vesicle Immunoisolation Components

1. Resuspension buffer: 50 mM HEPES, pH 7.5, 0.25 M sucrose, 1.5 mM $MgCl_2$, 0.2 mM EDTA, 150 mM NaCl, protease inhibitors.
2. Wash buffer: 50 mM HEPES, pH 7.5, 0.24 M sucrose, 1.5 mM $MgCl_2$, 0.2 mM EDTA, 150 mM NaCl.
3. 50 mM NH_4HCO_3.
4. Phosphate buffered saline (PBS): 137 mM NaCl, 2.7 mM KCl, 10 mM $Na_2HPO_4{\cdot}H_2O$, KH_2PO_4, pH 7.4.
5. Protein-A agarose beads.

6. Anti-GFP rabbit IgG (1 mg mL^{-1}) (e.g., anti-GFP IgG from Invitrogen) (*see* **Note 6**).
7. Rabbit IgG (e.g., from Santa Cruz Biotech. Inc.) (*see* **Note 7**).
8. Tube rotator.
9. Tabletop centrifuge with cooling capacity.

3 Methods

3.1 Plant Preparation

For proteomic analysis of TGN compartments, more than 12 g of tissue is required. Plants should be grown about 2–2.5 weeks in liquid media to yield sufficient root tissue.

1. 2–2.5 weeks before the vesicle isolation, sterilize seeds and stratify them at 4 °C overnight.
2. Grow 50–100 seeds in 200 mL liquid MS in a 500 mL Erlenmeyer flask while shaking at 150 rpm under a short day cycle at 22 °C.

3.2 Golgi/TGN Fractionation by Sucrose Density Gradient Ultracentrifugation

Initially, plant extracts from liquid grown plantlets are fractionated using discontinuous sucrose gradient centrifugation to enrich in SYP61 vesicles. All steps after harvesting tissues should be performed on ice. Figure 2 illustrates the sucrose fractionation procedure.

1. Rinse plants carefully in deionized water and pat dry with paper towels in a large Petri dish. Weigh plants.
2. Slice plants with a razor blade in the Petri dish set on ice. Transfer the finely sliced tissues into a cold mortar on ice.
3. Add ice-cold VIB to a final volume:weight ratio of 2:1 (e.g., 2 mL of VIB buffer for 1 g of plant tissue) and grind the plant tissue as gently as possible to a rough pulp (*see* **Note 8**). Place funnel with Miracloth over 50 mL conical centrifuge tube to filter the plant extract and centrifuge at 1,000 × *g* at 4 °C for 20 min.
4. Meanwhile, using a 10 mL pipette, add 8 mL of 38 % sucrose to 32 mL centrifuge tube. Load gently the supernatant from **step 3** (S1 fraction) on top of the sucrose cushion (*see* **Note 9**). Centrifuge at 100,000 × *g* at 4 °C for 1.5 h using a SW28 rotor or equivalent.
5. Place the tube on ice, and remove the plant extract liquid above the formed green interface band, without disturbing it. Carefully, using a 25 mL pipette, add 15 mL of 33 % sucrose on top of the green band, and then add 5 mL of 8 % sucrose. Centrifuge at 100,000 × *g* at 4 °C in SW28 rotor or equivalent for 2 h (*see* **Note 10**).
6. Using a 5 mL pipette, slowly remove 3–4 mL of the top layer (8 % sucrose). Using a Pasteur pipette, collect the interface

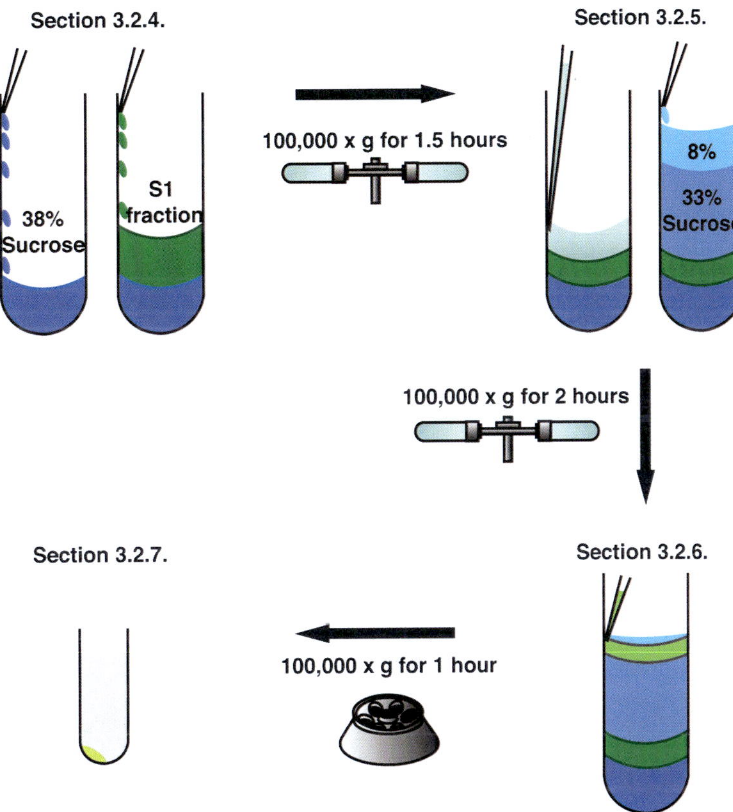

Fig. 2 Enrichment of Golgi/*trans*-Golgi Network compartments by sucrose gradient ultracentrifugation. Schematic illustration of the sequential sucrose gradient ultracentrifugation to isolate Golgi/TGN compartments. Briefly the S1 fraction of plant homogenates is loaded on a 38 % sucrose cushion and centrifuged at 100,000 × *g* for 1.5 h (Subheading 3.2, **step 4**). The upper phase is removed and a discontinuous gradient is formed by adding 33 and 8 % sucrose layers (Subheading 3.2, **step 5**). The interface between 8 and 33 % M sucrose fractions is collected (Subheading 3.2, **step 6)** and vesicles are retained for immunoisolation by centrifugation at 100,000 × *g* for 1 h (*see* Subheading 3.2, **step 7**)

band between the 8 and 33 % sucrose layers into an ice-chilled 30 mL centrifuge tube and add 0.5× volume of 50 mM HEPES (pH 7.5).

7. Centrifuge at 100,000 × *g* at 4 °C using a fixed angle 70Ti rotor or equivalent for 1 h. Decant the supernatant and keep the pellet at 4 °C over night (*see* **Note 11**).

3.3 Immunoisolation of SYP61 Vesicles

SYP61 vesicles are isolated by immunopurification from the fraction obtained in the previous section (Fig. 3).

1. Couple the GFP antibody to protein-A agarose beads. First, resuspend protein-A agarose beads in PBS and place 25 μL into a 1.5 mL microfuge tube. Add 500 μL of ice-cold PBS,

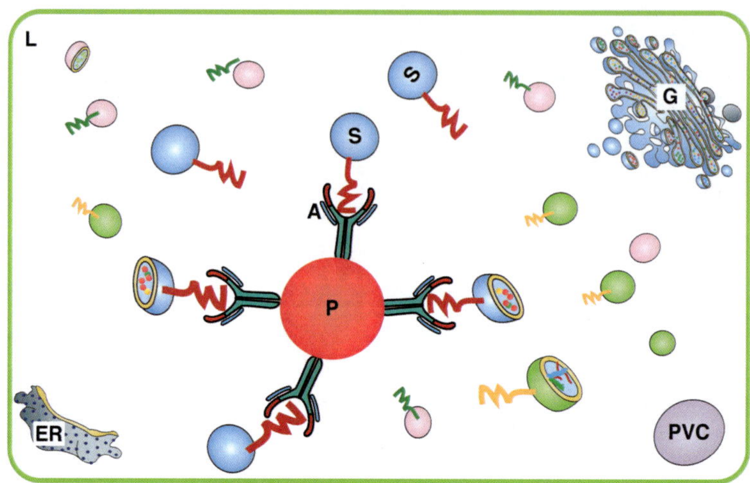

Fig. 3 Schematic diagram of vesicle immunoisolation. Vesicles are isolated from Golgi/TGN enriched fractions of cell lysates with the aid of an antibody against the target protein CFP-SYP61. Images are not to scale. Note that similar-sized vesicles containing different cargos are eliminated by immunopurification. Contamination of ER and pre-vacuolar compartment (PVC) is eliminated in this step, resulting in the purification of CFP-SYP61 vesicles (S). *P* Protein, *A* agarose bead, *A* antibody against the target protein (CFP-SYP61), *G* Golgi apparatus, *S* CFP-SYP61 vesicles, *PVC* pre-vacuolar compartment, *L* cell lysates

mix well by inverting the tube, and centrifuge at 10,000 × *g* for 30 s (*see* **Note 12**).

2. Discard supernatant using a pipette and add 2 μL (2 μg) of the anti-GFP antibody or 2 μg of the Rabbit IgG antibody for the control. Add cold PBS containing protease inhibitors to a final volume of 100 μL, mix well by inverting the tube, and incubate on a rotator for 2 h. Centrifuge at 10,000 × *g* for 30 s and discard the supernatant.
3. Equilibrate the antibody coupled agarose beads with 200 μL resuspension buffer on a rotator at 4 °C for 20 min. Centrifuge at 1,000 × *g* for 30 s. Carefully discard the supernatant.
4. Meanwhile, resuspend the vesicle pellets from Subheading 3.2, **step 7** in 400 μL of resuspension buffer and incubate with 25 μL of protein-A agarose beads (*see* **Note 13**). Gently mix the suspension for 20 min using a rotator at 4 °C and then centrifuge at 100 × *g* for 30 s. Collect the supernatant.
5. Add 300 μL of the supernatant collected from **step 4** with the antibody coupled agarose beads collected from **step 3** and mix for 1 h on a rotator at 4 °C. Centrifuge at 100 × *g* for 1 min (*see* **Note 14**).
6. Wash the pellet with 1 mL of wash buffer under gentle agitation at 4 °C for 2 min and centrifuge at 100 × *g* for 1 min. Repeat this step three times (*see* **Note 15**).

7. Wash a fourth time with 1 mL of 50 mM NH_4HCO_3 to equilibrate the sample (vesicles attached to beads) for proteomic analysis (*see* **Note 16**).

4 Notes

1. This protocol uses CFP-SYP61 tagged plants for vesicle isolation. The N-terminus CFP fusion to SYP61 allows for interaction with the antibody during isolation, while the SYP61 C-terminus facilitates the attachment to the vesicle membranes. Special consideration should be given to the "bait" protein that will be used for isolation, in particular the accessibility to the antibody.
2. It is important to start with sufficient plant material. Only a selected fraction from the sucrose gradient will be used for vesicle isolation. Note that more than 12 g of plant tissues are required.
3. Growth in liquid media yields more root tissues. However, when the target protein is highly abundant in specific tissues or developmental stages, collect the appropriate tissues to obtain a better yield.
4. The MS media used in this protocol are pH-adjusted. When other MS media are used, adjust the pH to 5.8–6.0.
5. The sucrose gradient procedure is adapted from ref. 19. If necessary, the sucrose gradient can be adjusted with more layers or different sucrose densities to enrich for the specific target vesicles.
6. Due to its readily availability and reactivity with CFP, the GFP antibody is used.
7. Rabbit IgG is used as a control, since the GFP antibody is raised in rabbit. Similarly the control must be chosen based on the nature of the antibody.
8. In this step, it is very important to keep the plant material cold, but not frozen. Ground tissue should be maintained as slurry.
9. A sharp interface between the sucrose cushion and supernatant should be visible. Carefully set up the centrifugation without disturbing the sharp interface.
10. The duration of centrifugation might vary depending on the model of the centrifuge in use. If the interfaces between the sucrose layers are not sharp, increase the centrifugation time. We found that the centrifugation force can be increased to $140{,}000 \times g$ in **steps 9** and **10.**
11. If time permits, the next steps can be performed immediately.
12. Instead of agarose beads other types, such as magnetic beads, can be used.

13. This step is necessary to minimize nonspecific binding to the protein-A agarose beads.
14. After centrifugation, the supernatant can be stored for determining the vesicle purification efficiency by Western blot analysis.
15. During the last washing step retain a small fraction ~10 % of the suspension to test the isolation efficiency. Centrifuge at 100×*g* for 1 min. Resuspend in 100 μL PBS and mix with appropriate SDS protein loading buffer to prepare the sample for SDS-PAGE and Western blot analysis.

    This step is required for analysis of the immunoisolation efficiency. Different fractions from the original enriched vesicle fraction (Subheading 3.3, **step 4**), the flow through (Subheading 3.3, **step 5**), and the immunoisolated fraction (Subheading 3.3, **step 6**) are analyzed by Western blot. The presence of CFP-SYP61 can be tested using a monoclonal antibody against GFP. In addition, antibodies against subcellular markers for the endoplasmic reticulum (ER) marker, BiP [20], and the pre-vacuolar compartment (PVC) marker, SYP21 [21], can be used to test for nontargeted inclusions in the purified vesicles. The ultrastructure of isolated vesicles can be observed by transmission electron microscopy.
16. Proteomic analysis can be carried out using a variety of approaches, depending on expertise and availability of equipment. When the protein content is low, a gel-free strategy of direct on-beads digestion can be employed and the resulting tryptic peptides can be analyzed by MudPIT nano-LC-MS/MS [22]. The resultant MS/MS data can be investigated with the latest version of the *Arabidopsis* protein database using Mascot (Matrix Sciences). A label-free quantification approach can be used to identify the specific vesicle peptides compared to the controls [23–25].

Acknowledgements

This work was supported by the UCD Hellman Fellowship and the NSF-IOS 1258135 to GD.

References

1. Surpin M, Raikhel N (2004) Traffic jams affect plant development and signal transduction. Nat Rev Mol Cell Biol 5:100–109
2. Worden N, Park E, Drakakaki G (2012) Trans-Golgi network: an intersection of trafficking cell wall components. J Integr Plant Biol 54:875–886
3. Hicks GR, Raikhel NV (2010) Advances in dissecting endomembrane trafficking with small molecules. Curr Opin Plant Biol 13:706–713
4. Brandizzi F, Fricker M, Hawes C (2002) A greener world: the revolution in plant bioimaging. Nat Rev Mol Cell Biol 3:520–530

5. Sparkes I, Brandizzi F (2012) Fluorescent protein-based technologies: shedding new light on the plant endomembrane system. Plant J 70:96–107
6. Geldner N, Denervaud-Tendon V, Hyman DL, Mayer U, Stierhof YD, Chory J (2009) Rapid, combinatorial analysis of membrane compartments in intact plants with a multicolor marker set. Plant J 59:169–178
7. Drakakaki G, Robert S, Szatmari A-M, Brown MQ, Nagawa S, Van Damme D, Leonard M, Yang Z, Girke T, Schmid SL, Russinova E, Friml J, Raikhel NV, Hicks GR (2011) Clusters of bioactive compounds target dynamic endomembrane networks in vivo. Proc Natl Acad Sci U S A 108:17850–17855
8. Kleffmann T, Russenberger D, von Zychlinski A, Christopher W, Sjölander K, Gruissem W, Baginsky S (2004) The Arabidopsis thaliana chloroplast proteome reveals pathway abundance and novel protein functions. Curr Biol 14:354–362
9. Mosley AL, Florens L, Wen Z, Washburn MP (2009) A label free quantitative proteomic analysis of the Saccharomyces cerevisiae nucleus. J Proteomics 72:110–120
10. Carter C, Pan S, Zouhar J, Avila EL, Girke T, Raikhel NV (2004) The vegetative vacuole proteome of Arabidopsis thaliana reveals predicted and unexpected proteins. Plant Cell 16:3285–3303
11. Eubel H, Heazlewood JL, Millar AH (2006) Isolation and subfractionation of plant mitochondria for proteomic analysis. Methods Mol Biol 355:49–62
12. Eubel H, Meyer EH, Taylor NL, Bussell JD, O'Toole N, Heazlewood JL, Castleden I, Small ID, Smith SM, Millar AH (2008) Novel proteins, putative membrane transporters, and an integrated metabolic network are revealed by quantitative proteomic analysis of arabidopsis cell culture peroxisomes. Plant Physiol 148:1809–1829
13. Asakura T, Hirose S, Katamine H, Kitajima A, Hori H, Sato MH, Fujiwara M, Shimamoto K, Mitsui T (2006) Isolation and proteomic analysis of rice Golgi membranes: cis-Golgi membranes labeled with GFP-SYP31. Plant Biotechnol 23:475–485
14. Dunkley TPJ, Hester S, Shadforth IP, Runions J, Weimar T, Hanton SL, Griffin JL, Bessant C, Brandizzi F, Hawes C, Watson RB, Dupree P, Lilley KS (2006) Mapping the Arabidopsis organelle proteome. Proc Natl Acad Sci 103:6518–6523
15. Nikolovski N, Rubtsov D, Segura MP, Miles GP, Stevens TJ, Dunkley TPJ, Munro S, Lilley KS, Dupree P (2012) Putative glycosyltransferases and other plant golgi apparatus proteins are revealed by LOPIT proteomics. Plant Physiol 160:1037–1051
16. Parsons HT, Christiansen K, Knierim B, Carroll A, Ito J, Batth TS, Smith-Moritz AM, Morrison S, McInerney P, Hadi MZ, Auer M, Mukhopadhyay A, Petzold CJ, Scheller HV, Loque D, Heazlewood JL (2012) Isolation and proteomic characterization of the Arabidopsis Golgi defines functional and novel components involved in plant cell wall biosynthesis. Plant Physiol 159:12–26
17. Drakakaki G, van de Ven W, Pan S, Miao Y, Wang J, Keinath NF, Weatherly B, Jiang L, Schumacher K, Hicks G, Raikhel N (2012) Isolation and proteomic analysis of the SYP61 compartment reveal its role in exocytic trafficking in Arabidopsis. Cell Res 22:413–424
18. Parsons HT, Drakakaki G, Heazlewood JL (2012) Proteomic dissection of the Arabidopsis Golgi and trans-Golgi network. Front Plant Sci 3:298
19. Drakakaki G, Zabotina O, Delgado I, Robert S, Keegstra K, Raikhel N (2006) Arabidopsis reversibly glycosylated polypeptides 1 and 2 are essential for pollen development. Plant Physiol 142:1480–1492
20. Morris JA, Dorner AJ, Edwards CA, Hendershot LM, Kaufman RJ (1997) Immunoglobulin binding protein (BiP) function is required to protect cells from endoplasmic reticulum stress but is not required for the secretion of selective proteins. J Biol Chem 272:4327–4334
21. da Silva CA, Marty-Mazars D, Bassham DC, Sanderfoot AA, Marty F, Raikhel NV (1997) The syntaxin homolog AtPEP12p resides on a late post-Golgi compartment in plants. Plant Cell 9:571–582
22. Fonslow BR, Carvalho PC, Academia K, Freeby S, Xu T, Nakorchevsky A, Paulus A, Yates JR (2011) Improvements in proteomic metrics of low abundance proteins through proteome equalization using ProteoMiner prior to MudPIT. J Proteome Res 10:3690–3700
23. Silva JC, Denny R, Dorschel CA, Gorenstein M, Kass IJ, Li GZ, McKenna T, Nold MJ, Richardson K, Young P, Geromanos S (2005) Quantitative proteomic analysis by accurate mass retention time pairs. Anal Chem 77:2187–2200
24. Weatherly DB, Atwood JA 3rd, Minning TA, Cavola C, Tarleton RL, Orlando R (2005) A Heuristic method for assigning a false-discovery rate for protein identifications from Mascot database search results. Mol Cell Proteomics 4:762–772
25. Cutillas PR, Vanhaesebroeck B (2007) Quantitative profile of five murine core proteomes using label-free functional proteomics. Mol Cell Proteomics 6:1560–1573

Chapter 15

Analysis of Global Ubiquitylation and Ubiquitin-Binding Domains Involved in Endosomal Trafficking

Kamila Kalinowska and Erika Isono

Abstract

Ubiquitylation is a reversible posttranslational modification that regulates various cellular pathways. Ubiquitylation of a plasma membrane protein was shown to serve as a signal for endocytosis of plasma membrane proteins in yeast and mammals as well as in plants. As more and more plant plasma membrane proteins are reported to be regulated through their ubiquitylation status, methods to analyze ubiquitylation and ubiquitin binding would be useful for the characterization of proteins involved in endocytosis of ubiquitylated cargo proteins.

Key words Ubiquitin, K48, K63, Western blot, Binding assay

1 Introduction

Posttranslational attachment of ubiquitin (Ub) to its substrate proteins is a highly conserved process that controls various cellular pathways [1]. The attachment of a single ubiquitin molecule, or monoubiquitylation, is sufficient as a signal in some of these pathways. However, ubiquitin molecules also form chains between the carboxyl group of the C-terminal glycine and one of seven lysine residues (K6, K11, K27, K29, K33, K48, and K63) within the ubiquitin molecule or with the N-terminal amino group (linear chain) [2]. Among the different chain types, K48- and K63 chains are the best characterized. These chains differ not only in their structural features [3, 4] but also in their cellular functions. K48-linked chains are mainly mediators of proteasomal degradation [5], whereas K63-linked chains are mostly involved in nonproteasomal pathways [6]. Both monoubiquitylation and K63-linked ubiquitylation were implied in endocytic trafficking of plasma membrane proteins [7, 8] and further studies have confirmed that many, if not all, plasma membrane proteins deploy ubiquitylation as an endocytosis signal.

Marisa S. Otegui (ed.), *Plant Endosomes: Methods and Protocols*, Methods in Molecular Biology, vol. 1209,
DOI 10.1007/978-1-4939-1420-3_15, © Springer Science+Business Media New York 2014

To ensure the delivery of ubiquitylated plasma membrane proteins, a series of multi-protein complexes are involved in the sorting of ubiquitylated cargo to the intraluminal vesicles of the multivesicular body (MVB). Endosomal Sorting Complex Required for Transport (ESCRT) machinery is essential for the MVB sorting pathway and is comprised of ESCRT-I, II, and III in plants [9]. The ESCRT-0 components characterized in yeast and mammalian systems seem to be absent in plants. ESCRT-0, ESCRT-I, and ESCRT-II have subunits with ubiquitin-binding domains. These ubiquitin-binding domains serve for anchoring the ubiquitylated membrane cargos to the endosomal membrane and guide them through the endocytic degradation route.

Ubiquitylation of target proteins is catalyzed by the activity of E1/E2/E3 enzymes. Ubiquitylation is a reversible process in which both deubiquitylating enzymes (DUBs) as well as ubiquitylating enzymes can regulate the stability of their substrate proteins. Associated Molecule with SH3 Domain of STAM (AMSH) proteins were shown to be DUBs that can interact with ESCRT-0 and ESCRT-III subunits and are implicated in the regulation of ESCRT-dependent endocytic degradation [10–13]. Inhibition of AMSH or ESCRT-III function in plants leads to the accumulation of ubiquitylated cargos, showing their importance in the endocytic degradation pathway [14].

Monoubiquitin, polyubiquitin, and ubiquitylated proteins can be detected using anti-ubiquitin antibodies. The global ubiquitin profile can be analyzed in any given material, e.g., specific tissues, mutants, or samples treated with specific drugs. It is also possible to characterize the status of chosen ubiquitin linkages in the sample material. One way to do so is by mass spectrophotometric analysis, though analysis of ubiquitin chain types using this method has multiple difficulties [15]. An alternative method is the use of available chain-type-specific antibodies, for example anti-K48 chain or anti-K63 chain antibody as described below [16].

Many proteins that are involved in the sorting of ubiquitylated membrane cargos have domains or motifs that have affinity towards ubiquitin or ubiquitin chains. The presence of such a ubiquitin-binding domain is thought to ensure the secure transport of the cargo proteins along the endocytic degradation route. To date different ubiquitin-binding domains are identified, e.g., ubiquitin interaction motif (UIM), ubiquitin-associated (UBA) domain, CUE domain, ubiquitin E2 enzyme variant (UEV) domain, GGAs and TOM1 (GAT) domain, GLUE domain, Vps27p, Hrs, and STAM (VHS) domain, and novel zinc finger (NZF) domain [17]. The presence of a ubiquitin-binding motif can be analyzed by pull-down assays with ubiquitin-conjugated matrices as described in the following or by making use of commercially available ubiquitin or ubiquitin chains. Ubiquitin-binding motifs are also a good tool for purification of ubiquitylated proteins. Tandem-repeated ubiquitin-binding entities (TUBEs) show high affinity towards ubiquitin chains and when conjugated to a matrix, can be used to capture ubiquitylated proteins from protein extracts [18, 19].

2 Materials

2.1 Plant Material and Total Protein Extraction

1. Growth medium: 4.3 g/l Murashige and Skoog medium containing Gamborg B5 vitamins, 10 g/l sucrose, 500 mg/l 4-morpholineethnesulfonic acid sodium salt (MES), 5.5 g/l plant agar. Adjust pH to 5.8 using potassium hydroxide (KOH).
2. Protein extraction buffer: 50 mM Tris–HCl pH 7.5, 150 mM NaCl, 0.5 % (v/v) Triton X-100. Store at room temperature. Add 1/50 volume of 50× stock of complete EDTA-free protease inhibitor cocktail (Roche) directly before use. Prepare the 50× protease inhibitor stock by dissolving one tablet in 1 ml protein extraction buffer; store stock at −20 °C.
3. Motor-driven homogenizer with a cooling jacket with 5-ml glass tubes and tightly fitting pestles (*see* **Note 1**).
4. 5× Laemmli buffer [20]: 250 mM Tris–HCl pH 6.8, 10 % (w/v) SDS, 50 % (w/v) glycerol, 0.05 % bromophenol blue (BPB), 5 % β-mercaptoethanol. Store at room temperature.

2.2 SDS-Polyacrylamide Gel Electrophoresis (PAGE)

1. Resolving (separating) gel buffer: 1.5 M Tris–HCl pH 8.8 and 0.4 % (w/v) SDS. Store at room temperature.
2. Stacking gel buffer: 0.5 M Tris–HCl pH 6.8 and 0.4 % (w/v) SDS. Store at room temperature.
3. 30 % acrylamide/bisacrylamide solution (37.5:1), 10 % (w/v) ammonium persulfate (APS), and *N,N,N,N'*-tetramethyl ethylenediamine (TEMED). Store at 4 °C.
4. 10× SDS running buffer: 250 mM Tris, 1.92 M glycin, 0.4 % (w/v) SDS. Dilute 10× stock in deionized water to prepare 1× running buffer.
5. Prestained molecular mass marker.
6. Electrophoresis system, e.g., Mini-PROTEAN III (Bio-Rad).
7. Purified Ubiquitin and polyubiquitin chains (Ub_{2-7}).

2.3 Western Blotting

1. Semidry transfer buffer: 25 mM Tris pH 8.3, 192 mM glycine, 20 % (v/v) methanol, 0.04 % (w/v) SDS. Store at room temperature.
2. Semidry transfer apparatus.
3. Four filter papers (1.2 mm) cut in the size 0.5 cm larger than the protein gel.
4. PVDF membrane cut in the size of protein gel.

5. 100 % methanol.
6. 10× Tris-buffered saline (TBS): 0.5 M Tris–HCl pH 7.5, 1.5 M NaCl, 10 mM $MgCl_2$.
7. Tris-buffered saline with Tween-20 (TBST): use 10× TBS stock to prepare 1× working solution. Add Tween-20 to 0.05 % (v/v). Store at room temperature.
8. Anti-Ub Antibody (e.g., P4D1 from Santa Cruz Biotechnology). Used for detection of mono- as well as all types of polyubiquitin and ubiquitin conjugates.
9. Anti-Ub (Lys48-Specific) antibody (e.g., clone Apu2 from Millipore). Used for detection of K48-linked ubiquitin chains.
10. Anti-Ub (Lys63-Specific) antibody (e.g., clone Apu3 Millipore). Used for detection of K63-linked ubiquitin chains.
11. Anti-mouse HRP-conjugated.
12. Anti-rabbit HRP-conjugated.
13. Enhanced chemiluminescent (ECL) reagents, e.g., SuperSignal West Femto Chemiluminescent Substrate from Thermo Scientific.
14. Development: Imaging device (e.g., LAS 4000 or LAS 4000 mini system from Fuji-Film).

2.4 NuPAGE SDS-PAGE and Transfer

1. NuPAGE 4–12 % Bis-Tris Gel.
2. 20× MES SDS Running Buffer: 1 M MES, 1 M Tris base, 2 % (w/v) SDS, 20 mM EDTA. Store 20× stock at 4 °C. Use the 20× stock to prepare 1× running buffer.
3. 4× NuPAGE SDS Sample Buffer: 564 mM Tris base, 416 mM Tris hydrochloride, 8 % (w/v) SDS, 40 % (w/v) glycerol, 2.04 mM EDTA, 0.88 mM SERVA Blue G250, 0.70 mM Phenol Red. Store at 4 °C.
4. Prestained molecular mass markers.
5. Gel apparatus, e.g., XCell SureLock Mini-Cell Electrophoresis System (Life Technologies).
6. 20× NuPAGE Transfer Buffer: 0.5 M bicine, 0.5 M Bis-Tris (free base), 20 mM EDTA. Store at 4 °C. Prepare 2× transfer buffer with 10 % (v/v) methanol.

Other materials for transfer and blotting are identical as in Subheading 2.3.

2.5 Recombinant Protein Purification from Bacteria

1. Buffer A: 50 mM Tris, 100 mM NaCl, 10 % (w/v) glycerol. Adjust the pH to 7.5 using HCl. Cool the buffer down to 4 °C for at least several hours. When the buffer temperature reaches 4 °C, readjust the pH to 7.5 using HCl.

2. Buffer A supplemented with 0.2 % Triton X-100 and 1× complete EDTA-free protease inhibitor (Roche), prepare directly before use.
3. Ultrasonic homogenizer (*see* **Note 2**).
4. Refrigerated centrifuge.
5. Magnetic beads (Nickel magnetic beads for His-tagged proteins and Glutathione magnetic beads for GST-fusion proteins) (*see* **Note 3**).
6. Rotator with 1.5-ml tube holders.
7. Magnetic separation racks (for 50- and 1.5-ml tubes), e.g., from New England BioLabs.
8. Protein molecular weight standards.

2.6 Ubiquitin In Vitro Binding Assay

1. 10× Phosphate Buffered Saline (PBS): 1.37 M NaCl, 27 mM KCl, 100 mM Na_2HPO_4, 20 mM KH_2PO_4; pH 7.4. Use the 10× stock to prepare 1× solution with ultrapure water.
2. 1× PBS with 1 mg/ml Bovine Serum Albumin (BSA).
3. Ubiquitin, agarose immobilized (e.g., from Enzo Life Sciences) (*see* **Note 4**).
4. Mini spin-columns, e.g., Mini Bio-Spin Chromatography Columns (Bio-Rad).
5. Rotator with 1.5-ml tube holders.
6. Heating block for 1.5-ml tubes.

2.7 Coomassie Brilliant Blue (CBB) Staining

1. Staining Solution: 40 % (v/v) ethanol, 7 % (v/v) acetic acid, 0.25 % (w/v) Coomassie Brilliant Blue (CBB).
2. Destaining solution: 40 % (v/v) ethanol, 7 % (v/v) acetic acid.

3 Methods

3.1 Plant Material and Total Protein Extraction

1. Grow *Arabidopsis thaliana* seedlings on solid growth medium for 7 days at 21 °C under continuous light. Collect 100 mg of seedlings in 1.5-ml microcentrifuge tube and freeze immediately in liquid nitrogen.
2. Cool the refrigerated centrifuge to 4 °C. Cool the protein extraction buffer on ice and add protease inhibitors. Place a cooled homogenization glass tube in the ice-filled cooling jacket. Transfer the frozen seedlings into the glass tube and add 200 μl protein extraction buffer. Homogenize each sample for 1 min at 1,600 U/min. Repeat the homogenization twice, spin the sample down, and cool it down between two homogenization rounds.

3. Transfer the plant extract to a new 1.5-ml tube and centrifuge for 20 min at 10,000 × g at 4 °C. Collect the supernatant in a new 1.5-ml tube. Take a small sample for protein concentration quantification, e.g., using the Bradford assay reagent.
4. Transfer 160 μl extract into a new 1.5-ml tube and add 40 μl 5× Laemmli buffer. Boil the sample for 5 min at 95 °C in a heating block.
5. Depending on the type of experiment, the samples should be loaded on different PAGE gels. For an anti-(poly)ubiquitin blot, load the samples on a 10 % SDS-PAGE gel. To visualize monoubiquitin or free polyubiquitin chains, load the samples on a NuPAGE 4–12 % Bis-Tris gel.

3.2 SDS-Polyacrylamide Gel Electrophoresis (PAGE)

1. Use a minigel system with 0.75-mm thick gels for the SDS-PAGE. Clean the plates with ethanol before assembling.
2. Prepare a 10 % resolving gel by mixing in a small beaker or 50-ml tube: 1.7 ml water, 1 ml resolving gel buffer, 1.3 ml 30 % acrylamide/bisacrylamide solution, 25 μl 10 % APS, and 3 μl TEMED. Pour the gel immediately after mixing the components and leave 1–1.5 cm space for the stacking gel. Carefully overlay the gel with water. Leave the gel for 15–30 min to polymerize.
3. Discard the water overlay. Remove rests of the overlay solution thoroughly with a paper towel.
4. Prepare the stacking gel by mixing 1.2 ml water, 0.5 ml stacking gel buffer, 0.3 ml 30 % acrylamide/bisacrylamide solution, 25 μl 10 % APS, and 3 μl TEMED. Mix the components and pour the gel. Insert the comb immediately avoiding formation of air bubbles. Leave the gel for 15 min to polymerize. Ready gel can be used directly or can be stored for several days at 4 °C wrapped in wet paper towels and plastic wrap.
5. Remove the comb carefully and assemble the gel apparatus. Fill the inner and outer chambers with 1× running buffer.
6. Load the 4 μl prestained molecular mass marker. Load 20 μg of total protein samples (*see* **Note 5**).
7. Run the gel at 20 mA (constant current) until the dye reaches the bottom of the gel.

3.3 Western Blotting

1. Activate the PVDF membrane for 30 s in 100 % methanol before immersing it in semidry transfer buffer.
2. Soak the filter papers in the semidry transfer buffer and assemble a stack in the following order (from bottom to top): two filter papers, PVDF membrane, gel, two filter papers. Eliminate air bubbles by rolling a glass tube over the transfer package after adding each filter paper.

3. Blot the gel for 1.5 h at a constant current of 2 mA/cm^2 membrane.
4. After the transfer, place the membrane in the blocking buffer and incubate for 15 min at room temperature on a shaker. Alternatively, incubate the membrane overnight at 4 °C on a shaker.
5. Prepare a proper dilution of the primary antibody in blocking solution. For the anti-Ub (P4D1), anti-Ub (Lys48-Specific, clone Apu2), and anti-Ub (Lys63-Specific, clone Apu3) antibodies we recommend 1:1,000 dilution.
6. Incubate the primary antibody with the membrane at room temperature with shaking for at least 1 h or overnight at 4 °C.
7. Remove the solution with the primary antibody. Wash the membrane for 15 min with TBST buffer. Repeat the step three times, using fresh TBST buffer each time.
8. Prepare a proper dilution of the secondary antibody in TBST.
9. Incubate secondary antibody with the membrane at room temperature with shaking for at least 45 min or overnight at 4 °C.
10. Wash the membrane as in **step 7**.
11. Take the membrane from the washing solution and remove excess liquid. Incubate the membrane with the ECL solution (600 μl is usually sufficient for a 6.5 × 8.0 cm membrane) for 5 min. Remove excess ECL solution and expose the membrane in a CCD camera system. Optimal exposure time varies between experiments. For the results of a typical polyubiquitin blot, *see* Fig. 1a.

3.4 NuPAGE SDS-PAGE and Transfer

1. Prepare 800 ml 1× running buffer. Unpack the precasted NuPAGE 4–12 % Bis-Tris Gel, remove the comb, and prepare the gel according to the manufacturer's instruction.
2. Assemble the gel apparatus and the gel. Fill the apparatus with 1× running buffer.
3. Load the samples on the gel:
   (a) For an anti-monoubiquitin blot, load 30 μg of total protein from seedling total protein extracts.
   (b) For ubiquitin-chain western blots, load 250 ng of ubiquitin or polyubiquitin chains (Ub_{2-7}).
4. Run the gel at 200 V for 35 min.
5. Disassemble the apparatus and incubate the gel for 15 min with the 2× transfer solution, containing 10 % (v/v) methanol.
6. Prepare a transfer stack as in the Subheading 3.3. Transfer the protein on the membrane for 25 min at 15 V.

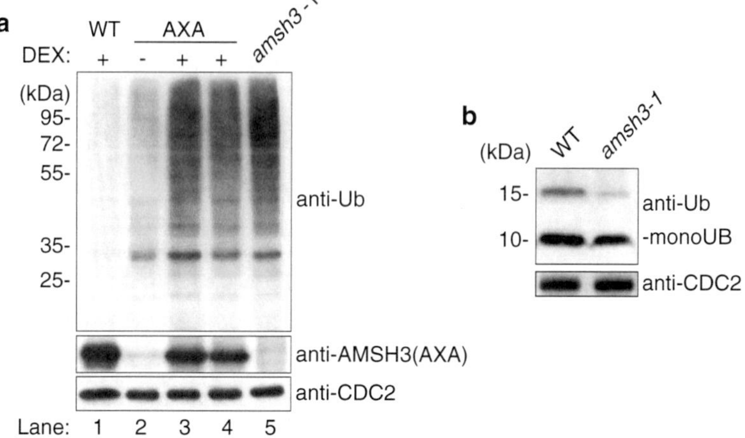

Fig. 1 Western blot of poly- and monoubiquitin profiles. **(a)** Western blot with anti-Ub, anti-AMSH3, and anti-CDC2 antibodies. Plant total extracts of DEX-induced plant lines overexpressing either a wild-type (*lane 1*) or a dominant-negative form of the DUB AMSH3 (*lanes 3* and *4*), and the *amsh3-1* knock-out mutant (*lane 5*) were subjected to the western blots. Note that accumulation of ubiquitin conjugates was observed in DEX-induced dominant-negative lines (*lanes 3* and *4*) and in knock-out mutants (*lane 5*) when compared to the wild type (*lane 1*) or to non-induced plants (*lane 2*). CDC2 was used as a loading control. **(b)** Monoubiquitin western blot of the wild type and knock-out mutant *amsh3-1*. The level of free ubiquitin is similar between the wild type and the *amsh3-1* mutant, which suggests that the pool of free ubiquitin is not affected in *amsh3-1*. CDC2 was used as a loading control. The position of free ubiquitin is indicated on the *right* of the panel

7. To detect free monoubiquitin in total plant extracts, boil the membrane in ultrapure water on a heating block for 10 min.
8. Membrane blocking, incubation with antibodies, and washing steps should be performed as described in Subheading 3.3. A typical monoubiquitin blot is shown in Fig. 1b, and the specificity of the K48- and K63-specific antibodies is shown in Fig. 2.

3.5 Recombinant Protein Purification from Bacteria

1. Cool down a refrigerated centrifuge for 50-ml tubes to 4 °C.
2. Place a frozen or fresh *E. coli* pellet on ice in a 50-ml falcon tube. The tube should contain bacterial pellets from 200–500 ml culture.
3. Prepare buffer A with 0.2 % Triton X-100 and 1× protease inhibitor. Add 20 ml buffer to the pellet. Place the tube in a beaker filled with ice.
4. Place the beaker with the tube in an ultrasonic homogenizer chamber.

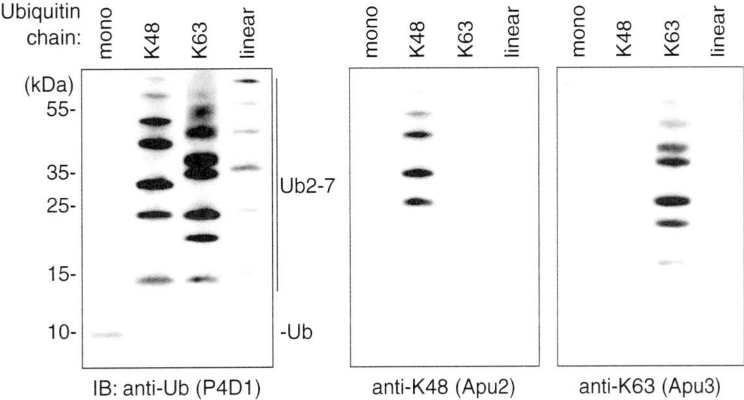

Fig. 2 Antibodies with different specificity towards ubiquitin linkages. The *left panel* shows a western blot using a global anti-ubiquitin antibody (P4D1). The *middle* and *right panels* present western blots conducted using a K48-(Apu2) and a K63-(Apu3) linkage specific antibody, respectively. Note that on the *left panel*, all chain types and monoubiquitin were detected whereas in the *middle* and *right panels* only specific linkages were visualized

5. Sonicate samples (five cycles at 20 % output) for 15 min. The solution should be clear afterwards. If it still looks cloudy, repeat the sonication for another 15 min.
6. Centrifuge the post-sonication solution at 9,300 × *g* for 10 min at 4 °C. The supernatant will then contain soluble protein of interest and should be kept. The pellet containing undisrupted *E. coli* cells and cell debris can be thrown away.
7. Take 100 μl of magnetic beads (*see* **Note 6**), transfer them to a 1.5-ml tube, and wash with 1 ml buffer A. Place the tubes in a magnetic rack to collect the beads. Remove the supernatant. Repeat twice. After washing, resuspend the beads in 500 μl buffer A.
8. Transfer the supernatant of the bacterial lysate to a new 50-ml tube. Using a tip with a cut-off end, add the washed beads to the supernatant.
9. Incubate the protein solution with the beads for 2 h at 4 °C with rotation.
10. Collect magnetic beads using a magnetic rack for 50-ml tubes and discard supernatant (you may take a small sample to test how much protein was not bound to the beads). It is recommended to work in a cold room to prevent protein degradation.
11. Discard supernatant. Wash the beads twice with 20 ml buffer A.
12. Discard the washing buffer, leaving ca. 500 μl buffer in the tube. Using a pipette and a tip with a cut-off end, transfer the beads into a 1.5-ml tube.

13. Wash the beads three times with 500 µl buffer A containing 0.2 % Triton and three times with buffer A only. Each time collect the beads using a magnetic rack and discard the supernatant.
14. Add 200 µl 1× PBS to the beads. Elute the protein using an appropriate competitor (imidazole for His-tagged and glutathione for GST-tagged protein). If necessary, the competitor can be removed by dialysis or desalting columns.
15. Take 6 µl of the purified protein, add 1.5 µl 5× Laemmli buffer, and incubate for 5 min at 95 °C. Analyze the purity and protein concentration on a CBB-stained SDS-PAGE gel using protein standards with known concentration, e.g., BenchMark Protein Ladder (Life Technologies).

3.6 Ubiquitin In Vitro Binding Assay

1. Cool down a refrigerated centrifuge for 1.5-ml tubes to 4 °C.
2. Using a tip with a cut-off end, transfer 20 µl of ubiquitin beads to a minicolumn and place the column on a 1.5-ml tube (*see* **Note 7**).
3. Wash the beads by adding 500 µl 1× PBS and centrifuge at 700 × *g* for 10 s at 4 °C. Discard the flow-through.
4. Close the bottom of the minicolumn. Add 400 µl of 1× PBS with 1 mg/ml BSA to block the beads.
5. Incubate the beads for 20 min at 4 °C with rotation.
6. Unplug the column and place it on a new 1.5-ml tube. Centrifuge the column at 700 × *g* for 10 s at 4 °C. Discard the flow-through.
7. Wash the column with 500 µl 1× PBS and centrifuge at 700 × *g* for 10 s at 4 °C. Discard the flow-through.
8. Plug the column. Add 8 µg of the protein or protein domain of interest (*see* **Note 8**) and add 1× PBS (*see* **Note 9**) to a final volume of 400 µl.
9. Using a tip with a cut-off end, take 40 µl as an input sample. Add 4.5 µl 5× Laemmli buffer, and incubate for 5 min at 80 °C. Let the sample cool to room temperature and then centrifuge it at 700 × *g* for 10 s. Keep the supernatant as the input sample.
10. Incubate the protein or domain of interest with Ub agarose for 4 h at 4 °C with rotation (*see* **Note 10**).
11. Unplug the column and place it on a new 1.5-ml tube. Centrifuge the column at 700 × *g* for 10 s at 4 °C. Discard the flow-through.
12. Wash the column at least four times with 500 µl 1× PBS. Centrifuge the column as described above.
13. Dry the column filter by centrifuging it for 5 s. Plug the column.

14. Add 40 μL 1× Laemmli on a minicolumn. Incubate the column on a heating block for 5 min at 80 °C and let it cool to room temperature.
15. Unplug the column and place on a new 1.5-ml tube. Centrifuge the column at 700 × *g* for 10 s at room temperature. Keep the flow-through.
16. Load 20 μl of input and post-binding material on a NuPAGE SDS-PAGE gel for analysis (*see* **Note 11**).

3.7 Coomassie Brilliant Blue (CBB) Staining

1. Disassemble the SDS-PAGE apparatus.
2. Pre-incubate the gel in destaining solution on a shaker for 15 min at room temperature, in order to remove excess SDS from the gel.
3. Discard the destaining solution. Pour the staining solution containing Coomassie Brilliant Blue (CBB) over the gel and incubate it on a shaker for 60 min at room temperature.
4. Discard the staining solution. Wash the gel gently several times with tap water. Pour the destaining solution over the gel and incubate it on a shaker at room temperature until the gel background is reduced to a satisfactory extend (Fig. 3). Presence of the recombinant protein in the pulldown fraction is indicative of ubiquitin-binding properties.

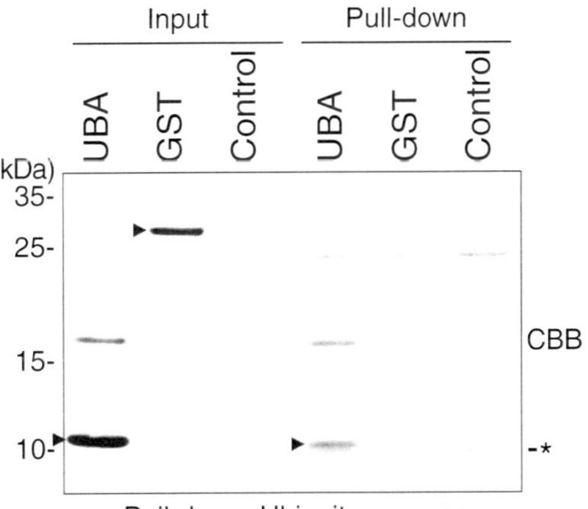

Fig. 3 Ubiquitin in vitro binding assay. The UBA domain of NBR1 (UBA) with a His-tag and a GST tag alone as a negative control was incubated with ubiquitin agarose. The input amount is shown on the left. The lane "Control" represents ubiquitin beads alone, without addition of any other protein. The right side of the blot shows the result of the binding assay. UBA domain of NBR1 bound to the ubiquitin agarose, whereas GST did not. In all three samples, two unspecific bands are visible and the asterisk shows co-eluted monoubiquitin. Positions of UBA and GST are indicated with *arrowheads* in the respective *lanes*

4 Notes

1. Alternatively, other homogenizing methods or equipment can be used, e.g., Tissue Lyser II (Qiagen) with glass beads. Note that the use of glass beads might result in less efficient protein extraction and higher material loss depending on the sample material.
2. *E. coli* cells can be disrupted also by other means, e.g., French press or bacterial lysis solutions.
3. Magnetic beads have an advantage over conventional agarose or sepharose beads, in that they usually result in higher protein purity and are easier to handle. However, they are usually more expensive than conventional matrices, and therefore the use should be decided according to the amount of beads required in an experiment. In case that conventional centrifugation-based methods are used, it is recommended to use minicolumns, e.g., Mini Bio-Spin Chromatography Columns (BioRad) to incubate and wash the agarose or sepharose beads.
4. Ubiquitin chains agarose, e.g., tetraubiquitin (K48- or K63-linked) agarose is also commercially available.
5. Proper loading is very important. If subtle differences are to be detected, it is recommended not to load too much protein per lane. Also, equal loading of samples is absolutely necessary to compare the amount between different genotypes or treatments.
6. The amount of the beads to be used depends on the volume of the cell culture, expression levels and solubility of the recombinant protein, and the binding capacity of the beads. Typically, for a 200–500 ml culture, 40–250 μl beads are used.
7. It is possible to wash the ubiquitin beads, block the beads, incubate the protein with the beads, and wash the beads after binding in normal 1.5-ml tubes, without using minicolumns in the whole procedure. Beads should be then spun each time when necessary at $800 \times g$ for 1 min at 4 °C. Be careful while removing supernatant with a pipette not to suck in the beads. The ubiquitin-binding affinity varies depending on the protein and domains; there are domains that bind to polyubiquitin more efficiently than to monoubiquitin. In this case, commercial polyubiquitin chains in combination with anti-ubiquitin antibody-conjugated agarose can be used for the binding assay.
8. This protocol is suitable for recombinant proteins which are available in the soluble fraction in abundance (μg order), as the assay will be analyzed by CBB staining. For proteins that cannot by purified in large quantities, less protein (as little as 500 ng)

can be used and the binding can be analyzed by detecting the protein or the tag with antibodies on a western blot.

9. In case there are problems with unspecific binding to the ubiquitin agarose, 0.1–0.5 % (v/v) Triton X-100 can be added to the binding buffer, as well as during washing steps.
10. The optimal incubation time of a protein of interest with ubiquitin agarose depends on the individual binding affinity and kinetics. For weak interactions, prolonged incubation time up to overnight can be advantageous.
11. For small proteins or domains (<15 kDa), precast gradient gels might be suitable, whereas normal SDS-PAGE gel can be used for analysis of larger proteins.

References

1. Kerscher O, Felberbaum R, Hochstrasser M (2006) Modification of proteins by ubiquitin and ubiquitin-like proteins. Annu Rev Cell Dev Biol 22:159–180
2. Xu P, Duong DM, Seyfried NT, Cheng D, Xie Y, Robert J, Rush J, Hochstrasser M, Finley D, Peng J (2009) Quantitative proteomics reveals the function of unconventional ubiquitin chains in proteasomal degradation. Cell 137(1):133–145
3. Varadan R, Assfalg M, Haririnia A, Raasi S, Pickart C, Fushman D (2004) Solution conformation of Lys63-linked di-ubiquitin chain provides clues to functional diversity of polyubiquitin signaling. J Biol Chem 279(8): 7055–7063
4. Cook WJ, Jeffrey LC, Carson M, Chen Z, Pickart CM (1992) Structure of a diubiquitin conjugate and a model for interaction with ubiquitin conjugating enzyme (E2). J Biol Chem 267(23):16467–16471
5. Chau V, Tobias JW, Bachmair A, Marriott D, Ecker DJ, Gonda DK, Varshavsky A (1989) A multiubiquitin chain is confined to specific lysine in a targeted short-lived protein. Science 243(4898):1576–1583
6. Spence J, Sadis S, Haas AL, Finley D (1995) A ubiquitin mutant with specific defects in DNA repair and multiubiquitination. Mol Cell Biol 15(3):1265–1273
7. Galan JM, Haguenauer-Tsapis R (1997) Ubiquitin lys63 is involved in ubiquitination of a yeast plasma membrane protein. EMBO J 16(19):5847–5854
8. Shih SC, Sloper-Mould KE, Hicke L (2000) Monoubiquitin carries a novel internalization signal that is appended to activated receptors. EMBO J 19(2):187–198
9. Winter V, Hauser MT (2006) Exploring the ESCRTing machinery in eukaryotes. Trends Plant Sci 11(3):115–123
10. McCullough J, Clague MJ, Urbe S (2004) AMSH is an endosome-associated ubiquitin isopeptidase. J Cell Biol 166(4):487–492
11. Isono E, Katsiarimpa A, Muller IK, Anzenberger F, Stierhof YD, Geldner N, Chory J, Schwechheimer C (2010) The deubiquitinating enzyme AMSH3 is required for intracellular trafficking and vacuole biogenesis in Arabidopsis thaliana. Plant Cell 22(6):1826–1837
12. McCullough J, Row PE, Lorenzo O, Doherty M, Beynon R, Clague MJ, Urbe S (2006) Activation of the endosome-associated ubiquitin isopeptidase AMSH by STAM, a component of the multivesicular body-sorting machinery. Curr Biol 16(2):160–165
13. Katsiarimpa A, Anzenberger F, Schlager N, Neubert S, Hauser MT, Schwechheimer C, Isono E (2011) The Arabidopsis deubiquitinating enzyme AMSH3 interacts with ESCRT-III subunits and regulates their localization. Plant Cell 23(8):3026–3040
14. Katsiarimpa A, Kalinowska K, Anzenberger F, Weis C, Ostertag M, Tsutsumi C, Schwechheimer C, Brunner F, Huckelhoven R, Isono E (2013) The deubiquitinating enzyme AMSH1 and the ESCRT-III subunit VPS2.1 are required for autophagic degradation in Arabidopsis. Plant Cell 25(6):2236–2252
15. Kirkpatrick DS, Denison C, Gygi SP (2005) Weighing in on ubiquitin: the expanding role of mass-spectrometry-based proteomics. Nat Cell Biol 7(8):750–757
16. Newton K, Matsumoto ML, Wertz IE, Kirkpatrick DS, Lill JR, Tan J, Dugger D, Gordon N, Sidhu SS, Fellouse FA, Komuves

L, French DM, Ferrando RE, Lam C, Compaan D, Yu C, Bosanac I, Hymowitz SG, Kelley RF, Dixit VM (2008) Ubiquitin chain editing revealed by polyubiquitin linkage-specific antibodies. Cell 134(4):668–678

17. Dikic I, Wakatsuki S, Walters KJ (2009) Ubiquitin-binding domains—from structures to functions. Nat Rev Mol Cell Biol 10(10): 659–671

18. Hjerpe R, Aillet F, Lopitz-Otsoa F, Lang V, England P, Rodriguez MS (2009) Efficient protection and isolation of ubiquitylated proteins using tandem ubiquitin-binding entities. EMBO Rep 10(11):1250–1258

19. Kim DY, Scalf M, Smith LM, Vierstra RD (2013) Advanced proteomic analyses yield a deep catalog of ubiquitylation targets in Arabidopsis. Plant Cell 25(5):1523–1540

20. Laemmli UK (1970) Cleavage of structural proteins during the assembly of the head of bacteriophage T4. Nature 227(5259): 680–685

Chapter 16

Analysis of Endocytosis and Ubiquitination of the BOR1 Transporter

Koji Kasai, Junpei Takano, and Toru Fujiwara

Abstract

Endocytosis and membrane trafficking are the major factors controlling the abundance of plasma membrane proteins, such as transporters and receptors. We have found that *Arabidopsis* borate transporter BOR1 is polarly localized to the inner (stele-facing) plasma membrane domain of various root cells under boron limitation, and when boron is supplied in excess, BOR1 is rapidly transferred to the vacuole for immediate degradation. The BOR1 polarity and degradation are controlled by membrane trafficking including endocytosis. In this chapter, we describe methods for observation of endocytic trafficking of BOR1, and detection of BOR1 ubiquitination that is required for vacuolar sorting for degradation.

Key words Endocytosis, Membrane trafficking, BOR1, Polar localization, Boron-induced degradation, Vacuolar sorting, Ubiquitination

1 Introduction

Boron is essential for plant growth, but toxic in excess [1]. *Arabidopsis* borate transporter BOR1 is a key protein maintaining boron homeostasis [2, 3]. BOR1 has borate exporter activity in yeast cells, and is involved in boron xylem loading and translocation to shoots.

Under low-boron conditions, BOR1 is accumulated in the plasma membrane, at the inner side (stele-facing) of various cells at the root tip region and endodermal cells of the root hair zone [4, 5]. It is reasonable to assume that the inward polarity is important for the efficient loading of limited amount of boron into the xylem. In the predicted cytosolic loop region of BOR1, there are three putative tyrosine-based sorting signal sequences (YXXØ: Y, tyrosine; X, any amino acid; Ø, a bulky hydrophobic residue). The tyrosine-based sorting signal is recognized by μ-subunit of the clathrin-adaptor protein (AP) complex for targeting membrane proteins into clathrin-coated vesicles at the plasma membrane and

Marisa S. Otegui (ed.), *Plant Endosomes: Methods and Protocols*, Methods in Molecular Biology, vol. 1209, DOI 10.1007/978-1-4939-1420-3_16,

the trans-Golgi network (TGN), and is involved in polar membrane trafficking [6]. Mutations in all of the three tyrosine residues, Y373, Y398, and Y405, of BOR1 cause a loss of polar localization in root tip cells, indicating a possible role of membrane trafficking in the establishment of BOR1 polarity via tyrosine-based sorting signals [4]. Another boron-transporting protein, the boric acid channel NIP5;1, is localized on the outer side (soil-facing) of root epidermal cells under low-boron conditions [4, 7]. The outward polarity of NIP5;1 and inward polarity of BOR1 establish an efficient radial boron transport route to the stele. The membrane trafficking mechanisms underlying NIP5;1 polar localization are not known.

Under high-boron conditions, BOR1 is immediately degraded via endocytosis and vacuolar trafficking [8]. The high-boron induced dot-like structures containing BOR1-GFP were partly co-localized with endocytic marker FM4-64 and endosomal Rab-GTPase Ara7 fused with mRFP [8, 9]. BOR1-GFP also accumulated in the brefeldin A (BFA)-induced endosomal aggregations. In the presence of high concentration of boron, GFP fluorescence derived from BOR1-GFP was observed in the vacuole. Immunogold electron microscopy showed that the internalized BOR1-GFP is transported to the TGN/early endosomes and then, into inner vesicles of multivesicular bodies (MVBs) [10]. These results demonstrate that BOR1 protein availability is regulated by endocytosis and postendocytic trafficking. The boron-induced degradation of BOR1 is deemed necessary to avoid toxicity of excess boron accumulation.

Ubiquitination is known to play an important role in the degradation of proteins [11]. In mammals, mono- and multi-monoubiquitination are crucial for endocytic trafficking of membrane proteins to lysosome [12]. Monoubiquitination also functions as a sorting signal of membrane proteins to MVBs, and is required for binding to the ESCRT protein sorting complex [13]. The ESCRT-related proteins in *Arabidopsis*, CHMP1A and CHMP1B, are necessary for MVB sorting of the auxin transporters, PIN1, PIN2, and AUX1 [14]. Several plant plasma membrane proteins, such as the PIN2 auxin efflux transporter, and the PIP2 water channel, have been reported to be ubiquitinated [15, 16]. However, in these cases, the role of the ubiquitination in protein trafficking is still unclear. For boron-induced BOR1 degradation, mono- or diubiquitination is required [17]. The substitution of lysine at residue 590 with alanine (K590A) in BOR1 completely blocks both ubiquitination and degradation of BOR1. K590 residue of BOR1 is not required for endocytic internalization of BOR1, but is necessary for BOR1 translocation to MVBs, indicating that boron-induced ubiquitination is a critical signal for sorting of internalized BOR1 to MVBs and subsequent vacuolar degradation. A similar study on the *Arabidopsis* IRT1 high-affinity iron transporter was reported by Barberon et al. [18]. They showed that both endocytosis and MVB sorting of IRT1 require monoubiquitination at two lysine residues.

This chapter describes the procedures for preparation of low-boron medium, observation of BOR1 trafficking by GFP imaging using confocal microscopy, and detection of boron-induced ubiquitination of BOR1 including hydroponic culture, microsome isolation, immunoprecipitation, and immunoblot analysis.

2 Materials

Prepare all solutions using MilliQ water (Millipore Corp., Bedford, MA).

2.1 MGRL Medium

1. 200× MGRL nutrients stock solutions [19], *see* Table 1. Store at room temp.
2. 0.3 M boric acid solution.
3. Sucrose.
4. Gellan gum (*see* **Note 1**).
5. 1 L polycarbonate bottles.
6. Sterile disposable rectangular petri dish (140 × 100 × 14.5 mm).

2.2 Plant Materials

1. *Arabidopsis thaliana* transgenic line harboring BOR1 promoter:BOR1-GFP [4].
2. 70% (v/v) and 99.5% (v/v) Ethanol.

Table 1
Constituents of stock solutions for MGRL medium

Stock solutions	Nutrients	per 1 L	Final conc. (1×)
200 × Pi	$NaH_2PO_4 \cdot 2H_2O$	47.2 g	1.51 mM
	$Na_2HPO_4 \cdot 12H_2O$	18.4 g	0.257 mM
200 × Mg	$MgSO_4 \cdot 7H_2O$	74.0 g	1.50 mM
200 × Ca, K, N	$Ca(NO_3)_2 \cdot 4H_2O$	94.4 g	2.00 mM
	KNO_3	60.7 g	3.00 mM
200 × micro –Fe, –B	$MnSO_4$	311 mg	10.3 μM
	$ZnSO_4 \cdot 7H_2O$	57.5 mg	1.00 μM
	$CuSO_4 \cdot 5H_2O$	50.0 mg	1.00 μM
	$CoCl_2 \cdot 6H_2O$	6.20 mg	130 nM
	$(NH_4)_6Mo_7O_{24} \cdot 4H_2O$	5.90 mg	24.0 nM
200 × Fe	Fe(III)-EDTA·$3H_2O$	4.21 g	50 μM
300 mM boric acid	H_3BO_3	18.5 g	–

3. Wooden toothpick.
4. Surgical tape (Micropore, 3M, Germany).
5. Microcentrifuge tube (1.5 ml).

2.3 Confocal Microscopy

1. Slide glasses and cover glasses (Thickness No.1, 24 × 36 mm).
2. Cover glass chambers or glass base dish (Thickness No.1S, Asahi Glass Co., Ltd., Tokyo, Japan).
3. FM4-64 (Molecular Probes, Carlsbad, CA, USA): Prepare 10 mM stock solution in water. Store at −20 °C. Dilute the stock solution in MGRL liquid medium containing 0.3 μM boric acid to a concentration of 25 μM before use.
4. Brefeldin A (BFA): Prepare 50 mM stock solution in DMSO. Store at −20 °C. Dilute the stock solution in MGRL liquid medium containing 0.3 μM boric acid to a concentration of 50 μM. Dilute the stock solution in MGRL liquid medium containing 0.3 μM boric acid to a concentration of 50 μM before use.
5. Cycloheximide (CHX): Prepare 25 mM stock solution in water before use. Dilute the stock solution in MGRL liquid medium containing 0.3 μM boric acid to a concentration of 50 μM.
6. Wortmannin (Sigma, St. Louis, MO, USA): Prepare 50 mM stock solution in DMSO. Store at −20 °C. Dilute the stock solution in MGRL liquid medium containing 0.3 or 100 μM boric acid to a concentration of 16.5 μM.
7. Razor blade and tweezers.
8. PCR tube (200 μl).
9. Confocal laser scanning microscope with 488 nm excitation line and filter sets for detection of GFP.

2.4 Detection of Ubiquitination

2.4.1 Hydroponic Culture

1. Formed polyethylene sheets (15 × 100 × 4 mm).
2. Stapler.
3. Light-shielded plastic container (90 × 150 × 90 mm). Surface of plastic container is covered with black vinyl tape.
4. Air pump.

2.4.2 Preparation of Microsomal Fractions

1. Homogenization buffer: 250 mM Tris–HCl (pH 7.8), 25 mM ethylenediaminetetraacetic acid (EDTA), 290 mM sucrose, 75 mM 2-mercaptoethanol, 10 mM *N*-etylmaleimide, 1 mM phenylmethylsulfonyl fluoride (PMSF), and one tablet/50 ml of protease inhibitor cocktail (complete, EDTA-free; Roche, Indianapolis, IN, USA). Prepare before use.
2. Storage buffer: 50 mM Tris–HCl (pH 8.0), 150 mM NaCl, 10 mM *N*-etylmaleimide, 1 mM PMSF, and one tablet/50 ml of protease inhibitor cocktail (complete, EDTA-free; Roche). Prepare before use.

3. Conical tube (50 ml).
4. Scissors.
5. PT-2100 POLYTRON mixer (Kinematica Inc., Bohemia, NY, USA).
6. Centrifuge tube (5 ml).
7. Conical tube (15 ml).
8. Ultracentrifuge tube (6 ml, 6PC Thick-walled tube, Hitachi-koki, Tokyo, Japan).
9. Ultracentrifuge himac CS150GXL with S80AT rotor or S55A2 rotor (Hitachi-koki, Tokyo, Japan).
10. Microcentrifuge tube (1.5 ml Eppendorf tube).
11. Pestle that fits into the bottom of 1.5 ml microcentrifuge tube.
12. Quick Start Protein Assay Kit I (Bio-Rad, Hercules, CA, USA)

2.4.3 Immuno-precipitation

1. 2× solubilization buffer: 50 mM Tris–HCl (pH 7.5), 50 mM dithiothreitol, 5 % glycerol, 5 % sodium dodecyl sulfate (SDS), 5 mM EDTA, 0.03% bromophenol blue, one tablet/10 ml of protease inhibitor cocktail (complete mini, EDTA-free; Roche).
2. Agarose conjugated anti-GFP monoclonal antibody (50 % gel slurry; MBL, Nagoya, Japan).
3. Syringe and 27 G needle.

2.4.4 SDS-PAGE and Immunoblot Analysis

1. 30% acrylamide/Bis solution (29.76:0.24, *see* **Note 2**): 29.76 g of acrylamide and 0.24 g of Bis in water. Filter through a 0.45 μm filter membrane, and store at 4 °C in the dark.
2. 10 % Ammonium persulfate (APS).
3. *N*, *N*, *N*, *N*′- tetramethylethylenediamine (TEMED).
4. 1.5 M Tris–HCl (pH 8.8).
5. 0.5 M Tris–HCl (pH6.8).
6. 10 % sodium dodecyl sulfate (SDS).
7. 5 M NaCl.
8. Gel electrophoresis system (BE-220, BIO CRAFT, Tokyo, Japan).
9. SDS-PAGE running buffer: 25 mM Tris, 192 mM glycine, 0.1 % SDS.
10. Blotting buffer for membrane protein: 0.1 M Tris, 0.192 M glycine, 5 % methanol, 0.02 % SDS.
11. PVDF membrane (Immobilon-P transfer membrane, Millipore Corp., Bedford, MA).
12. Methanol.
13. Filter paper (Semidry gel pad, 85×90 mm).

14. Semidry blotting system (BE-330, BIO CRAFT, Tokyo, Japan).
15. TBST: 0.02 M Tris–HCl (pH 8.0), 0.15 M NaCl, 0.1 % Tween20.
16. ECL Advance Blocking Agent (GE Healthcare, *see* **Note 3**). Prepare 2 % blocking solution in TBST before use.
17. Anti-Ubiquitin monoclonal antibody (P4D1; Santa Cruz Biotechnology, Santa Cruz, CA). Use at 0.5 μg/ml in 2 % ECL advance blocking reagent.
18. Anti-GFP mouse monoclonal antibody GF200 (Nakalai Tesque, Kyoto, Japan). Use at 2.2 μg/ml in 2 % ECL advance blocking reagent.
19. Horseradish peroxidase-conjugated goat antibody to mouse IgG (GE Healthcare). Use at a dilution of 1:25,000 in 2 % ECL advance blocking reagent.
20. ECL Advance Western Blotting Detection kit (GE Healthcare, *see* **Note 2**).
21. CCD camera-based quantitative biomolecular imager (LAS4000, GE healthcare).

3 Methods

3.1 MGRL Medium

1. Eight hundred milliliters of MGRL medium is made for ten plates of solid MGRL medium. Use 1 L polycarbonate bottles. Do not use glass bottles (*see* **Note 4**).
2. Add 4 ml each of 200× MGRL nutrients stock solutions (200× Pi, 200× Mg, 200× Ca, K, N, 200× micro –Fe, –B, and 200× Fe. *see* Table 1) into about 700 ml of MilliQ water.
3. Add boric acid at final concentration of 0.3–100 μM.
4. Add 16 g sucrose, stirred in magnetic stirrer until fully dissolved, and then make up to 800 ml.
5. Slowly add 12 g of gellan gum to the medium, stirring the medium using magnetic stirrer.
6. Autoclave the medium for 15 min at 121 °C.
7. Distribute approximately 75 ml to sterile disposable rectangular petri dish, wait until the medium solidifies, and then, let the plates rest for 30 min on a clean bench.
8. MGRL liquid medium. Add 1/200 volume of 200× MGRL nutrients stock solutions 200× Pi, 200× Mg, 200× Ca, K, N, 200× micro –Fe, –B, and 200× Fe) into MilliQ water, and supplement with 0.3–100 μM of boric acid. Do not add sucrose.

3.2 Growth of Arabidopsis thaliana for GFP Imaging

1. Surface sterilization of *Arabidopsis* seeds. Pretreat the seeds in 70 % ethanol for 1–2 min, and then soak the seeds in 99.5 % ethanol for 2 min in 1.5 ml microcentrifuge tube. Discard ethanol by decantation and then remove the remaining ethanol as much as possible using the micropipette. Allow to air-dry for a few hours.
2. Sow the seeds using wet wooden toothpick onto MGRL solid medium containing 0.3–3 μM of boric acid (low-boron conditions). Seal the dish by using surgical tape. After a 2–4-day cold acclimation at 4 °C, place the plates vertically at 22 °C under a 16 h light:8 h dark cycle for 4–7 day (*see* **Note 5**).

3.3 Confocal Laser Scanning Microscopy

3.3.1 Standard Method

1. Cut off approximately 1 cm of root tip of transgenic BOR1 promoter:BOR1-GFP plant using razor blade.
2. The root tip is placed on a slide glass using tweezers with a drop of MGRL liquid medium, and a cover glass is placed on top of it.
3. Optionally, the plasma membrane can be stained using the fluorescent dye, FM4-64. Dilute 10 mM stock solution of FM4-64 in MGRL liquid medium to a final concentration of 25 μM, and incubate root tip in this solution for 2 min using 200 μl PCR tube.
4. For imaging BOR1-GFP, a laser scanning confocal microscope with a 488 nm excitation line and a 40× water immersion objective can be used. We collect emitted fluorescence with a 505–530-nm band-pass filter for GFP and a 650-nm long-pass filter for FM4-64. Any alternative that allows imaging of GFP and FM4-64 is possible.

3.3.2 Polar Localization and Boron-Induced Degradation of BOR1-GFP

1. To observe BOR1-GFP polar localization, find an optical section through the center of root (Fig. 1a).
2. To observe boron-induced degradation of BOR1-GFP, incubate the root tip in MGRL liquid medium supplemented with 100 μM boric acid (high-boron condition) for 0.5 or 2 h at 22 °C in the light using a 200 μl PCR tube. Do not close the lid to avoid hypoxia. If incubation is achieved in the dark by wrapping with aluminum foil, GFP signal can be detected in the vacuole (Fig. 1a, b, *see* **Note 6**).

3.3.3 BFA and Wortmannin Treatment

Membrane trafficking inhibitors such as BFA and wortmannin are useful for studying membrane protein trafficking. BFA blocks translocation of membrane proteins from early endosomes (EEs) to MVBs and to the plasma membrane but does not block endocytosis. BFA causes accumulation of internalized BOR1-GFP in aggregates known as BFA bodies, indicating that BOR1-GFP is internalized by endocytosis (Fig. 1c). Wortmannin is an inhibitor

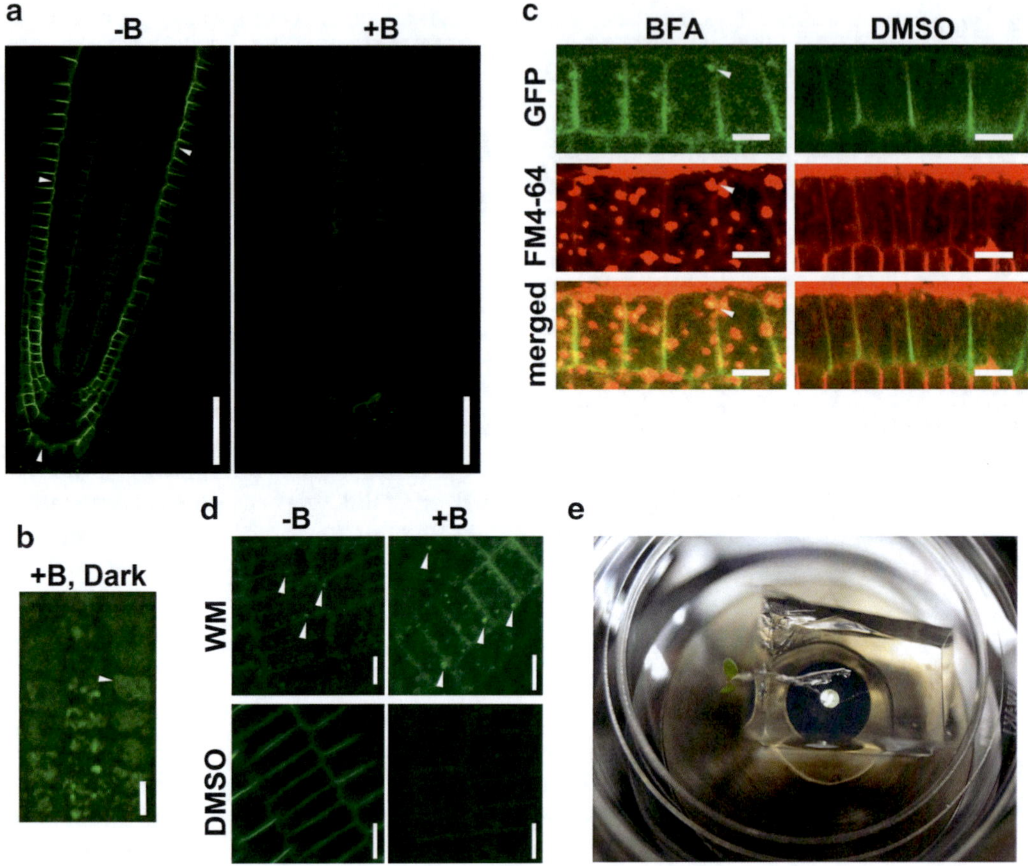

Fig. 1 Subcellular localization of BOR1-GFP observed by confocal microscopy. (**a**) BOR1-GFP in the root tip region. Under low-boron condition (–B, 3 μM boric acid), BOR1-GFP is localized in the plasma membrane, and inward polarity is observed as indicated by *arrows*. Two-hour incubation with high-boron medium (+B, 100 μM boric acid) causes BOR1-GFP degradation. Scale bars, 50 μm. (**b**) Distribution of GFP signals derived from BOR1-GFP in the epidermal cells of root tips after incubation with 100 μM boric acid medium in the dark for 2 h. *Arrow* indicates GFP signal detected in the vacuole. Scale bars, 10 μm. (**c**) Distribution of BOR1-GFP after treatment with BFA or DMSO. The GFP and FM4-64 signal are shown in *green* and *red*, respectively, and in the merged images, overlapping signals of GFP and FM4-64 appear in *yellow*. *Arrowheads* indicate co-localization of BOR1-GFP and FM4-64 in the BFA body. Scale bars, 10 μm. (**d**) Distribution of BOR1-GFP after treatment with wortmannin or DMSO. Wortmannin treatment was performed with low-boron medium (–B, 3 μM boric acid) or high-boron medium (+B, 100 μM boric acid). Arrowheads indicate example of BOR1-GFP signal in the dot structure representing enlarged MVBs. Scale bars, 10 μm. (**e**) Recommended setting for time-lapse imaging using an inverted microscope. The seedling was placed on a glass bottom dish and the root was covered by a block of solid medium

of phosphatidyl-inositol 3-kinase and induces homotypic fusions of the prevacuolar compartment (PVC)/MVBs. Localization of BOR1-GFP in PVC/MVBs can be visualized by wortmannin treatment (Fig. 1d).

1. BFA treatment. Incubate a root tip with 50 μM CHX for 30 min (*see* **Note 7**), and with 25 μM FM4-64 for 2 min, and

then with 50 μM CHX and 50 μM BFA. All incubations are performed at 22 °C in the light using 200 μl PCR tube. Do not close the lid. DMSO is used as a solvent control.

2. Wortmannin treatment. Incubate a root with 25 μM FM4-64 for 2 min followed by incubation with 0.3 or 100 μM boric acid containing MGRL medium supplemented with 16.5 μM wortmannin for 90 min at 22 °C in the light using 200 μl PCR tube. Do not close the lid. DMSO is used as a solvent control.

3.3.4 Time-Lapse Imaging Using Inverted Microscopes

For time-lapse imaging with healthy plants, intact seedlings with appropriate growth conditions can be set on inverted microscopes. Here we describe the methods to observe boron-dependent endocytosis of BOR1-GFP [20].

1. Grow plants on vertically placed solid MGRL medium containing 0.3–3 μM boric acid, 1 % sucrose, and 1.5 % gellan gum for 4–5 days.
2. Transfer the seedlings onto a cover-glass chamber or a glass-base dish using tweezers with a drop of MGRL liquid medium.
3. Cover the roots of the seedlings with a block of MGRL medium containing 100 μM boric acid (high-boron condition) and 0.5–1.5 % gellan gum (Fig. 1e). Inhibitors can be included in the medium.
4. Image roots in a confocal laser scanning system and a 40× water immersion objective.

3.4 Detection of Boron-Induced Ubiquitination of BOR1-GFP

3.4.1 Hydroponic Culture

1. Sow 25 surface-sterilized seeds on MGRL solid medium containing 30 μM boric acid and let plants grow vertically at 22 °C under a 16 h light:8 h dark cycle for 16 days.
2. Remove carefully 12–13 plants from the plate and place them between two formed-polyethylene-sheets (15 × 100 × 4 mm); attach the two polyethylene sheets with a stapler (Fig. 2a, b).
3. Float two sets of the sheets (in total 25 plants) on 400 ml of MGRL liquid medium in a light-shielded plastic container (Fig. 2b).
4. Grow plants hydroponically with MGRL liquid medium containing 30 μM B for 4 days and then with 3 μM B MGRL liquid medium for additional 4 days under a 10 h light:14 h dark cycle at 22 °C. The liquid medium should be continuously aerated by an air pump.
5. To induce ubiquitination of BOR1-GFP, treat plants with 100 μM boric acid (high-boron condition) in MGRL liquid medium for 0.5, 1, and 2 h. Plants incubated for 2 h with 3 μM boric acid (low-boron condition) MGRL liquid medium are used as negative control.

3.4.2 Preparation of Microsomal Fractions

1. Distribute 5 ml homogenization buffer in a 50 ml conical tube, and chill it on ice.
2. Using scissors, harvest roots from hydroponically grown plants treated with 100 μM B MGRL liquid medium for 0.5 h. Remove excess liquid medium with a paper towel.
3. Immerse roots (approximately 0.5–1 g fresh weight) rapidly in 5 ml of ice-cold homogenization buffer using tweezers; homogenize using a PT-2100 POLYTRON mixer at 26,000 rpm for 5 s at 4 °C. Repeat homogenization step four times with intervals of 15 s.
4. Centrifuge the homogenate at 8,000 × *g* for 10 min at 4 °C.
5. While waiting for centrifugation to be completed, prepare centrifugal filter for removing debris. Make a hole at the bottom and the lid of a 5 ml centrifuge tube using an eyeleteer, and fill the bottom of the tube with appropriate quantities of absorbent cotton. The filter tube is inserted into a 15 ml conical tube (Fig. 2c).

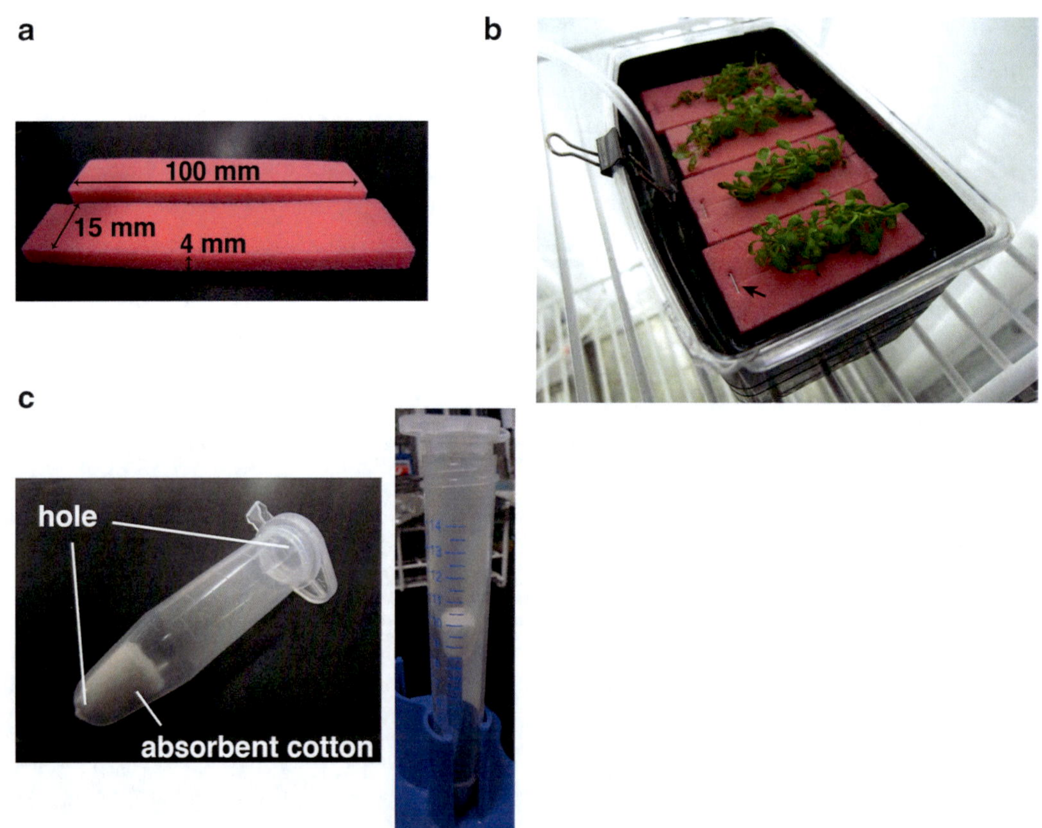

Fig. 2 Experimental apparatus used for the hydroponic culture and microsome isolation. (**a**) Formed-polyethylene-sheets (15 × 100 × 4 mm). (**b**) A hydroponic culture system. Plants placed between two formed-polyethylene-sheets are fixed using stapler as shown by an *arrow*. (**c**) A centrifugal filter for removing debris

6. Transfer the supernatant into the filter tube and centrifuge at 200 × *g* for 5 min at 4 °C to remove debris.
7. Transfer the filtrate into a 6 ml ultracentrifuge tube and centrifuge at 100,000 × *g* for 15 min at 4 °C using ultracentrifuge (himac CS150GXL with S80AT rotor).
8. Remove the supernatant, and wash the pellet briefly with 1 ml of ice-cold storage buffer by pipetting. Add 100 µl of ice-cold storage buffer and resuspend the microsome-pellet by pipetting with a micropipette. Transfer the suspension into a 1.5 ml microcentrifuge tube; suspend microsomes completely using a pestle that fits into the bottom of 1.5 ml microcentrifuge tube.
9. Determine protein concentration using the Quick Start Protein Assay Kit I (Bio-Rad). Freeze samples in liquid nitrogen and store at –80 °C.

3.4.3 Immunoprecipitation of BOR1-GFP

1. Transfer microsomal protein fractions containing 50 µg proteins in a 1.5 ml Eppendorf 3810× microcentrifuge tube (*see* **Note 8**), and centrifuge at 100,000 × *g* for 15 min at 4 °C using ultracentrifuge (Himac CS150GXL with S55A2 rotor).
2. Remove the supernatant completely using micropipette and resolve the pellet in 300 µl of lysis buffer using a pestle that fits into the bottom of 1.5 ml microcentrifuge tube. Incubate the tube with gentle agitation for 30 min at 4 °C
3. Thoroughly suspend the vial of agarose conjugated anti-GFP monoclonal antibody to make a uniform suspension of the resin; add 20 µl of the resin slurry to a sample and incubate for 1.5 h at 4 °C with gentle shaking. After incubation, the resin is pelleted by centrifugation at 2,500 × *g* for 10 s at 4 °C, and the supernatant is removed using a syringe and a 27 G needle (*see* **Note 9**).
4. Add 200 µl of lysis buffer to the resin and mix by tapping, then centrifuge at 2,500 × *g* for 10 s at 4 °C and remove the supernatant using a syringe and a 27 G needle. Repeat this washing step for five times.
5. Resuspend the agarose in 20 µl of 2× solubilization buffer, and incubate at 65 °C for 10 min (*see* **Note 10**). Centrifuge at 10,000 × *g* for 5 min, and use 18 µl of the supernatant for immunoblot analysis.

3.4.4 Immunoblot Analysis

1. Preparation of SDS-PAGE separation gel (8 % acrylamide). To a 50 ml conical tube, add 3.45 ml H_2O, 2 ml of 30 % acrylamide/Bis solution (29.76:0.24, *see* **Note 2**), 1.9 ml of 1.5 M Tris–HCl (pH.8.8), 75 µl of 10 % SDS, 75 µl of 10 % APS,

19.5 μl of 5 M NaCl (*see* **Note 2**), and 4.5 μl of TEMED. Pour the 5.5 ml of gel solution to a gel cassette of BIO CRAFT Gel electrophoresis system (BE-220), and gently overlay with water. Allow the gel to polymerize for 30 min at room temp. Remove the overlaying water using filter paper.

2. Preparation of SDS-PAGE stacking gel. Mix 2.1 ml H_2O, 0.5 ml of 30 % acrylamide/Bis solution (29.76:0.24), 0.38 ml of 1 M Tris–HCl (pH 6.8), 30 μl of 10 %SDS, 30 μl of 10 % APS, 7.8 μl of 5 M NaCl, and 3 μl of TEMED. Pour 2 ml of the stacking gel solution to the cassette, insert the comb, and allow the gel to polymerize for 30 min to 1 h at room temp.
3. After polymerization of the gel, remove the comb, fill tank with SDS-PAGE running buffer, and load 18 μl of sample in each well. Run gel electrophoresis at 20 mA until the dye reaches the bottom of the gel.
4. Remove the stacking gel using spatula and equilibrate separating gel in blotting buffer for 5 min.
5. Cut PVDF membrane to the size of the gel, soak it in methanol for 1–2 min, and then soak the membrane and two pieces of filter papers in blotting buffer.
6. Place one piece of wet filter paper onto the platinum anode of semidry blotting system (BE-330, BIO CRAFT), and lay the gel on the filter paper.
7. Place the wet PVDF membrane on the gel. Make sure that there are no air bubbles between the filter and the gel.
8. Place another piece of wet filter paper onto the PVDF filter, and smooth out any air bubbles by rolling a pipette across its surface.
9. Place the cathode plate of semidry blotting system on the top of the stack.
10. Run the semidry transfer system at 5 V for 30 min followed by 10 V for 2.5 h.
11. Block the membrane in the 2 % blocking solution for 1 h at room temp or overnight at 4 °C.
12. Incubate the membrane with anti-ubiquitin monoclonal antibody for 1 h at room temp.
13. Wash the membrane with TBST three times, 10 min each.
14. Incubate the membrane with horseradish peroxidase-conjugated goat antibody to mouse IgG.
15. Wash the membrane as in **step 13**.
16. Incubate the membrane with ECL advance reagents for 1–5 min at room temp, and detect the chemiluminescence with an LAS-4000 imaging system (Fig. 3).

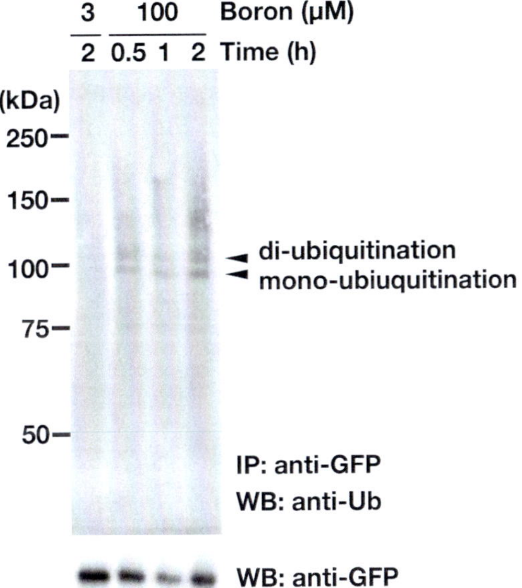

Fig. 3 Detection of boron-induced ubiquitination of BOR1-GFP. Samples were prepared from the roots of transgenic plants expressing BOR1-GFP, which were treated with 3 or 100 μM boron for the indicated times. The BOR1-GFPs were immunoprecipitated with anti-GFP antibody, and ubiquitination was assessed by immunoblotting with anti-ubiquitin antibody (anti-Ub) or anti-GFP antibody. Arrowhead indicates the presumed band sizes for the mono-ubiquitination and di-ubiquitination, respectively

4 Notes

1. We use gellan gum (Wako Chemical, Osaka, Japan) to make low-boron medium. Agar contains significant amount of boron.
2. The Laemmli's method is the most widely used system of SDS-PAGE [21]. For separation of the membrane proteins, we use a modified Laemmli's system [22]. In this system, the ratio of acrylamide:Bis is 29.76:0.24, and NaCl is added to give a concentration of 13 mM. This system provides sharper bands of the membrane proteins.
3. ECL Advance Western Blotting Detection kit (GE Healthcare) is now discontinued. But, similar detection kits obtained from other companies should be able to use. ECL Advance Blocking Agent is now available as ECL Prime Blocking Reagent (Cat. No. RPN418).
4. We use plastic bottles to make low-boron medium to avoid contamination from boron in glass bottle. Most of the glass bottles for scientific research are made from borosilicate glass.
5. When the boron level in the medium is properly controlled, *Arabidopsis* plants show growth retardation by boron deficiency.

Compare the growth of plants under low (0–1 μM) and a high (30–100 μM) boron conditions.

6. The vacuole-targeted GFP is rapidly degraded in the vacuole of plants in light conditions; in contrast, it is stably accumulated in dark conditions [23, 24]. Alternatively, concanamycin A, an inhibitor of vacuolar-type proton-ATPase, can be used to inhibit GFP degradation in the vacuole [8].
7. BFA treatment was performed in the presence of CHX to prevent accumulation of newly synthesized protein. Thus, the GFP signals in the BFA-induced compartments are not derived from newly synthesized proteins, but are derived from the plasma membrane via endocytosis.
8. We found that 1.5 ml Eppendorf tube 3810× (Eppendorf, Hamburg, Germany) can be used for ultracentrifugation at 100,000 × *g* for 15 min in our centrifugation system.
9. The agarose resin cannot pass through the 27 G needle.
10. We do not boil the SDS-PAGE samples of membrane proteins, because it is known that boiling causes aggregation for some membrane proteins [25].

Acknowledgement

This work was supported by a Grant-in-Aid for JSPS Fellows (no. 19-7094 to K.K.) from the Japan Society for the Promotion of Science, by a Grant-in-Aid for Young Scientists (Wakate B-237800060, to K.K.) from the Ministry of Education, Culture, Sports, Science and Technology, Japan, by a Grant-in-Aid for Scientific Research (to T.F.), and a Grant-in-Aid for Scientific Research Priority Areas (to T.F.) from the Ministry of Education, Culture, Sports, Science and Technology, Japan, by the NEXT program (to J.T.) from the Japanese Society for the Promotion of Science.

References

1. Warrington K (1923) The effect of boric acid and borax on the broad bean and certain other plants. Ann Bot 37:629–672
2. Noguchi K, Yasumori M, Imai T, Naito S, Matsunaga T, Oda H, Hayashi H, Chino M, Fujiwara T (1997) *bor1-1*, an *Arabidopsis thaliana* mutant that requires a high level of boron. Plant Physiol 115:901–906
3. Takano J, Noguchi K, Yasumori M, Kobayashi M, Gajdos Z, Miwa K, Hayashi H, Yoneyama T, Fujiwara T (2002) Arabidopsis boron transporter for xylem loading. Nature 420:337–340
4. Takano J, Tanaka M, Toyoda A, Miwa K, Kasai K, Fuji K, Onouchi H, Naito S, Fujiwara T (2010) Polar localization and degradation of Arabidopsis boron transporters through distinct trafficking pathways. Proc Natl Acad Sci U S A 107:5220–5225
5. Takano J, Miwa K, Fujiwara T (2008) Boron transport mechanisms: collaboration of channels and transporters. Trends Plant Sci 13: 451–457
6. Mellman I, Nelson WJ (2008) Coordinated protein sorting, targeting and distribution in

polarized cells. Nat Rev Mol Cell Biol 9: 833–845

7. Takano J, Wada M, Ludewig U, Schaaf G, von Wirén N, Fujiwara T (2006) The Arabidopsis major intrinsic protein NIP5;1 is essential for efficient boron uptake and plant development under boron limitation. Plant Cell 18: 1498–1509
8. Takano J, Miwa K, Yuan L, von Wirén N, Fujiwara T (2005) Endocytosis and degradation of BOR1, a boron transporter of Arabidopsis thaliana, regulated by boron availability. Proc Natl Acad Sci U S A 102: 12276–12281
9. Ueda T, Uemura T, Sato MH, Nakano A (2004) Functional differentiation of endosomes in Arabidopsis cells. Plant J 40: 783–789
10. Viotti C, Bubeck J, Stierhof YD, Krebs M, Langhans M, van den Berg W, van Dongen W, Richter S, Geldner N, Takano J, Jürgens G, de Vries SC, Robinson DG, Schumacher K (2010) Endocytic and secretory traffic in *Arabidopsis* merge in the trans-golgi network/early endosome, an independent and highly dynamic organelle. Plant Cell 22:1344–1357
11. Guerra DD, Callis J (2012) Ubiquitin on the move: the ubiquitin modification system plays diverse roles in the regulation of endoplasmic reticulum- and plasma membrane-localized proteins. Plant Physiol 160:56–64
12. Katzmann DJ, Babst M, Emr SD (2001) Ubiquitin-dependent sorting into the multivesicular body pathway requires the function of a conserved endosomal protein sorting complex, ESCRT-I. Cell 106:145–155
13. Mukhopadhyay D, Riezman H (2007) Proteasome-independent functions of ubiquitin in endocytosis and signaling. Science 315: 201–205
14. Spitzer C, Reyes FC, Buono R, Sliwinski MK, Haas TJ, Otegui MS (2009) The ESCRT-related CHMP1A and B proteins mediate multivesicular body sorting of auxin carriers in *Arabidopsis* and are required for plant development. Plant Cell 21:749–766
15. Abas L, Benjamins R, Malenica N, Paciorek T, Wisniewska J, Moulinier-Anzola JC, Sieberer T, Friml J, Luschnig C (2006) Intracellular trafficking and proteolysis of the *Arabidopsis* auxin-efflux facilitator PIN2 are involved in root gravitropism. Nat Cell Biol 8:249–256
16. Lee HK, Cho SK, Son O, Xu ZY, Hwang I, Kim WT (2009) Drought stress-induced Rma1H1, a RING membrane-anchor E3 ubiquitin ligase homolog, regulates aquaporin levels via ubiquitination in transgenic *Arabidopsis* plants. Plant Cell 21:622–641
17. Kasai K, Takano J, Miwa K, Toyoda A, Fujiwara T (2011) High boron-induced ubiquitination regulates vacuolar sorting of the BOR1 borate transporter in *Arabidopsis thaliana*. J Biol Chem 286:6175–6183
18. Barberon M, Zelazny E, Robert S, Conéjéro G, Curie C, Friml J, Vert G (2011) Monoubiquitin-dependent endocytosis of the iron-regulated transporter 1 (IRT1) transporter controls iron uptake in plants. Proc Natl Acad Sci U S A 108:E450–E458
19. Fujiwara T, Hirai MY, Chino M, Komeda Y, Naito S (1992) Effects of sulfur nutrition on expression of the soybean seed storage protein genes in transgenic petunia. Plant Physiol 99:263–268
20. Yoshinari A, Kasai K, Fujiwara T, Naito S, Takano J (2012) Polar localization and endocytic degradation of a boron transporter, BOR1, is dependent on specific tyrosine residues. Plant Signal Behav 7:46–49
21. Laemmli UK (1970) Cleavage of structural proteins during the assembly of the head of bacteriophage T4. Nature 227:680–685
22. Ito K, Bassford PJ Jr, Beckwith J (1981) Protein localization in *E. coli*: is there a common step in the secretion of periplasmic and outer-membrane proteins? Cell 24:707 717
23. Tamura K, Shimada T, Ono E, Tanaka Y, Nagatani A, Higashi SI, Watanabe M, Nishimura M, Hara-Nishimura I (2003) Why green fluorescent fusion proteins have not been observed in the vacuoles of higher plants. Plant J 35:545–555
24. Kleine-Vehn J, Leitner J, Zwiewka M, Sauer M, Abas L, Luschnig C, Friml J (2008) Differential degradation of PIN2 auxin efflux carrier by retromer-dependent vacuolar targeting. Proc Natl Acad Sci U S A 105:17812–17817
25. Rayson BM (1989) Rates of synthesis and degradation of Na+-K+-ATPase during chronic ouabain treatment. Am J Physiol 256:C75–C80

Chapter 17

Ubiquitination of Plant Immune Receptors

Jinggeng Zhou, Ping He, and Libo Shan

Abstract

Ubiquitin is a highly conserved regulatory protein consisting of 76 amino acids and ubiquitously expressed in all eukaryotic cells. The reversible ubiquitin conjugation to a wide variety of target proteins, a process known as ubiquitination or ubiquitylation, serves as one of the most important and prevalent posttranslational modifications to regulate the myriad actions of protein cellular functions, including protein degradation, vesicle trafficking, and subcellular localization. Protein ubiquitination is an ATP-dependent stepwise covalent attachment of one or more ubiquitin molecules to target proteins mediated by a hierarchical enzymatic cascade consisting of an E1 ubiquitin-activating enzyme, E2 ubiquitin-conjugating enzyme, and E3 ubiquitin ligase. The plant plasma membrane resident receptor-like kinase Flagellin Sensing 2 (FLS2) recognizes bacterial flagellin and initiates innate immune signaling to defend against pathogen attacks. We have recently shown that two plant U-box E3 ubiquitin ligases PUB12 and PUB13 directly ubiquitinate FLS2 and promote flagellin-induced FLS2 degradation, which in turn attenuates FLS2 signaling to prevent excessive or prolonged activation of immune responses. Here, we use FLS2 as an example to describe a protocol for detection of protein ubiquitination in plant cells in vivo and in test tubes in vitro. In addition, we elaborate the approach to identify different types of ubiquitin linkages by using various lysine mutants of ubiquitin. The various in vivo and in vitro ubiquitination assays will provide researchers with the tools to address how ubiquitination regulates diverse cellular functions of target proteins.

Key words Ubiquitin, Ubiquitination, Polyubiquitin chain, Immunoblot, Immunoprecipitation (IP)

1 Introduction

Ubiquitin (originally named as ubiquitous immunopoietic polypeptide) is a highly phylogenetically conserved, 76-amino acid, 8.6 kDa protein that exists in all eukaryotic cells [1–3]. One or more ubiquitin molecules could be covalently attached to a wide range of proteins via its C-terminus or one of its seven lysine (K) residues, which was known as ubiquitination or ubiquitylation [1, 4]. Protein ubiquitination is catalyzed in an ATP-dependent manner by a stepwise enzymatic reaction cascade consisting of an E1 ubiquitin-activating enzyme, E2 ubiquitin-conjugating enzyme, and E3 ubiquitin ligase [1, 2, 5, 6] (Fig. 1a). A single ubiquitin via its C-terminal glycine residue can be attached to a target protein,

Marisa S. Otegui (ed.), *Plant Endosomes: Methods and Protocols*, Methods in Molecular Biology, vol. 1209,
DOI 10.1007/978-1-4939-1420-3_17, © Springer Science+Business Media New York 2014

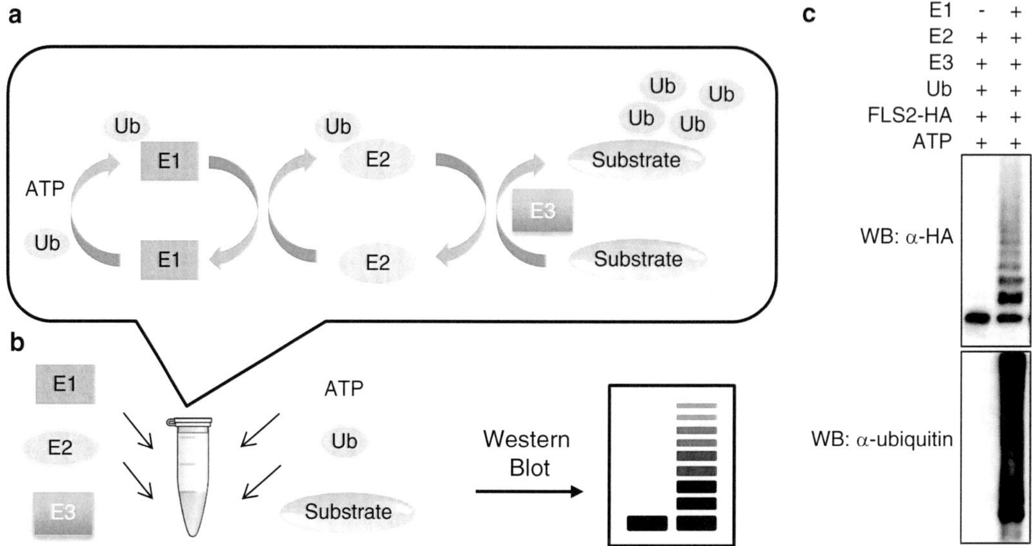

Fig. 1 In vitro ubiquitination assay. (**a**) Protein ubiquitination is catalyzed in an ATP-dependent manner by hierarchical stepwise enzymatic reactions consisting of an *E1* ubiquitin-activating enzyme, *E2* ubiquitin-conjugating enzyme, and *E3* ubiquitin ligase. (**b**) In an in vitro ubiquitination assay, the substrate proteins are mixed with *E1*, *E2*, *E3*, ubiquitin, and ATP in a 1.5 mL tube for a reaction and subjected for an immunoblot with an antibody specifically recognizing the substrates. The ubiquitinated proteins are typically observed as strong smears or ladders with the molecular weight increase at about 8.6 kDa per band. (**c**) Immunoblot of ubiquitinated proteins. Ubiquitinated FLS2-HA was detected by anti-HA immunoblot; total ubiquitinated proteins were shown in anti-ubiquitin immunoblot

which is known as monoubiquitination [4]. Monoubiquitination can occur at multiple lysine residues of a target protein, which is known as multi-monoubiquitination [4, 7]. Further successive addition of many self-linked ubiquitin molecules to form a linear or branched ubiquitin chain is termed as polyubiquitination [5]. Any of the seven lysine residues of ubiquitin could be used in the polyubiquitin chain synthesis, resulting in a highly complex chain [4]. Notably, ubiquitination is a reversible process in the cells as many enzymes could cleave ubiquitin from mono-, multi-, and polyubiquitinated proteins [2, 3, 8].

Different types of ubiquitination modifications likely lead to distinct fates of the targeted proteins. Mono- or multi-monoubiquitination of cell surface-resident receptors often serves as a signal for receptor endocytosis and vesicle trafficking, which lead to either subsequent degradation in lysosomes, or recycling to the cell surface [1, 9, 10]. For polyubiquitination, distinct polyubiquitin chains apparently play different roles in various signaling pathways [11]. Two well-studied examples are the K48- and K63-linked chains [2, 12]. The former type of polyubiquitin chain often directs the target proteins for proteasome-dependent proteolytic degradation [10], whereas the proteins tagged with K63-linked chain are often associated with DNA damage responses and signaling processes, such as activation of the transcription factor NF-κB [13]. In addition, protein ubiquitination has been shown to modulate the protein subcellular dynamics [9]. The unanchored polyubiquitination chains also appear to function as immune elicitors to activate innate immune signaling cascade [14]. Thus, protein ubiquitination constitutes one of the key posttranslational modification processes that modulate the myriad actions of protein cellular functions in diverse signaling pathways.

The plant plasma membrane resident receptor-like kinase (RLK) Flagellin Sensing 2 (FLS2) recognizes bacterial flagellin and functions as an immune receptor to initiate defense responses against pathogen attacks [15, 16]. Upon flagellin perception, FLS2 instantaneously heterodimerizes with another RLK BAK1, which likely results in the subsequent phosphorylation of FLS2 and BAK1 complex [16, 17]. The activated BAK1 phosphorylates a cytosolic kinase BIK1, which in turn leads to the release of BIK1 from FLS2/BAK1 complex to propagate intracellular immune responses [18]. Once immune receptor is activated, it is essential to down-regulate immune signaling to prevent excessive or prolonged activation of immune responses. It has been observed that flagellin induces FLS2 translocation into intracellular vesicles, followed by degradation in a proteasome-dependent pathway [19]. We have recently identified two plant U-box E3 ubiquitin ligases PUB12 and PUB13 that directly ubiquitinate FLS2 and promote flagellin-induced FLS2 degradation, which in turn attenuates FLS2 signaling (Fig. 2). BAK1 directly phosphorylates PUB12 and PUB13 and is required for flagellin-induced FLS2-PUB12/13 association. Interestingly, FLS2 kinase activity and PEST domain, which are required for internalization, are dispensable for PUB12/13-mediated FLS2 ubiquitination, suggesting different mechanisms or E3 ligases mediate FLS2 internalization and degradation [20]. Here, we use FLS2 as an example to describe a protocol to detect protein ubiquitination in vivo in plant cells and in vitro in test tubes.

To establish an in vivo plant ubiquitination assay, we amplified UBQ10 from Col-0 cDNA library with polymerase chain reaction (PCR) and cloned it into a plant expression vector under the control of a constitutive Cauliflower Mosaic Virus 35S promoter. The UBQ10 was tagged with 2× FLAG epitope at its N-terminus (FLAG-UBQ), whereas FLS2 was tagged with 2× HA epitope at its C-terminus in a plant expression vector (FLS2-HA). The epitope tag of ubiquitin should be placed upstream of N-terminus of the open reading frame, leaving the C-terminus intact for subsequent isopeptide linkage to substrates. The epitope tag of FLS2 was placed at its C-terminus to avoid the potential influence of protein posttranslational modifications signal at its N-terminus. The freshly isolated *Arabidopsis* mesophyll protoplasts were transfected with

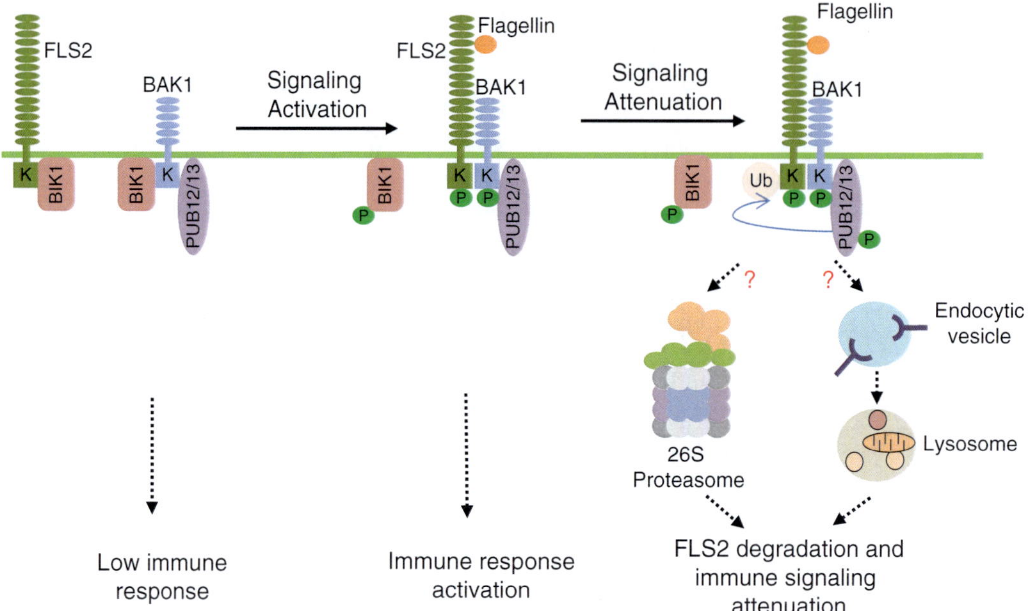

Fig. 2 Activation and attenuation of FLS2 signaling. Flagellin induces complex formation and transphosphorylation to activate FLS2 signaling and immune responses. The activated PUB12/13 ubiquitinate FLS2 leading to its degradation and attenuate immune signaling

FLAG-UBQ and FLS2-HA and treated with specific elicitors such as flg22 (a conserved 22-amino acid peptide of flagellin). The ubiquitinated FLS2-HA proteins could be detected by anti-HA immunoblotting after immunoprecipitation with anti-FLAG antibodies or anti-FLAG immunoblotting after immunoprecipitation with anti-HA antibodies (Fig. 3). The polyubiquitinated proteins are typically observed as strong smears or ladders with the molecular weight increase at about 8.6 kDa per band. In the case of the proteins with monoubiquitination modification, one extra band that is about 8.6 kDa bigger than the predicated size is often observed. Considering the abundance of endogenous ubiquitins which could also attach to FLS2 while are not able to be detected by anti-FLAG antibodies, the co-transfection of FLAG-UBQ with FLS2-HA greatly reduces the attachment of endogenous ubiquitins for the easy immunoprecipitation and immunoblotting with anti-HA or anti-FLAG antibodies. This assay does not require the knowledge of specific E3 ligases and is particularly sensitive to detect the dynamic changes of the target protein ubiquitination modification upon ligand treatment as it reveals all the potential ubiquitination modification events mediated by various endogenous E3 ligases to the target proteins. An in vitro ubiquitination assay could provide the direct evidence for the E3 ligase activity and the ubiquitination of target proteins. For in vitro assay that is optimized to detect Arabidopsis protein ubiquitination, we use Arabidopsis E1

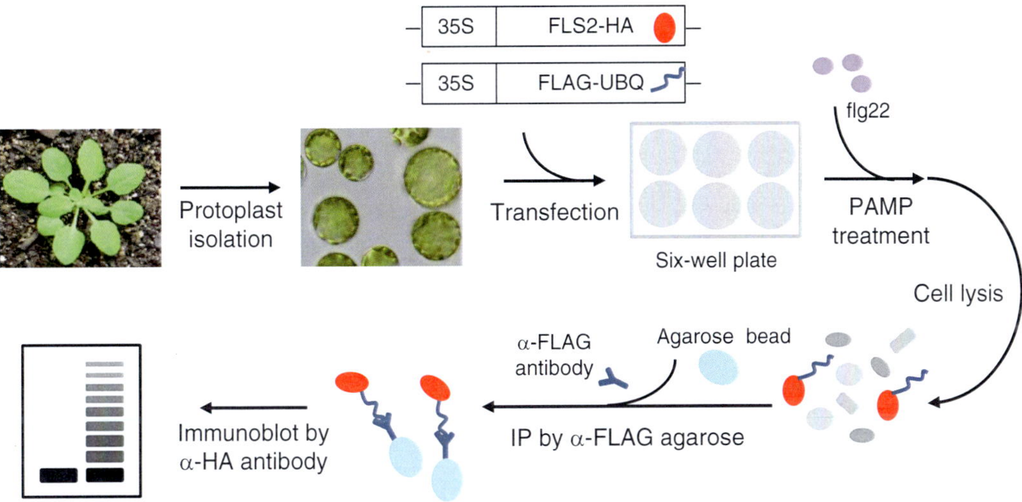

Fig. 3 In vivo ubiquitination of FLS2 assay. The *Arabidopsis* protoplasts co-transfected with FLAG-UBQ and FLS2-HA were treated with 1 μM flg22 for 30 min. The ubiquitinated FLS2-HA proteins could be detected by anti-HA immunoblot after immunoprecipitation (IP) with anti-FLAG antibodies

(AtUBA1, AT2g30110) and E2 (AtUBC8, AT5g41700) that were purified from *E. coli* as 6× His fusion protein. The AtUBC8 was chosen as an E2 because it has a broad activity towards different E3 ligases in in vitro ubiquitination assay [21, 22]. In the case of FLS2 ubiquitination assay, we use glutathione S-transferase (GST)-fused PUB13 (AtPUB13, At3g46510) as an E3 and maltose binding protein (MBP)-fused FLS2 cytosolic domain (FLS2CD) as a substrate (Fig. 1b, c). In addition, to explore the different types of ubiquitin linkage, we could perform the in vivo and in vitro ubiquitination assays with different versions of ubiquitin mutants that carry series of lysine (K) to Arginine (R) mutations. The polyubiquitin ladder would not be formed if the relevant ubiquitin lysine is mutated (Fig. 4).

2 Materials

2.1 In Vivo Ubiquitination Assay

1. FLAG-UBQ construct: Arabidopsis *UBQ10* gene tagged with 2× FLAG epitope at its N-terminus in a plant expression vector under the control of a constitutive Cauliflower Mosaic Virus 35S promoter.
2. Target gene constructs: PCR amplify the coding region of FLS2 or other target genes into a plant expression vector with 2× HA epitope at the C-terminus under the control of constitutive Cauliflower Mosaic Virus 35S promoter.
3. Protoplast isolation: refer the protocol reported by He et al. [23].

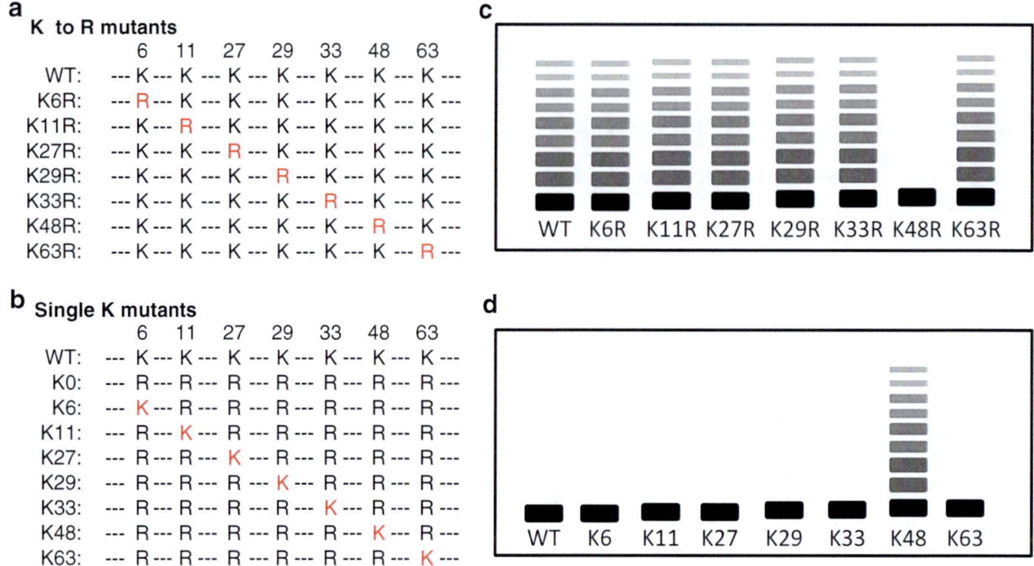

Fig. 4 Identification of the types of polyubiquitination chains. (**a**) K to R point mutants of ubiquitin. Only K residues of ubiquitin are shown and the numbers indicate the positions of Ks in ubiquitin. Each mutant carries a K to R mutation in an indicated position. (**b**) Single K mutants of ubiquitin. All the K residues are mutated into Rs except the indicated positions. (**c**) and (**d**) Identification of polyubiquitin chain structure in vitro; here depicts only one type of the polyubiquitin chains. In the case shown here, only K48 ubiquitin is involved in chain synthesis. Therefore, K48R ubiquitin cannot be used as a substrate (**c**), while among the single-lysine ubiquitins, only K48 ubiquitin is compatible for the polyubiquitin chain synthesis (**d**)

4. 2 mL round-bottom tubes.
5. Clinical centrifuge.
6. 40 % (w/v) polyethylene glycol (PEG) solution: 40 % (w/v) of PEG4000, 0.1 M $CaCl_2$, and 0.2 M mannitol.
7. W5 solution: 154 mM NaCl, 125 mM $CaCl_2$, 5 mM KCl, 2 mM MES pH 5.7.
8. MMg solution: 0.4 M mannitol, 15 mM $MgCl_2$, 4 mM MES pH 5.7.
9. WI solution: 0.5 M mannitol, 20 mM KCl, 4 mM MES pH 5.7.
10. Six-well tissue culture plates.
11. Flg22, the conserved 22-amino acid peptide of flagellin.

2.2 In Vitro Ubiquitination Assay

1. 5× UBQ buffer: 100 mM Tris–HCl pH 7.5, 25 mM $MgCl_2$, 2.5 mM DTT, 10 mM ATP.
2. Enzymes: His_6-AtUBA1 (E1, AT2g30110), His_6-AtUBC8 (E2, AT5g41700), GST-AtPUB13 (E3, AT3g46510), 1 μg/μL for each protein stored in 25 % glycerol (v/v) solution, at −20 °C (*see* **Note 1**).

3. MBP-FLS2CD-HA (cytosolic domain of FLS2 tagged with HA at its C terminus), 1 μg/μL stored in 25 % (v/v) glycerol solution, at −20 °C (*see* **Note 1**).
4. His_6-ubiquitin (BostonBiochem).
5. His_6-ubiquitin K to R mutants:

   His_6-ubiquitin K6R (BostonBiochem).

   His_6-ubiquitin K11R (BostonBiochem).

   His_6-ubiquitin K27R (BostonBiochem).

   His_6-ubiquitin K29R (BostonBiochem).

   His_6-ubiquitin K33R (BostonBiochem).

   His_6-ubiquitin K48R (BostonBiochem).

   His_6-ubiquitin K63R (BostonBiochem).

   His_6-ubiquitin K6 only (BostonBiochem).

   His_6-ubiquitin K11 only (BostonBiochem).

   His_6-ubiquitin K27 only (BostonBiochem).

   His_6-ubiquitin K29 only (BostonBiochem).

   His_6-ubiquitin K33 only (BostonBiochem).

   His_6-ubiquitin K48 only (BostonBiochem).

   His_6-ubiquitin K63 only (BostonBiochem).

   His_6-ubiquitin no lysines K0 (BostonBiochem).

2.3 Immunoprecipitation and Immunoblot

1. 4× SDS-PAGE sample buffer: 0.2 M Tris–HCl pH 6.8, 80 g/L SDS, 40 % (v/v) glycerol, 50 mM EDTA pH 8.0, 0.8 g/L Bromophenol Blue, and 4 % (v/v) β-Mercaptoethanol.
2. Sodium dodecyl sulfate-polyacrylamide gel (SDS-PAGE), containing 7.5 % (w/v) polyacrylamide-co-methylene-bis-acrylamide.
3. Antibody: anti-HA-peroxidase (HRP), anti-FLAG-M2 peroxidase (HRP) (Sigma), anti-ubiquitin, anti-Rabbit IgG-peroxidase (HRP), anti-FLAG M2 affinity gel (Sigma).
4. Immunoblot PVDF membrane for protein blotting.
5. Immunoprecipitation buffer (IP buffer): 150 mM NaCl, 50 mM Tris–HCl pH 7.5, 5 mM EDTA pH 8.0, 1 % (v/v) Triton X100, 1 mM DTT, 2 mM NaF, 2 mM Na_3VO_4, 0.5 % (v/v) protease inhibitor cocktail.
6. 50 mM Tris–HCl pH 7.5.
7. SDS-PAGE running buffer: 3 g/L Tris base and 14.4 g/L glycine and 1 g/L SDS.
8. Transfer buffer: 3 g/L Tris base, 14.4 g/L glycine, 20 % (v/v) methanol, and 0.1 g/L SDS, final pH 8.3.

9. Phosphate-buffered saline (PBS) buffer: 8 g/L NaCl, 0.2 g/L KCl, 2.14 g/L Na_2HPO_4, and 0.2 g/L KH_2PO_4.
10. PBST buffer: add 0.1 % (v/v) Tween-20 into PBS solution before use.
11. Blocking solution: 5 % (w/v) milk dissolved in PBST solution.
12. Supersignal west pico chemiluminescent substrate (Thermo).

3 Methods

3.1 In Vivo Ubiquitination Assay

1. Add 50 μL of the mixed FLAG-UBQ and FLS2-HA DNA with the concentration of 1.8 μg/μL into a 2 mL round-bottom tube (*see* **Note 2**).
2. Add 500 μL of protoplasts with the density of 2×10^5 cells/mL suspended in MMg solution (*see* **Note 3**).
3. After adding protoplasts, immediately add 550 μL of freshly prepared 40 % PEG, gently mix the DNA and protoplasts by tip-taping the tube.
4. Incubate at room temperature (23 °C) for 8–10 min.
5. Add 1.5 mL of W5 solution and mix well to stop the transfection.
6. Spin at 100 × *g* with a clinical centrifuge for 1 min and suck out the solution without disturbing the protoplast pellets.
7. Resuspend the protoplasts in 100 μL of WI solution gently.
8. Transfer the protoplasts into a six-well tissue culture plate with 1 mL of WI (*see* **Note 4**).
9. Incubate the protoplasts 8–12 h at the room temperature (23 °C).
10. Add flg22 to a final concentration of 1 μM and incubate 30 min at the room temperature.
11. Harvest the protoplasts by centrifugation at 100 × *g* for 1 min and suck out the supernatant.
12. Add 300 μL of IP buffer, vortex for 5–10 seconds, and chill on ice for 5 min.
13. Remove the cellular debris by centrifugation at 13,000 rpm for 5 min at 4 °C. Keep the supernatant for the following immunoprecipitation (IP).
14. Take out 15 μL of the IP samples, add 5 μL of 4× SDS sample loading buffer, heat at 95 °C for 5 min, and store as protein loading controls.
15. Add 20 μL of anti-FLAG agarose into the remaining IP samples (*see* **Note 5**).

16. Incubate at 4 °C for 2–4 h on a rocker.
17. Spin down the agarose beads at 100 × *g* for 1 min, and suck out the supernatant completely.
18. Wash the agarose beads with 500 μl ice-cold IP buffer (*see* **Note 6**).
19. Repeat **steps 17** and **18**.
20. Wash one more time with 500 μl ice-cold 50 mM Tris–HCl pH 7.5, and suck out the supernatant completely.
21. Add 5 μL of 4× SDS-PAGE sample buffer and heat at 95 °C for 10 min.
22. Store the samples at −20 °C until ready for immunoblotting.

3.2 In Vitro Ubiquitination Assay

1. Prepare each reaction in a 1.5 mL tube by mixing the following:

1 μg	E1
1 μg	E2
1 μg	E3
1 μg	Ubiquitin or various ubiquitin mutants
0.5 μg	Substrate protein tagged with HA epitope
6 μL	5× UBQ buffer (*see* **Note 7**)

Add water to a final volume of 30 μL.

No E1 is added in the reaction as a negative control (Fig. 4b) (*see* **Note 8**).

2. Spin down at 250 × *g*. Incubate the reaction tubes on a roller at 100 rpm for 2 h at 28 °C (*see* **Note 9**).
3. Add 10 μL of 4× SDS-PAGE sample buffer and heat the samples at 95 °C for 5 min to stop the reaction.
4. Freeze and store the samples at −20 °C until ready for immunoblotting.

3.3 Immunoblot Detection

1. Load 10 μL of the reaction samples from in vivo or in vitro ubiquitination assays onto a 7.5 % acrylamide SDS-PAGE gel. Include pre-stained protein molecular weight standards in one lane as reference to estimate the molecular weight of the target proteins.
2. Load the same samples to another SDS-PAGE as **step 1**. Both gels are treated exactly the same from **steps 3** to **6**.
3. Use a constant voltage of 80 V for 20–40 min until the protein dyes reach the resolving gel, and run another 1–2 h at 150 V until the target proteins approach the bottom of the gel (can be estimated according to the predicated molecular weight).

4. Fully wet the PVDF membranes by immersion in methanol for 30 s, and then in transfer buffer for 15–30 min to facilitate the binding of ubiquitinated proteins to the membrane.
5. Electro-transfer proteins from the SDS-PAGE to a PVDF membrane at the constant current of 200 mA for 2–3 h in transfer buffer (*see* **Note 10**).
6. Incubate the membrane in a small tray with blocking buffer at room temperature for 1 h on an orbital rocker. Alternatively, membrane can be blocked overnight in a cold room.
7. Incubate one membrane with anti-HA-HRP antibodies for 2 h or longer on an orbital rocker in a cold room (*see* **Note 11**).
8. Incubate the other membrane with anti-ubiquitin antibody. All the ubiquitinated proteins, including E1, E2, E3, and substrate proteins, could be potentially detected by anti-ubiquitin antibody. Strong smears are often seen (**step 12**), indicating a successful ubiquitination reaction (Fig. 1c). Since the ubiquitin in the in vivo ubiquitin assay is tagged with FLAG epitope, anti-FLAG-HRP antibody can also be used to assess the reaction efficiency (Fig. 3).
9. Wash the membrane with PBST buffer for three times, 10 min for each time.
10. If anti-ubiquitin antibody is used in **step 8**, incubate the membrane with anti-rabbit IgG-HRP secondary antibody with a dilution of 1:5,000 in blocking buffer for 30–60 min at room temperature and then repeat **step 9**. For the membrane that is incubated with a HRP-conjugated primary antibody, **step 10** should be skipped (*see* **Note 12**).
11. Immerse the membrane in a small volume of "supersignal west pico chemiluminescent substrate." After 1 min, immediately seal the membrane inside a small transparent plastic bag (*see* **Note 13**).
12. Stick the membrane to the inside of a cassette and expose the membrane to a film in the dark room.

4 Notes

1. N-terminal His_6-tagged *At*UBA1 (E1) and *At*UBC8 (E2) proteins were expressed in *Escherichia coli* and purified through standard nickel-agarose beads. N-terminal GST-tagged *At*PUB13 protein was expressed from *E.coli* and purified through standard glutathione beads. Salts remained in the proteins will interfere with the in vitro ubiquitylation reaction. It is recommended to desalt the enzymes by flowing through the 30 KD centrifugal filters (Millipore). Proteins could be adjusted for storage with a concentration of up to

5 μg/μL at −80 °C. Keep the proteins as small aliquots and avoid repetitive freeze-thaw cycles.

2. The quality of DNA is critical for protoplast transfection. Low-quality DNA could result in reduced protein expression level or even dead protoplasts. It is highly recommended to use CsCl gradients for plasmid DNA isolation. The ratio of FLAG-UBQ and the target DNAs can vary from 2:1 to 1:2, depending on the target protein expression level.
3. The amount of protoplasts used for transfection can be scaled up or down with the same DNA/protoplast ratio depending on the target protein expression and ubiquitination level and the purpose of individual experiments. A large scale of protoplasts, such as 10–20 mL with a concentration of 2×10^5cells/mL, can be used to express proteins for mass spectrometry analysis of protein ubiquitination sites and chain linkages.
4. To prevent sticking of protoplasts to the bottom of the plates, the plates can be pretreated with 5 % calf serum for 1 s before use.
5. Anti-FLAG agarose needs to be prewashed with IP buffer to remove storage solution which may interfere with protein binding.
6. At least 500 μL IP buffer is recommended for each wash, and any trace amount of supernatant should be completely removed with a fine tip. This will reduce the nonspecific protein contamination. DTT, NaF, Na_3VO_4, and protease inhibitor cocktail can be eliminated in the IP washing buffer.
7. Do not frequently freeze and thaw the 5× UBQ buffer as ATP will lose its activity.
8. In addition to wild-type ubiquitin protein, His_6-ubiquitin lysine mutants can be used to identify the type of polyubiquitin chains (Fig. 4a, b). For instance, if only K48 is used for polyubiquitin chain synthesis, ubiquitin K48R mutation will block polyubiquitination and the ladder collapses to a monoubiquitination band (Fig. 4c). At the same time, among the single lysine ubiquitin proteins, ubiquitin K48, but not the others, is competent for the polyubiquitin chain synthesis shown as a ladder in the corresponding lane (Fig. 4d).
9. The incubation time could be extended up to 12 h to enhance protein ubiquitination efficiency. However, the prolonged incubation of proteins at 28 °C may affect protein stability. Typically, incubation times no longer than 12 h are recommended.
10. Both "semidry" and "wet transfer" are acceptable, but the wet transfer tends to have relatively higher transfer efficiency, especially for high molecular weight proteins. So we recommend transferring ubiquitinated proteins with "wet transfer"

method described by Towbin [24]. Moreover, autoclaving the membranes in deionized water for 20 min will denature the ubiquitinated proteins, thereby increasing the reactivity to anti-ubiquitin antibodies, some of which only target denatured ubiquitins. Heat activation may not be applied to all anti-ubiquitin antibodies [25].

11. Antibody can be diluted in blocking buffer at a ratio of 1:2,000–1:5,000. For the in vitro ubiquitination assay, the incubation time of a primary antibody can be as short as 30 min.
12. Secondary antibody can be diluted in blocking buffer at a ratio of 1:5,000–1:10,000. High concentration and too long incubation time may enhance the background.
13. "Supersignal west femto chemiluminescent substrate" (Thermo) can be used to enhance immunoblot signal. We recommend mixing "Supersignal west femto chemiluminescent substrate" with "Supersignal west pico chemiluminescent substrate" (Thermo) at a ratio of no higher than 1:10 to avoid too strong bleaching signal.

Acknowledgements

We thank Dr. Dongping Lu for the initial work to establish various ubiquitination assays. The work was supported by the funds from NIH R01GM092893 to P.H and R01GM097247 to L.S.

References

1. Bonifacino JS, Weissman AM (1998) Ubiquitin and the control of protein fate in the secretory and endocytic pathways. Annu Rev Cell Dev Biol 14:19–57
2. Smalle J, Vierstra RD (2004) The ubiquitin 26S proteasome proteolytic pathway. Annu Rev Plant Biol 55:555–590
3. Vierstra RD (2009) The ubiquitin-26S proteasome system at the nexus of plant biology. Nat Rev Mol Cell Biol 10(6):385–397
4. Peng J, Schwartz D, Elias JE et al (2003) A proteomics approach to understanding protein ubiquitination. Nat Biotechnol 21(8): 921–926
5. Finley D (2009) Recognition and processing of ubiquitin-protein conjugates by the proteasome. Annu Rev Biochem 78:477–513
6. Tian G, Finley D (2012) Cell biology: destruction deconstructed. Nature 482(7384): 170–171
7. Hicke L (2001) Protein regulation by monoubiquitin. Nat Rev Mol Cell Biol 2(3):195–201
8. Lee MJ, Lee BH, Hanna J et al (2010) Trimming of ubiquitin chains by proteasome-associated deubiquitinating enzymes. Mol Cell Proteomics 10(5), R110003871
9. Shih SC, Sloper-Mould KE, Hicke L (2000) Monoubiquitin carries a novel internalization signal that is appended to activated receptors. EMBO J 19(2):187–198
10. Thrower JS, Hoffman L, Rechsteiner M et al (2000) Recognition of the polyubiquitin proteolytic signal. EMBO J 19(1):94–102
11. Volk S, Wang M, Pickart CM (2005) Chemical and genetic strategies for manipulating polyubiquitin chain structure. Methods Enzymol 399:3–20
12. Jacobson AD, Zhang NY, Xu P et al (2009) The lysine 48 and lysine 63 ubiquitin conjugates are processed differently by the 26 s proteasome. J Biol Chem 284(51):35485–35494
13. Sun L, Chen ZJ (2004) The novel functions of ubiquitination in signaling. Curr Opin Cell Biol 16(2):119–126

14. Xia ZP, Sun L, Chen X et al (2009) Direct activation of protein kinases by unanchored polyubiquitin chains. Nature 461(7260):114–119
15. Gomez-Gomez L, Boller T (2000) FLS2: an LRR receptor-like kinase involved in the perception of the bacterial elicitor flagellin in Arabidopsis. Mol Cell 5(6):1003–1011
16. Chinchilla D, Zipfel C, Robatzek S et al (2007) A flagellin-induced complex of the receptor FLS2 and BAK1 initiates plant defence. Nature 448(7152):497–500
17. Heese A, Hann DR, Gimenez-Ibanez S et al (2007) The receptor-like kinase SERK3/BAK1 is a central regulator of innate immunity in plants. Proc Natl Acad Sci U S A 104(29): 12217–12222
18. Lu D, Wu S, Gao X et al (2010) A receptor-like cytoplasmic kinase, BIK1, associates with a flagellin receptor complex to initiate plant innate immunity. Proc Natl Acad Sci U S A 107(1): 496–501
19. Robatzek S, Chinchilla D, Boller T (2006) Ligand-induced endocytosis of the pattern recognition receptor FLS2 in Arabidopsis. Genes Dev 20(5):537–542
20. Lu D, Lin W, Gao X et al (2011) Direct ubiquitination of pattern recognition receptor FLS2 attenuates plant innate immunity. Science 332(6036):1439–1442
21. Kraft E, Stone SL, Ma L et al (2005) Genome analysis and functional characterization of the E2 and RING-type E3 ligase ubiquitination enzymes of Arabidopsis. Plant Physiol 139(4): 1597–1611
22. Stone SL, Hauksdottir H, Troy A et al (2005) Functional analysis of the RING-type ubiquitin ligase family of Arabidopsis. Plant Physiol 137(1):13–30
23. He P, Shan L, Sheen J (2007) The use of protoplasts to study innate immune responses. Methods Mol Biol 354:1–9
24. Towbin H, Staehelin T, Gordon J (1979) Electrophoretic transfer of proteins from polyacrylamide gels to nitrocellulose sheets: procedure and some applications. Proc Natl Acad Sci U S A 76(9):4350–4354
25. Swerdlow PS, Finley D, Varshavsky A (1986) Enhancement of immunoblot sensitivity by heating of hydrated filters. Anal Biochem 156(1):147–153

Chapter 18

Ubiquitylation-Mediated Control of Polar Auxin Transport: Analysis of *Arabidopsis* PIN2 Auxin Transport Protein

Johannes Leitner and Christian Luschnig

Abstract

Reversible, covalent modification by the small protein ubiquitin acts in a variety of pathways controlling protein fate in virtually all aspects of cellular function. For example, ubiquitylation of plasma membrane proteins modulates their intracellular sorting and turnover, thereby decisively influencing crosstalk between cells and their environment. In recent years, experimental work performed with the model plant *Arabidopsis thaliana* demonstrated ubiquitylation of a number of plasma membrane proteins, including the auxin efflux carrier protein PIN2. By using solubilized membrane protein immunoprecipitation assays, we established quantitative approaches, suitable for analysis of PIN2 ubiquitylation and variations therein. Applicability of this robust approach is not restricted to PIN auxin carriers, but could be extended to analysis of further plant membrane proteins that are controlled by variations in their ubiquitylation status.

Key words Membrane protein, Ubiquitylation, Polar auxin transport, Immunoprecipitation, PIN2, *Arabidopsis*

1 Introduction

In recent years, several reports described the identification of plant plasma membrane proteins that are subject to reversible ubiquitylation [1–6]. Some of these plant membrane proteins appear to undergo dynamic adjustments in their ubiquitylation, which appears to be involved in control of their intracellular sorting and proteolytic degradation [3, 4, 6, 7]. These findings established a role for membrane protein ubiquitylation in plants, analogous to the situation described for animals and fungi, acting in the regulation of nutrient and hormonal transport as well as in responses to external, biotic or abiotic signals [8].

Determination of protein ubiquitylation typically involves selective immunoprecipitation of the protein of interest, followed by SDS-PAGE, Western blotting, and probing with ubiquitin-specific antibodies. Analysis of membrane protein ubiquitylation requires selective enrichment of solubilized membrane protein

Marisa S. Otegui (ed.), *Plant Endosomes: Methods and Protocols*, Methods in Molecular Biology, vol. 1209, DOI 10.1007/978-1-4939-1420-3_18,

fractions, often times involving laborious ultracentrifugation steps, requiring large amounts of material, and causing loss of considerable amounts of membrane fractions during preparation [9]. This later point could be particularly problematic, when it comes to determination of variations in posttranslational protein modifications like ubiquitylation, as such modifications might strongly depend on the intracellular localization of the protein of interest. As a result, the output of such experiments could be biased, owing to disproportionate enrichment of membrane protein fractions from different cellular membranes.

We made use of a modified membrane protein extraction protocol, optimized for plant extracts and apparently less susceptible to loss of microsomal membrane protein pools [9]. Solubilized membrane proteins were then subject to immunoprecipitation with affinity-purified PIN2 antibody, followed by detection of PIN2 ubiquitylation using different, commercially available anti-ubiquitin antibodies [10, 11]. This experimental setup allowed us to determine ubiquitylation of cellular PIN2 in quantitative terms, and demonstrated variations in PIN2 total ubiquitylation levels in response to intrinsic or environmental stimuli [6, 7].

2 Materials

Prepare all solutions using ultrapure water (ddH_2O).

2.1 Seed Sterilization Solution and Plant Growth Medium PN(+/–)S (Plant Nutrition +/– Sucrose)

1. Seed sterilization solution: 6 % sodium hypochlorite, 0.1 % (v/v) Triton X-100, use immediately or store at 4 °C for up to 4 days.
   Prepare stock solutions (2–8) in advance, before preparing PN (Plant nutrition medium; [12].
2. 50 % (w/v) sucrose in ddH_2O, autoclave and store at room temperature (RT).
3. 0.1 % (w/v) agarose in ddH_2O, autoclave and store at RT.
4. 500 mM KNO_3 in ddH_2O, autoclave and store at RT.
5. 200 mM $MgSO_4$ in ddH_2O, autoclave and store at RT.
6. 200 mM $Ca(NO_3)_2$ in ddH_2O, autoclave and store at RT.
7. 250 mM KPO_4 buffer (pH 5.5): 235 mM KH_2PO_4 and 15 mM K_2HPO_4 in ddH_2O, autoclave and store at RT.
8. 1,000× micronutrients mix: 70 μM H_3BO_3, 14 μM $MnCl_2$, 0.5 μM $CuSO_4$, 1 μM $ZnSO_4$, 0.2 μM Na_2MoO_4, 10 μM NaCl, 0.01 μM $CoCl_2$ in ddH_2O, autoclave and store at RT.
9. PN(–S) plant growth medium: 50 μM Na_2EDTA, 50 μM $FeSO_4$, 5 mM KNO_3, 2 mM $MgSO_4$, 2 mM $Ca(NO_3)_2$, 2.5 mM KPO_4 buffer, 1× micronutrients, bring pH to 5.5 with 1 M KOH, add 1 % (w/v) agar, autoclave, and store at RT.

Optional: before pouring plates, add 1 % (v/v) sucrose to obtain PN(+S). For liquid PN(−S), leave out agar before autoclaving and do not add sucrose before use.

2.2 Extraction Buffer for Total Membrane Extraction

Cool microcentrifuge to 4 °C prior to protein extraction.

Preparation of 1× Extraction buffer:

Prepare a stock solution of 1.75× Extraction buffer and freeze 1 ml aliquots at −20 °C. Prior to extraction, dilute 1.75× Extraction buffer by adding sodium molybdate, phenylarsine oxide (PAO), water, and *N*-Ethylmaleimide (NEM). After adding DTE, okadaic acid, benzamidine, Pefabloc-SC, Aprotinin/Leupeptin, E64, Pepstatin, PMSF, borate acid, and ascorbic acid the 1× Extraction buffer is ready to use [6, 9].

Prepare solutions 4–25 in advance.

1. 1.75× Extraction buffer: 50 mM Tris–HCl (pH 7.3), 20 mM Tris–HCl (pH 8.8), 3.54 mM D-Sorbitol, 0.35 % (w/v) casein, 10 mM EDTA, 10 mM EGTA, 40 mM β-GP, 20 mM KPO_4 buffer (pH 7.4), and 50 mM NaF; vortex, aliquot, and freeze at −20 °C. Add 2 mM sodium molybdate and 2 mM PAO before preparing 1× Extraction buffer.
2. 1× Extraction buffer: Dilute 1.75× Extraction buffer with ddH_2O and add 20 mM NEM. After vortexing, add 10 mM boric acid, 2.5 mM benzamidine, 1 mM DTE, 2 mM Pefabloc-SC, 2 nM okadaic acid, 2× Aprotinin/Leupeptin, 2× E64, 2× Pepstatin, 2× PMSF, and 10 mM ascorbic acid and vortex. Now, the buffer is ready to use.
3. 200 mM NEM in ddH_2O (*see* **Note 1**).
4. 10 % (w/v) Polyvinylpolypyrrolidone (PVPP) in ddH_2O, keep at 4 °C (*see* **Note 2**).
5. 1 M Tris–HCl (pH 7.3) in ddH_2O, autoclave and store at RT.
6. 1 M Tris–HCl (pH 8.8) in ddH_2O, autoclave and store at RT.
7. 4.88 M D-sorbitol in ddH_2O, keep at 4 °C.
8. 7 % casein in ddH_2O, store at −20 °C.
9. 0.5 M EDTA (pH 8.0) in ddH_2O, autoclave and store at RT.
10. 0.5 M EGTA (pH 8.0) in ddH_2O, autoclave and store at RT.
11. 1 M β-Glycerophosphate(Disodium salt) (β-GP) in ddH_2O, store at 4 °C.
12. 1 M KPO_4 buffer (pH 7.4): mix 80.2 parts of 1 M K_2HPO_4 with 19.8 parts of 1 M KH_2PO_4, store at RT.
13. 2.5 M NaF in ddH_2O, store at 4 °C (*see* **Note 3**).
14. 0.5 M Na_2MoO_4 (sodium molybdate) in ddH_2O, store at 4 °C.
15. 0.5 M PAO in DMSO, store at 4 °C.
16. 2.7 M boric acid (pH 8.0) in ddH_2O, store at 4 °C.

17. 100 mM benzamidine in 50 % glycerol, store at −20 °C.
18. 0.5 M DTE in ddH_2O, store at −20 °C.
19. 200 mM Pefabloc-SC in ddH_2O, store at −20 °C.
20. 2 μM okadaic acid in DMSO, store at 4 °C.
21. 1,000× Aprotinin/Leupeptin: 10 mg Aprotinin and 10 mg Leupeptin in 10 ml ddH_2O. Vacuum-dry aliquots and store at −20 °C. Dissolve dry pellet in ddH_2O. Store at −20 °C.
22. 1,000× E64: 10 mg E64 in 4 ml MeOH. Vacuum-dry aliquots and store at 4 °C. Dissolve dry pellet in DMSO. Store at −20 °C.
23. 1,000× Pepstatin: 10 mg Pepstatin in 5 ml MeOH. Vacuum-dry aliquots and store at 4 °C. Dissolve dry pellet in DMSO. Store at −20 °C.
24. 1,000× PMSF: 2 g PMSF in 10 ml MeOH. Vacuum dry aliquots and store at 4 °C. Dissolve dry pellet in DMSO. Store at −20 °C.
25. 1 M ascorbic acid in ddH_2O, store at −20 °C.

2.3 Immunoprecipitation Buffer (RIPA Buffer) and Immunoprecipitation Wash Buffer (RIPA Wash Buffer)

1. RIPA buffer: 50 mM Tris–HCl (pH 7.4), 150 mM NaCl, 1 % (v/v) NP-40, 1 % (w/v) sodium deoxycholate, 0.1 % (w/v) SDS, 20 mM NEM, 1× Roche complete Mini Protease Inhibitor Cocktail, 1× PMSF in ddH_2O.
2. RIPA wash buffer: 50 mM Tris–HCl (pH 7.4), 150 mM NaCl, 1 % (v/v) NP-40, 1 % (w/v) sodium deoxycholate, 0.1 % (w/v) sodium dodecyl sulfate (SDS), 20 mM NEM in ddH_2O. Protease inhibitors are not added during washing steps.
3. 1 M Tris–HCl (pH 7.4) in ddH_2O, autoclave and store at RT.
4. 1 M NaCl in ddH_2O, autoclave and store at RT.
5. 10 % (w/v) sodium deoxycholate in ddH_2O, store at −20 °C.
6. 10 % (w/v) SDS in ddH_2O, store at RT.
7. 200 mM NEM in ddH_2O (*see* **Note 1**).
8. 10× Roche complete Mini Protease inhibitor tablets (EDTA-free) in ddH_2O, store at −20 °C.
9. 1,000× PMSF: 2 g PMSF in 10 ml MeOH. Vacuum dry aliquots and store at 4 °C. Dissolve dry pellet in DMSO. Store at −20 °C.
10. Protein A Sepharose: Protein A Sepharose 6 MB (GE Healthcare).
11. 20× Tris Buffered Saline (TBS) (pH 7.6): 400 mM Tris, 2.7 M NaCl in ddH_2O.
12. TBS containing 5 % (w/v) bovine serum albumin (BSA).

2.4 1× Sample Buffer for Protein Elution

1× sample buffer: 0.5 % (w/v) CHAPS, 3 % (w/v) SDS, 60 mM DTE, 125 mM Tris–HCl (pH 6.8), 30 % (v/v) glycerol, 0.5× complete Mini Protease Inhibitor Cocktail, and 1× PMSF. Add bromophenol blue with a toothpick, store at −20 °C.

1. 10 % (w/v) SDS in ddH_2O, store at RT.
2. 1 M Tris–HCl (pH 6.8) in ddH_2O, autoclave and store at RT.
3. 1000× PMSF: 2 g PMSF in 10 ml MeOH. Vacuum dry aliquots and store at 4 °C. Dissolve dry pellet in DMSO. Store at −20 °C.
4. 10× Roche complete Mini Protease inhibitor tablets (EDTA-free) in ddH_2O, store at −20 °C.
5. 5 % (w/v) CHAPS in ddH_2O, store at −20 °C.
6. 0.5 M DTE in ddH_2O, store at −20 °C.

2.5 SDS-UREA-PAGE and Immunoblotting

1. Bio-Rad Mini-PROTEAN Tetra Cell for SDS-UREA-PAGE.
2. Bio-Rad Mini-PROTEAN 10-well combs.
3. Bio-Rad Mini Trans-Blot Module for Western Blotting.
4. Stainless steel frame (*see* Subheading 3.12).
5. Thermo Scientific PageRuler Prestained Protein Ladder.
6. Resolving gel mixture: 463 mM Tris–HCl (pH 8.8), 0.1 % (w/v) SDS, Acrylamide/Bisacrylamide (29:1) solution (6 % (v/v) for ubiquitin detection, 10 % (v/v) for PIN2 detection), 0.04 % (w/v) APS, and 0.1 % (v/v) TEMED in ddH_2O.
7. Stacking gel mixture: 3 M Urea, 157 mM Tris–HCl (pH 6.8), 0.19 % (w/v) SDS, 5 % (v/v) Acrylamide/Bisacrylamide (29:1) solution, 0.05 % (w/v) APS, 0.125 % (v/v) TEMED in ddH_2O.
8. 1 M Tris–HCl (pH 8.8) in ddH_2O, autoclave and store at RT.
9. 1 M Tris–HCl (pH 6.8) in ddH_2O, autoclave and store at RT.
10. 10 % (w/v) SDS in ddH_2O, store at RT.
11. Ready-to-use 30 % Acrylamide/Bisacrylamide (29:1) solution.
12. 10 % APS in ddH_2O, store at −20 °C.
13. TEMED.
14. Nitrocellulose Membrane (Amersham Hybond ECL, GE Healthcare).
15. HRP Detection (Pierce ECL Western Blotting Substrate, Thermo Scientific).
16. X-Ray film (Mediphot X90/RP; Colenta).
17. 8 M Urea in ddH_2O: Wash 3 ml Amberlite mixed-bed resin in 200 ml ddH_2O for 30 min at RT on a shaker. Remove water, add 200 ml ddH_2O and wash for another 30 min. Remove water. Add 50 ml 8 M urea solution to the beads and shake for 30 min at RT. Aliquot solution and store at −20 °C.

18. SDS-PAGE Running buffer: 25 mM Tris, 192 mM glycine, 0.12 % (w/v) SDS in ddH_2O.
19. Western Blot Transfer buffer: 25 mM Tris, 192 mM glycine, 10 % (v/v) MeOH, 0.0375 % (w/v) SDS in ddH_2O. Store at 4 °C.
20. 20× TBS (pH 7.6): 400 mM Tris, 2.7 M NaCl in ddH_2O.
21. TBS containing 0.1 % Tween 20 (TBST).
22. Blocking solutions:
    For mouse anti-PIN2 antibody: 3 % (w/v) bovine serum albumin (BSA) in TBST.
    For mouse anti-ubiquitin antibodies: 5 % (w/v) skim milk in TBST.

2.6 Antibodies

1. Rabbit polyclonal anti-PIN2, affinity purified [2]. Used for immunoprecipitation of PIN2. Approximately 1 μg IgG for a single immunoprecipitation experiment.
2. Mouse polyclonal anti-PIN2, crude serum [2]. Used for detection of immunoprecipitated PIN2. 1:1,000 in 3 % (w/v) BSA/TBST.
3. Mouse monoclonal anti-ubiquitin (clone P4D1) from Santa Cruz. Used for detection of PIN2 ubiquitylation. 1:50 in 5 % (w/v) skim milk/TBST.
4. Mouse monoclonal anti-K63-linked ubiquitin (clone HWA4C4) from E-Bioscience. Used for detection of K63-linked poly-ubiquitin chains. 1:50 in 5 % (w/v) skim milk/TBST.
5. Mouse monoclonal anti–α-tubulin (clone B-5-1-2) from Sigma. Used for normalization of protein extracts before immunoprecipitation. 1:50,000 in 5 % (w/v) skim milk/TBST.
6. Secondary goat anti-mouse IgG (HRP-coupled) antibody from Dianova. Dilute 1:5,000 for detection of mouse anti-ubiquitin antibodies (in TBST), and 1:20,000 for detection of mouse anti-PIN2 and α-tubulin antibody (in 0.5 % (w/v) BSA/TBST).

3 Methods

3.1 Seed Batch and Propagation

Make sure to use a batch of seeds that germinates well and exhibits a homogenous growth pattern (i.e., the roots should be equally long).

For a single immunoprecipitation estimate roughly 12,000 seeds (results in approximately 200 mg of root tissue; *see* **Note 4**).

3.2 Seed Surface Sterilization

Examples are given for the purpose of a single immunoprecipitation experiment:

Seed sterilization and plating is done in a laminar flow hood.

1. Transfer 500 μl seeds into a 50 ml Falcon tube and sterilize in 20 ml of sterilization solution by vigorous vortexing for 10 min (*see* **Note 5**).
2. Allow the seeds to settle and carefully pour off the sterilization solution.
3. Wash seeds three times with 20 ml of ddH_2O for 5 min by vortexing (*see* **Note 6**).
4. Transfer surface-sterilized seeds for 72 h at 4 °C into darkness to promote simultaneous germination.

3.3 Seed Resuspension in 0.1 % Agarose

1. After 72 h, pour off water from the last washing step. Remove remaining water with a pipette (*see* **Note 7**).
2. Add 13 ml 0.1 % agarose and vortex vigorously to ensure equal distribution of the seeds.

Depending on whether seedlings are cultivated on solid or liquid medium, follow Subheadings 3.4 and 3.5 (for solid medium) or Subheadings 3.6 and 3.7 (for growth in liquid medium).

3.4 Plating Seeds on Solid PN(+S) Plates

1. Plate the surface-sterilized seeds (approximately 12,000) on four square plates (12 cm × 12 cm) with six rows per plate (Fig. 1, *see* **Note 8**).
2. Allow agarose to dry with the lids of the plates open (*see* **Note 9**).
3. Seal plates and arrange them vertically in a growth chamber (*see* **Note 10**).

3.5 Harvesting the Roots from Solid PN(+S) Plates

After usually 3–5 days after germination, depending on the growth conditions, harvest roots (*see* **Note 11**).

1. First cut roots horizontally at the root–hypocotyl junction with a razor blade.
2. Collect roots using Blotting Membrane tweezers (*see* **Note 12**).
3. Remove excess liquid from root tissue between two paper towels (*see* **Note 13**).
4. Use regular tweezers and transfer roots into a 2 ml Safe-Lock tube containing two metal beads with 4 mm diameter and immediately freeze in liquid nitrogen.

3.6 Seedling Growth in Liquid PN(−S) on Nylon Mesh in 6-Well Plates

When treating plantlets with growth regulators, etc. prior to root harvesting, seeds are germinated and grown under semi-sterile conditions on nylon mesh in 6-well plates filled with liquid PN(−S) (*see* **Note 14**).

Fig. 1 Example stencil for dense plating of *Arabidopsis thaliana* seeds on solid agar plates (12 cm × 12 cm) for root harvesting. Plating six rows per plate ensures high density of plants without adversely affecting plant growth. Root harvesting is eased because roots usually do not penetrate the agar

1. Fill wells with liquid PN(–S) and place nylon mesh on top (*see* **Note 15**).
2. Distribute 1 ml of seeds resuspended in 0.1 % agarose per well.
3. Prepare a styrofoam box or similar, filled with moistened paper towels and place the 6-well plates into the box. Cover the box with clear acrylic glass and transfer it into the growth chamber (*see* **Note 16**).
4. For treatment of roots, prepare a fresh 6-well plate filled with medium containing compounds of interest at required concentration. Transfer mesh with germinated seedlings and dip roots into the freshly prepared medium (*see* **Note 17**).

3.7 Harvesting Roots Grown on Nylon Mesh in Liquid PN(–S)

1. Lift the nylon mesh, turn it around, and place the side facing cotyledons on paper towels.
2. Cut the roots with a razor blade, collect tissue with tweezers, dry on filter paper, and freeze in liquid nitrogen immediately (*see* Subheading 3.5).

3.8 Grinding of Frozen Root Material

Before grinding root material, equilibrate PVPP solution, prepare 1× Extraction buffer, and cool a microcentrifuge to 4 °C (*see* Subheading 3.9).

1. Homogenize root material in a mixer mill (e.g., Retsch mixer mill MM2000 for 2 min, with the amplitude set at 60).
2. After milling, immediately freeze tubes in liquid nitrogen again (*see* **Note 18**).

3.9 Total Membrane Extraction from Arabidopsis thaliana Roots

The procedure is based on the method originally described in Abas and Luschnig (2010). Briefly, homogenized root material is resuspended in high-density 1× Extraction buffer and precleared with PVPP. Total membrane fractions are pelleted in a final centrifugation step at 16,000×*g*. All steps are carried out on ice and in a cooled (4 °C) microcentrifuge.

1. Prior to extraction of root material (see above), equilibrate a 10 % (w/v) PVPP suspension with 1× Extraction buffer. Per mg of root material, transfer 0.5 μl of 10 % (w/v) PVPP suspension into a fresh 1.5 ml tube by using a cut yellow tip attached to a pipette and add 500 μl 1× Extraction buffer (*see* **Note 19**).
2. Vortex briefly and centrifuge tube for a few seconds. Leave on ice for 10 min; decant and discard the 1× Extraction buffer afterwards. Keep the tube on ice.
3. Add 1× Extraction buffer (1.7 μl per mg of root material) to the homogenized plant material (from Subheading 3.8). The frozen root material causes the 1× Extraction buffer to freeze. Vortex until the buffer is completely liquefied; place tube on ice (*see* **Note 20**).
4. Remove metal beads from the tube with a stir bar retriever and transfer the homogenate to the tube with equilibrated PVPP. Optionally, vortex tube and put on ice for 30 s each to ensure efficient extraction.
5. After 5 min on ice, centrifuge the sample at 4 °C (2 min, 500×*g*).
6. Transfer the supernatant, containing membrane fraction into a fresh 1.5 ml tube.
7. Re-extract the PVPP/root material homogenate with the same amount of 1× Extraction buffer for 1 min.
8. Centrifuge at 4 °C (2 min, 500×*g*).
9. Combine supernatants and vortex.
10. Clear the supernatant from minor traces of PVPP by centrifugation at 4 °C (2 min, 500×*g*) and transfer supernatant into a fresh 1.5 ml tube.
11. Finally, transfer 200 μl aliquots of the supernatant into fresh 1.5 ml tubes and centrifuge at 4 °C for 2 h at 16,000×*g* to pellet membranes (*see* **Note 21**).
12. After centrifugation, discard the supernatant and freeze the pellets at −80 °C. Alternatively, pellets can be used immediately.

3.10 Immunoprecipitation of PIN2

Carry out all steps on ice or at 4 °C to avoid protein degradation.

1. Resuspend pelleted membrane fractions (*see* Subheading 3.9) in altogether 1 ml freshly prepared RIPA buffer.
2. Efficient resuspension of the pellet in RIPA buffer is crucial for efficient PIN2 immunoprecipitation. Therefore, first resuspend pellets with a blue tip attached to a pipette followed by resuspension with a capillary tip (the ones used for loading SDS-PAGE gels) attached to a pipette.
3. Incubate the sample at 4 °C on a Vibrax shaker (IKA) at 1,600 rpm for 1 h.
4. Pellet insoluble material by centrifugation for 50 s at 16,000 × *g* at 4 °C.
5. Transfer the supernatant into a fresh tube and add ~1 μg IgG (25–50 μl, depending on concentration) of affinity purified rabbit anti-PIN2 antibody.
6. Incubate tubes overnight at 4 °C on a rotator (Stuart) at 10 rpm.
7. Equilibrate 60 μl of Protein A Sepharose in 1 ml of TBS at 4 °C on a rotator at 10 rpm for 1 h, followed by three successive washing steps in 1 ml of TBS (*see* **Note 22**). For efficient washing, add TBS to the beads, invert several times, and keep on ice until the beads settle at the bottom of the tube.
8. Block the beads overnight in 1 ml of TBS supplemented with 5 % BSA (w/v).
9. The following day, wash blocked Protein A Sepharose beads three times in 1 ml of TBS.
10. Add 100 μl of a 50 % Protein A Sepharose slurry (i.e., 50 μl bed volume and 50 μl TBS) to tubes containing PIN2-IgG conjugates and incubate for 4 h on a rotator at 4 °C at 10 rpm.
11. Wash beads five times (for each washing step, add 1 ml of RIPA wash buffer and mix by inverting the tube for six times) (*see* **Note 23**).
12. After washing, remove RIPA wash buffer by first taking off most of the buffer with a blue tip attached to a pipette followed by removing the rest of RIPA wash buffer by using a yellow tip attached to a pipette (*see* **Note 24**).
13. For elution of antibody–protein complexes from Protein A Sepharose, add 50 μl of 1× sample buffer and incubate the sample in a heat block at 25 °C with shaking at 900 rpm for 25 min.
14. Centrifuge the tube for 1 min at 16,000 × *g* at RT and transfer the supernatant into a fresh 1.5 ml tube, using a yellow tip. At this stage, samples can either be subjected to SDS-UREA-PAGE or they can be stored at −80 °C.

3.11 SDS-UREA-PAGE of PIN2 Immunoprecipitates

For detection of PIN2 from immunoprecipitates, use 10 % resolving gels.

For detection of ubiquitin signals from immunoprecipitates, use 6 % resolving gels.

The stacking gels are 5 % and 3 M Urea.

1. Pour 6.5 ml of the resolving gel mixture (per gel) between assembled glass plates (gel dimensions: 5 cm × 8.5 cm × 0.15 cm) and overlay gels with isopropanol for 1 h.
2. After polymerization, remove isopropanol, rinse gels with ddH_2O, overlay with SDS-PAGE running buffer, and cover the whole gel-casting stand in plastic foil. Keep gels on the bench overnight to ensure complete polymerization (*see* **Note 25**).
3. On the next day, unwrap the casting stand, remove excess running buffer, rinse the gels with water, and pour 3 ml of stacking gel mixture on top of the resolving gel.
4. Insert the combs, avoiding air bubbles. Allow stacking gels to polymerize for 2 h.
5. Remove gels from the casting stand and assemble for electrophoresis. Fill the chamber with running buffer and load protein samples together with a pre-stained protein ladder. Run gels at 100 V until the bromophenol blue dye reaches the end of the stacking gels. Then set current to 150 V until the dye has migrated to the bottom of the resolving gels.

3.12 Western Blotting and Detection of PIN2 and Ubiquitin Signals

Proteins are transferred from polyacrylamide gels onto nitrocellulose membranes by wet transfer.

1. Prepare Western Blot Transfer buffer in advance and refrigerate at 4 °C.
2. Cut membranes and filter paper slightly bigger than the gel. For analysis of ubiquitylation of immunoprecipitated PIN2, cut membrane pieces with a dimension of 13 cm × 11 cm to facilitate further steps in membrane manipulation (see below).
3. After electrophoresis is completed, turn off power supply, disassemble apparatus, and remove stacking gels.
4. Soak resolving gels, membranes, filter papers, and fiber pads in transfer buffer for 10 min. Place the transfer cassettes into the transfer tank, ensuring that the membrane is positioned next to the anode. Fill the transfer tank with cold transfer buffer and cool the whole device in an ice/water bath during transfer.
5. Carry out transfer at 100 V for 1 h.
6. After transfer, rinse membranes designated for detection of PIN2 signals in ddH_2O and TBST five times each and wash for 3 min in TBST (*see* **Note 26**).

7. For detection of PIN2 ubiquitylation, rinse membrane in water immediately after transfer (without washing in TBST) and attach it to a stainless steel frame using small binder clips or similar (Fig. 2, *see* **Note 27**).
8. Transfer the frame with membrane assembled into a glass bowl with boiling ddH_2O for 20 min. Make sure that the membrane is entirely covered in water. This step facilitates recognition of ubiquitin epitopes.
9. After boiling, cut and discard wrinkled outer membrane areas (usually approximately 0.5 cm on each side).
10. Rinse and wash membranes in TBST (*see* **step 6**).
11. Block all membranes in blocking solution for 2 h at RT.
12. Rinse and wash membranes in ddH_2O and TBST five times each and wash for 3 min in TBST. Incubate with primary antibody overnight at 4 °C.
13. Rinse and wash membranes in ddH_2O and TBST as above and incubate with secondary antibody (coupled to horseradish peroxidase) for 2 h at RT.
14. Rinse and wash membranes in ddH_2O and TBST as described above.
15. Transfer membranes onto a square petri dish and apply HRP chemiluminescence substrate. Allow the substrate to react for 2 min, remove the membrane from the substrate and place it between two sheets of transparent plastic.
16. Remove excess substrate by gently pressing with a paper towel over the upper plastic sheet.
17. After 1–5 min exposure develop X-ray film.
18. Documentation of results (Fig. 3).

4 Notes

1. For each experiment, dilute NEM freshly in ddH_2O. Do not prepare a stock solution, as NEM is highly unstable. NEM powder is stored at 4 °C.
2. PVPP is insoluble in water. Vortex a 10 % (w/v) PVPP stock vigorously and take out amounts required.
3. The powder does not dissolve completely. Therefore, extensive vortexing is recommended before each use.
4. When using agravitropic roots for detection of PIN2 ubiquitylation, prepare approximately 30,000 seeds, as agravitropic roots tend to be more difficult to harvest.
5. The required amounts of seeds can easily be measured in a 1.5 ml tube.

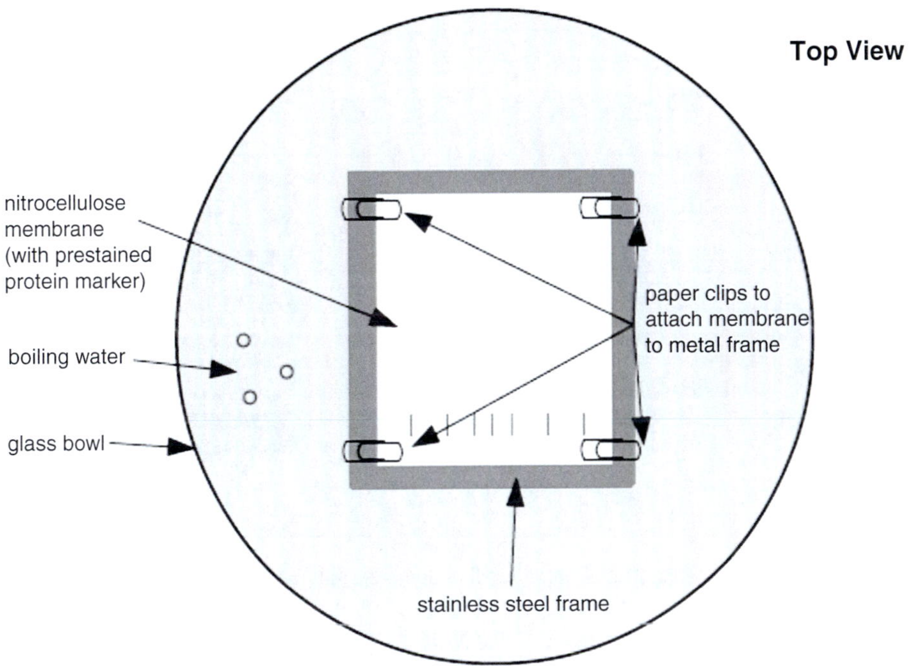

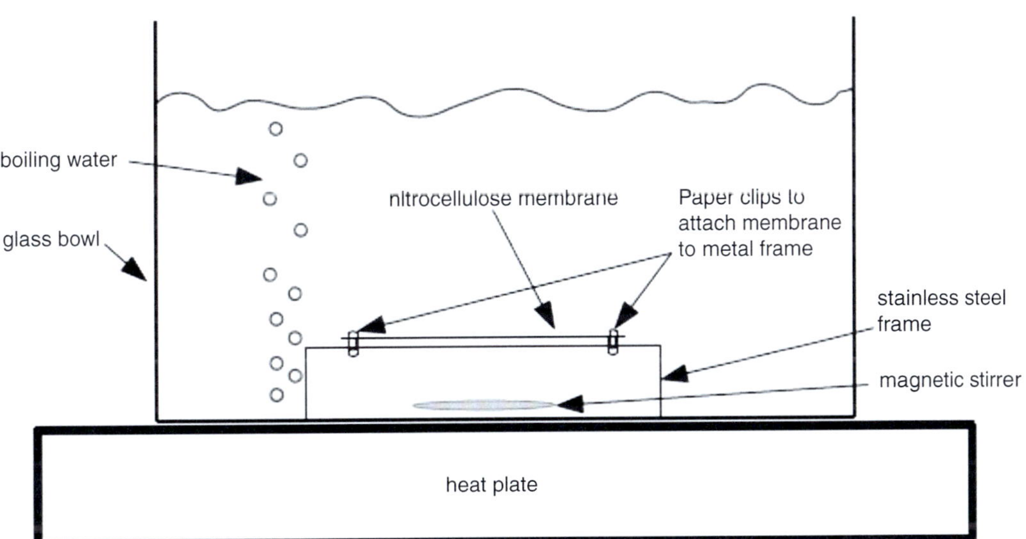

Fig. 2 Example device made of stainless steal for boiling nitrocellulose membranes in water after protein transfer to facilitate detection of ubiquitin epitopes in immunoprecipitated samples. *Top*: Assembled device with a membrane attached to the metal frame with four paper clips. *Bottom*: Side view of the metal frame placed in a glass bowl containing boiling water

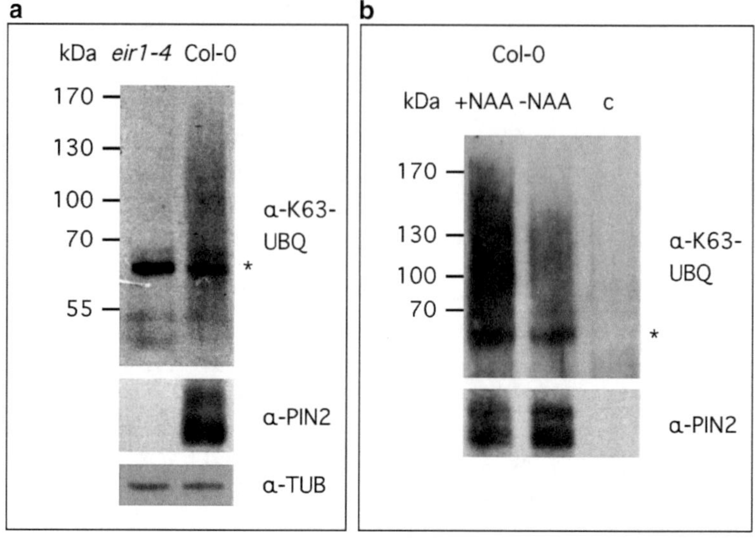

Fig. 3 Western blots on PIN2-IPs, demonstrating ubiquitylation of PIN2. (**a**) PIN2 IPs from *eir1-4* and Col-0, performed with rabbit anti-PIN2 and probed with non-discriminating ubiquitin antibody clone P4D1 (α-UBQ). A diffuse signal extending into the high molecular weight region is observed in Col-0 samples, reflecting ubiquitylation of PIN2. A distinct signal corresponding to IgG is detectable in precipitated samples (*asterisk*). Probing IPs with mouse anti-PIN2 antibody (α-PIN2, *middle panel*) indicates specificity of the immunoprecipitation. Probing solubilized membrane protein extracts for α-tubulin (α-TUB) before immunoprecipitation (*bottom panel*) demonstrates comparable amounts of protein in both samples. (**b**) 4-day-old seedlings were incubated with 20 μM NAA for 3 h (+NAA) and subjected to IP with α-PIN2. Untreated control seedlings (−NAA) were processed in parallel. After immunoprecipitation, PIN2 amounts were normalized by probing with mouse anti-PIN2 (α-PIN2, *bottom panel*), followed by another immunoprecipitation, which was probed for K63-linked ubiquitin chains (α-K63-UBQ, *top panel*). "c" indicates control immunoprecipitation, in which the primary rabbit anti-PIN2 antibody has been omitted. A distinct signal corresponding to IgG is detectable (*asterisk*)

6. Tap hard on top of the Falcon tube with the palm of your hand. This causes the clogged seeds to settle at the bottom of the tube.
7. Take care that seeds do not get lost during aspiration.
8. Plating less seeds per row may not give sufficient root material for detection of PIN2 ubiquitylation, whereas plating of too many seeds per row causes roots to penetrate the agar, which makes subsequent root harvesting difficult (a stencil for plating 6 rows on a 12 cm × 12 cm square petri dish is provided in Fig. 1. The square reflects the borders of a square petri dish, the six shorter, horizontal lines show the position, where seeds are to be plated. Do not plate to the outer edges of the plate, as this makes subsequent harvesting of the roots more tedious.

Laminating this stencil prolongs its life time, allows for spraying with 70 % ethanol before putting it into the laminar flow hood, and makes it easier to work with). Plate 500 μl of resuspended seeds per row. Take up seeds with a blue tip. Then manually take off the tip and carefully distribute the seeds by employing the capillary forces between the agar surface and the pipette tip. Make sure that the seeds are equally distributed.

9. Do not over-dry the plates as this may affect plant growth.
10. Using air-permeable Leukoplast instead of Parafilm for sealing plates, results in more efficient plant growth.
11. Try to avoid excessive manipulation of seedlings when testing for example variations in PIN2 ubiquitylation in response to gravistimulation.
12. If the roots tend to firmly stick to the agar, spray them with a little amount of water. This makes harvesting easier.
13. Do not apply too much pressure, as this may cause mechanical damage, possibly leading to the release of lytic enzymes from vacuoles.
14. Use a nylon mesh with 250 μM pore size. This allows the roots to grow into liquid medium, but prevents the seeds from dropping into the growth medium.
15. Fill up wells until surface tension leads to formation of a liquid cone. Then carefully place the mesh onto wells, avoiding the inclusion of air bubbles between mesh and liquid medium. Avoiding air bubbles is essential, as the seeds need to be in contact with the medium for efficient germination. Avoiding sucrose in the growth medium reduces chances of contamination.
16. Drill some holes into the acrylic glass to ensure airflow. Germination may take longer when seeds are placed in liquid PN(−S) compared to solid PN(+S). After germination, change medium every day to prevent bacterial and/or fungal contaminations.
17. Make sure that roots are in direct contact with the liquid medium.
18. Alternatively, roots can also be ground using precooled mortar and pestle.
19. Use a pair of scissors to cut approximately 1 cm at the end of the tip.
20. When using a bead mill, always wear protective glasses! Quickly open the tube after taking it out from liquid nitrogen. Otherwise, the tube is likely to burst. This would lead to drastic loss of root material.
21. Do not centrifuge more than 200 μl, as this may cause incomplete pelleting of membrane fractions due to reduced centrifugation forces [9].

22. Always pipette Protein A Sepharose with a cut yellow tip (*see* **Note 19**).
23. After inversion, let beads settle to the bottom of the tube before taking off RIPA wash buffer to minimize loss of Protein A Sepharose beads.
24. Firmly quench the end of a yellow tip with membrane tweezers to avoid aspiration of Protein A Sepharose beads.
25. Preparation of resolving gels should be done in parallel to overnight PIN2 immunoprecipitation.
26. The washing protocol is adopted from Wu and others [13] and reduces background when probing with anti-ubiquitin antibodies.
27. We constructed a frame made of stainless steel. The surface (14 cm × 12 cm) is used to attach the Nitrocellulose Membrane with four paper or binder clips, with the membrane side formerly attached to the resolving gel oriented towards the outer side. Clipping the membrane to the metal plate avoids curling of the membrane during boiling. Pillars (created by bending the shorter edges of the metal plate by 90°) of the device should be at least 1 cm to allow for positioning of a stir bar beneath the device. *See* Fig. 2 for a schematic presentation.

Acknowledgements

Establishment of this protocol was supported by a grant from the Austrian Science Funds to C.L. (FWF, P P25931).

References

1. Göhre V, Spallek T, Haweker H, Mersmann S, Mentzel T, Boller T, de Torres M, Mansfield JW, Robatzek S (2008) Plant pattern-recognition receptor FLS2 is directed for degradation by the bacterial ubiquitin ligase AvrPtoB. Curr Biol 18(23):1824–1832. doi:10.1016/j.cub.2008.10.063
2. Abas L, Benjamins R, Malenica N, Paciorek T, Wisniewska J, Moulinier-Anzola JC, Sieberer T, Friml J, Luschnig C (2006) Intracellular trafficking and proteolysis of the Arabidopsis auxin-efflux facilitator PIN2 are involved in root gravitropism. Nat Cell Biol 8(3):249–256. doi:10.1038/ncb1369
3. Kasai K, Takano J, Miwa K, Toyoda A, Fujiwara T (2011) High boron-induced ubiquitination regulates vacuolar sorting of the BOR1 borate transporter in Arabidopsis thaliana. J Biol Chem 286(8):6175–6183. doi:10.1074/jbc.M110.184929
4. Roberts D, Pedmale UV, Morrow J, Sachdev S, Lechner E, Tang X, Zheng N, Hannink M, Genschik P, Liscum E (2011) Modulation of phototropic responsiveness in Arabidopsis through ubiquitination of phototropin 1 by the CUL3-Ring E3 ubiquitin ligase CRL3(NPH3). Plant Cell 23(10):3627–3640. doi:10.1105/tpc.111.087999
5. Barberon M, Zelazny E, Robert S, Conejero G, Curie C, Friml J, Vert G (2011) Monoubiquitin-dependent endocytosis of the iron-regulated transporter 1 (IRT1) transporter controls iron uptake in plants. Proc Natl Acad Sci U S A 108(32):E450–E458. doi:10.1073/pnas.1100659108
6. Leitner J, Petrasek J, Tomanov K, Retzer K, Parezova M, Korbei B, Bachmair A, Zazimalova E, Luschnig C (2012) Lysine63-linked ubiquitylation of PIN2 auxin carrier protein governs hormonally controlled adaptation of

Arabidopsis root growth. Proc Natl Acad Sci U S A 109(21):8322–8327. doi:10.1073/pnas.1200824109

7. Leitner J, Retzer K, Korbei B, Luschnig C (2012) Dynamics in PIN2 auxin carrier ubiquitylation in gravity-responding Arabidopsis roots. Plant Signal Behav 7(10):1271–1273. doi:10.4161/psb.21715
8. Reyes FC, Buono R, Otegui MS (2011) Plant endosomal trafficking pathways. Curr Opin Plant Biol 14(6):666–673. doi:10.1016/j.pbi.2011.07.009
9. Abas L, Luschnig C (2010) Maximum yields of microsomal-type membranes from small amounts of plant material without requiring ultracentrifugation. Anal Biochem 401(2):217–227. doi:10.1016/j.ab.2010.02.030
10. Kahana A, Gottschling DE (1999) DOT4 links silencing and cell growth in Saccharomyces cerevisiae. Mol Cell Biol 19(10):6608–6620
11. Wang H, Matsuzawa A, Brown SA, Zhou J, Guy CS, Tseng PH, Forbes K, Nicholson TP, Sheppard PW, Hacker H, Karin M, Vignali DA (2008) Analysis of nondegradative protein ubiquitylation with a monoclonal antibody specific for lysine-63-linked polyubiquitin. Proc Natl Acad Sci U S A 105(51):20197–20202. doi:10.1073/pnas.0810461105
12. Haughn GW, Somerville C (1986) Sulfonylurea-resistant mutants of Arabidopsis thMiana. Mol Gen Genet 204:430–434
13. Wu M, Stockley PG, Martin WJ (2002) An improved western blotting technique effectively reduces background. Electrophoresis 23(15):2373–2376. doi:10.1002/1522-2683(200208)23:15<2373::Aid-Elps2373>3.0.Co;2-W

Chapter 19

Chemical Genomics Screening for Biomodulators of Endomembrane System Trafficking

Carlos Rubilar-Hernández, Glenn R. Hicks, and Lorena Norambuena

Abstract

Cell proteins traffic through complex and tightly regulated pathways. Although the endomembrane system is essential, its different pathways are still not well understood. In order to dissect protein trafficking pathways, chemical genomic screenings have been performed. This strategy has been utilized to successfully discover bioactive chemicals with a specific cellular action and in most cases, tunable and reversible effects. Once the bioactive chemical is identified, further strategies can be used to find the target proteins that are important for functionality of trafficking pathways. This approach can be combined with the powerful genetic tools available for model organisms. Drug-hypersensitive and drug-resistant mutant isolation can lead to the identification of cellular pathways affected by a bioactive chemical and reveal its protein target(s). Here, we describe an approach to look for hypersensitive and resistant mutants to a specific bioactive chemical that affects protein trafficking in yeast. This approach can be followed and adapted to any other pathway or cellular process that can be screened phenotypically, serving as a guide for novel screens in yeast. More importantly, information provided by this approach can potentially be extrapolated to other organisms like plants. Thus, the method described can be of broad utility to plant biologists.

Key words Bioactive compound, Carboxypeptidase Y, Chemical biology, Endomembrane, Endocytosis, Primary and secondary screening, Secretory route

1 Introduction

Although there is some variation, the endomembrane trafficking pathways in eukaryotic organisms are in general conserved. For example, yeast and plants share several mechanisms regarding protein and lipid trafficking [1] but plants have additional routes for specialized trafficking to two different types of vacuoles [2, 3]. Additionally, many genes encoding endomembrane trafficking components show higher levels of redundancy in plants than in yeast and mammals. Nevertheless, endomembrane system conservation is evident by the fact that plant endomembrane-related genes can complement many membrane trafficking components in the yeast *Saccharomyces cerevisiae* [4–6]. Thus, *S. cerevisiae* has been extensively used to perform wide-genome screening due to

Marisa S. Otegui (ed.), *Plant Endosomes: Methods and Protocols*, Methods in Molecular Biology, vol. 1209, DOI 10.1007/978-1-4939-1420-3_19,

the availability of genomics tools and the manageability of the experimental system. Plants are less manageable for these types of screens due to the fact that they are multicellular organisms with long life cycles and the potential for greater gene redundancy. Knowledge of trafficking pathways derived from yeast represents a very important starting point to explore the conservation or diversification of equivalent pathways in plants. One successful approach is to combine the experimental simplicity of yeast with the ability of chemical genomics approaches to discover new trafficking phenotypes [7, 8].

Chemical genomics uses small bioactive molecules as effectors of biological pathways [9–12]. Such biomodulators are useful for different purposes, such as identifying novel pathways and their gene targets and as reversible reagents for transiently perturbing biological processes, including endomembrane trafficking. Thus, small molecule biomodulators are powerful tools to dissect drug-sensitive pathways as a means to discover and characterize intracellular functional networks. Chemical biology can complement genetics for finding new pathways and addressing gene redundancy [13–16]. Since a biomodulator may target specific family members, it can overcome gene redundancy which is the strong advantage of this strategy and has resulted in significant discoveries [13, 17, 18].

In this chapter, we describe an experimental strategy to uncover endomembrane trafficking pathways affected by a particular bioactive compound and provide the basis to identify its protein target. This approach merges chemical biology with the use of genetics, taking advantage of mutant collections available in the yeast *Saccharomyces cerevisiae*. The screen we present is designed to identify small molecules that perturb vacuole trafficking by causing the aberrant extracellular secretion of the vacuole protease carboxypeptidase Y (CPY). However, with the design of other reporting systems, the approach can be used to discover compounds and new phenotypes related to other endomembrane compartments and functions. Thus, this chapter can serve as a guide for novel screens in yeast. Searching for drug-hypersensitive and drug-resistant mutants allows the identification of cellular pathways affected by a bioactive chemical as well as likely reveals proteins related to endomembrane system processes [8, 19–24].

2 Materials

2.1 Yeast Collection, Growth Culture Medium

1. YPD culture media: 2 % w/v peptone, 1 % w/v yeast extract, and 2 % w/v dextrose. Add distilled water and sterilize by autoclaving.
2. Yeast storage glycerol medium: Sterile 15 % v/v glycerol containing YPD culture media.

3. Yeast collection: *Saccharomyces cerevisiae* BY4742 parental strain haploid deletion library (MATα, *his3Δ1*, *leu2Δ0*, *lys2Δ0*, *ura3Δ0*; Open Biosystems, Huntsville, AL). This collection is a good choice to observe strong phenotypes in the screening because gene deletion results in a total loss of its function in every haploid strain. This collection comes in 96-well format plates (master plates) containing 4,800 strains. The company delivers the collection in YPD media with 15 % v/v glycerol (*see* **Note 1**).
4. Chemical compound stock: 100× chemical compound solubilized in 100 % dimethyl sulfoxide (DMSO).

2.2 CPY Secretion Assay Components

1. 10× TBS buffer (200 mM Tris–HCl, 1.5 M NaCl). Weigh 24.2 g Tris–HCl, 87.7 g NaCl and transfer them to a glass beaker. Add distilled water to 800 mL, adjust to pH 8.0 with 1 M HCl, and mix with a magnetic stirrer. Complete volume to 1 L with distilled water.
2. TTBS (1× TBS buffer, 0.05 % v/v Tween-20®). Dilute 100 mL of 10× TBS buffer to 1 L with distilled water and transfer to a glass beaker. Mix it with 500 μL Tween-20® (*see* **Note 2**).
3. Blocking solution: 5 % w/v nonfat dry milk in TTBS.
4. Primary antibody solution: Dilute 1 μL mouse monoclonal antibody against *Saccharomyces cerevisiae* CPY (e.g., from Invitrogen) in 5 mL TTBS (1:5,000 dilution, *see* **Note 3**). Store at −20 °C.
5. Secondary antibody solution: Dilute 1 μL goat monoclonal anti-mouse conjugated to alkaline phosphatase enzyme (e.g., from Reserve Ap™, KPL) in 5 mL TTBS (1:5,000 dilution, *see* **Note 3**). Store at −20 °C.
6. Alkaline phosphatase developing solution: 0.1 M Tris–HCl pH 9.5, 0.1 M NaCl, 5 mM $MgCl_2$.
7. Nitro blue tetrazolium stock solution (NBT, 50 mg/mL): Dissolve 500 mg NBT in 10 mL 85 % v/v dimethylformamide (DMF). Store in darkness at −20 °C.
   5-bromo-4-chloro-3-indolyl phosphate stock solutions (BCIP, 50 mg/mL): Dissolve 500 mg BCIP in 10 mL 100 % DMF. Wrap in aluminum foil and store at −20 °C.

2.3 Endocytic Trafficking Assay Components

1. *N*-(3-triethylammoniumpropyl)-4-(6-(4-(diethylamino)-phenyl)hexatrienyl) pyridinium dibromide (FM4-64) stock solution (16 mM FM4-64). Dissolve 100 μg FM4-64 (Invitrogen) in 10.2 μL distilled water. Wrap in aluminum foil and store at −20 °C.
2. Concanavalin A working solution (1 mg/mL). Dissolve Concanavalin A in distilled water. Store at −20 °C.

3 Methods

3.1 Primary Screening: Chemical Treatment

1. First, make a backup of the entire collection once it arrives (stocking plate, *see* **Note 1**). Use a disposable 96-pin replicator to collect a small aliquot (about 0.2–0.3 μL) from each well of a master plate (*see* **Note 1**). Add 150 μL YPD to each well using a multichannel micropipette. Cover the plates with a breath-easy membrane to allow air interchange. Incubate the plates for 2 days in a shaker at 28 °C in darkness.
2. Centrifuge both master and backup plates at 3,000 ×*g* for 10 min; remove the supernatant and add 150 μL of the yeast storage glycerol medium. Mix exhaustively to resuspend the pellet and freeze both master and backup plates at −80 °C.
3. To start the primary screening, prepare replicate plates from the backup plates. Defrost one or more of the backup plates and proceed according to **step 1**. After 2 days at 28 °C, these plates will be used to prepare the treatment plates. Dilute 3.5 μL of yeast culture with 75 μL of YPD culture medium in a new sterile 96-well plate. Add 75 μL of YPD culture medium containing 2× of the bioactive compound. The chemical concentration depends on the type of screening to be performed (*see* **Note 4**). In addition, a control treatment has to be performed with media containing 1 % DMSO as a final concentration. Cover the plates with a breath-easy membrane and wrap them with aluminum foil. Incubate the plates for 3 days at 28 °C.

3.2 Primary Screening: CPY Secretion Assay

The screening phenotype assay will measure the presence of Carboxypeptidase Y (CPY) on the yeast extracellular media by dot-blot [7].

1. Centrifuge the plates at 3,000 ×*g* for 10 min to separate media from cells.
2. Transfer the extracellular media from each treatment into a new plate using a multichannel micropipette. This step has to be performed carefully to avoid cell contamination.
3. Transfer the protein from the extracellular medium to a nitrocellulose membrane using a vacuum-based device apparatus (e.g., Bio-Dot from BioRad). To prepare the nitrocellulose membrane, size it to fit a 96-well plate and imbibe it in TTBS (*see* **Note 5**). After 20 min, put the membrane onto the vacuum apparatus following the manufacturer's instructions. To avoid bubbles while the samples are transferred, add 75 μL TTBS into each Bio-dot apparatus well. Add 75 μL of extracellular media recovered from treatment plates. When the samples are loaded, pay special attention to the positions of the treatment plate samples on the vacuum apparatus (*see* **Note 5**).

Transfer the 150 μL in each well to the nitrocellulose membrane by applying vacuum with a vacuum pump.

4. Remove the nitrocellulose membrane and wash it with TTBS to remove any bioactive chemical on the membrane that could interfere with antibody recognition.
5. Incubate the nitrocellulose membrane where yeast-secreted proteins are fixed with blocking solution for 2 h at room temperature. After this, wash the membrane with TTBS for 5 min three times.
6. Incubate the membrane with primary antibody solution for 2 h at room temperature (*see* **Notes 6** and 7). Then, remove the solution and wash the membrane as in **step 5**.
7. Incubate the membrane with secondary antibody solution at room temperature (*see* **Note** 7). After 2 h, wash the membrane as in **step 5**.
8. To develop the dot-blot, incubate the membrane into alkaline phosphatase developing solution containing 500 μg/mL NBT and 100 μg/mL BCIP (10 mL per membrane).
9. Rinse the membrane with 10 mM Ethylenediaminetetraacetic acid (EDTA) to stop the reaction when pink-purple background begins to appear.
10. Score the primary screening results (*see* **Note 8** and Fig. 1).

3.3 Secondary Screening

Identified hits from primary screening have to be validated in a secondary screening. The validation is essential to eliminate false-positives identified as hits in the primary screening. In addition, since the screening is done in a high-throughput manner, no duplicates are performed; therefore hits have to be retested. The bioactive compound effect should be dose-dependent. Therefore, the hit validation should be performed by retesting the strains at different concentrations of the chemical in the phenotype assay. Consistent with the primary screening, the secondary screening phenotype would be CPY secretion. However, primary and secondary screenings would depend upon the desired biomolecule activities and the nature of the available assay. After hit confirmation, a different phenotype may be analyzed following the same design. Due to the tight link between the secretory and endocytic routes, endocytosis may be analyzed as well (Subheading 3.3.2). The convenience of using a fluorescent tracer such as FM4-64 makes it a good choice to use in this mutant collection because it allows testing the endocytic route in a temporal in vivo assay. Therefore, it gives information about the kinetics of endocytosis as well as endosome compartment morphology in the mutant strains. Both parameters, endocytic kinetics and endosome morphology, must be contrasted with wild-type parental strain performance in regular conditions and also in the presence of the bioactive chemical (Fig. 3).

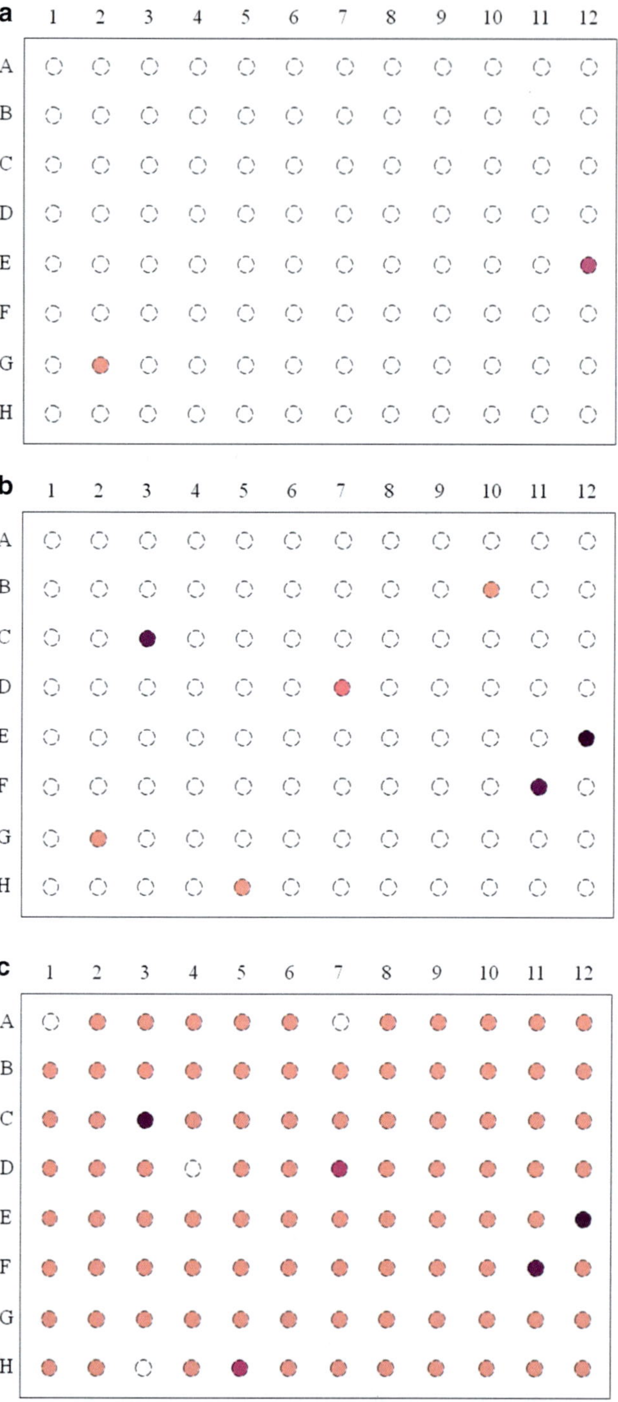

Fig. 1 Primary hypersensitive and resistance screening. CPY detection by dot-blot assay in extracellular media from mutant strains using antibodies against *Saccharomyces cerevisiae* CPY is shown. Stained spots indicate presence of CPY in the extracellular media from mutant strains incubated in YPD containing 1 % v/v DMSO (**a**, control condition) and a suitable bioactive chemical concentration for hypersensitive (**b**) and resistance screening (**c**). Strains in wells *E12* and

This assay may be performed for the primary screen as well; however, it is too laborious and time-consuming for screening the whole mutant collection. For protein endocytosis pathway would be more specific to use a fluorescent protein marker to select an endocytic pathway of interest between different compartments. The latter would need to have the marker expression in the entire collection.

3.3.1 Secondary Screening: Chemical Treatment

1. Grow wild-type and every positive mutant strain in 5 mL YPD culture media for 2 days at 28 °C.
2. Dilute every culture to 1:200 with YPD. Aliquot 75 μL of diluted cell culture into each well on a row (or column) of a sterile 96-well plate. The choice of rows and columns would depend on the number of bioactive chemical concentrations to be tested. Do not forget to include the wild-type strain since the CPY secretion phenotype will be compared to it. Always include a control without the chemical. Add 75 μL YPD containing 1 % v/v DMSO as non-drug control. In the remaining columns, add consecutively 75 μL YPD with appropriate 2× chemical compound concentration for dose testing (*see* **Note 9**). Therefore, each plate row corresponds to different yeast strains and each plate column represents different chemical concentrations (Fig. 2).
3. Cover the plate with a breath-easy membrane, wrap in aluminum foil, and incubate the plates for 3 days at 28 °C (*see* Subheading 3.1).
4. Perform the CPY secretion assay following the steps included in Subheading 3.2: CPY secretion assay method.
5. Analyze the secondary screening results (*see* **Note 10** and Fig. 2)

3.3.2 Secondary Screening: Endocytic Trafficking Assay (See **Note 11***)*

1. After chemical treatment (Subheading 3.3.1), take 30 μL of resuspended cell cultures and mix with 30 μL of 48 μM FM4-64 (24 μM FM4-64 final) at 4 °C for 30 min in darkness (wrapped in aluminum foil).
2. To remove the unbound FM4-64, spin cell suspension and replace FM4-64-containing medium with fresh YPD medium. Incubate at 28 °C in an incubator for different periods of time to evaluate FM4-64 trafficking.

Fig. 1 (continued) *G2* secrete CPY; therefore, they are considered CPY secretors. A mutant is identified as a primary hypersensitive hit if CPY secretion is induced by the bioactive chemical. In this case, the mutant hits are located in wells *B10*, *C3*, *D7*, *E12*, *F11*, and *H5*. On the other hand, a mutant is identified as a primary resistant hit if extracellular CPY secretion is not detected in this assay. Therefore, the mutant hits of this screening are placed in wells *A7* and *D4*. The footprinting wells *A1* and *H3* (*blank wells*) are negative controls in both assays. *Rows* and *columns* are indicated by *capital letters* and *numbers*, respectively

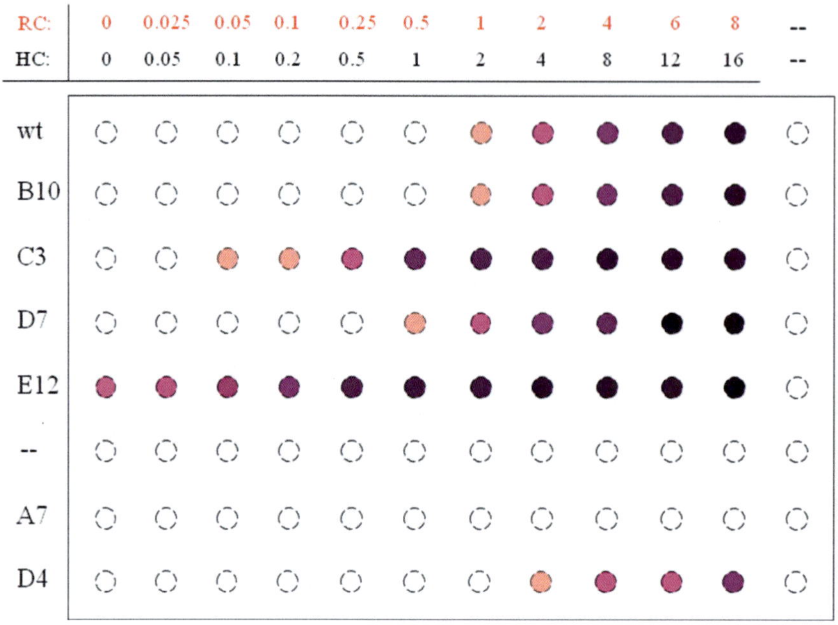

RC:	0	0.025	0.05	0.1	0.25	0.5	1	2	4	6	8	--
HC:	0	0.05	0.1	0.2	0.5	1	2	4	8	12	16	--

wt
B10
C3
D7
E12
--
A7
D4

Fig. 2 Secondary screening. Extracellular CPY detection assay is performed to retest culture media from primary hits. The bioactive chemical dose selection criteria are described in **Note 9**. Different drug concentrations for retesting bioactive chemical hypersensitivity (*HC*, in *black*) and resistance (*RC*, in *red*) are used to challenge primary screening hits. The final concentration is shown by two scales: folds of the primary hypersensitive screening concentration (*HC*) and primary resistance screening concentration (*RC*). Condition with 0 % compound contains 1 % v/v DMSO. Each column corresponds to one of the bioactive chemical concentrations applied in the assay. Each row corresponds to different strains. Row A contains the wild-type strain; rows B to E, the primary hypersensitive hits; and rows G and H the primary resistant hits. In this case, row F does not contain any yeast strain. The last column (twelfth column) contains only yeast culture media. The B10 strain behaves like the wild-type strain; therefore this strain is a false positive. The *C3*, *D7*, and *E12* (CPY secretor) strains are more sensitive than the wild-type strain validating them as hypersensitive hits. The *A7* and *D4* strains are more resistant to the chemical action than the wild-type strain. Therefore, *A7* and *D4* mutant strains are verified as resistant

3. To visualize endocytic trafficking set 5 μL cell suspension 15, 25, 40, and 60 min after FM4-64 treatment on a Concanavalin A-pretreated slide. Always evaluate FM4-64 distribution at beginning of the 28 °C incubation period (time 0). The slide is pretreated with 5 μL 1 mg/mL Concanavalin A for 30 s.
4. Evaluate FM4-64 intracellular localization using a confocal microscope (543 nm exciting laser; 560 nm band-pass emission filter, *see* **Note 12**). Image analysis can be done using image analysis software such as Magnification 1.5.1 (Orbicule® 2008) or ImageJ/Fiji software (http://fiji.sc/Fiji).
5. Perform a comparative analysis of the endocytic trafficking kinetics between mutant and wild-type cells in bioactive chemical as in control condition. Typical temporal intracellular dye distribution described in wild-type cells [25] is shown in Fig. 3.

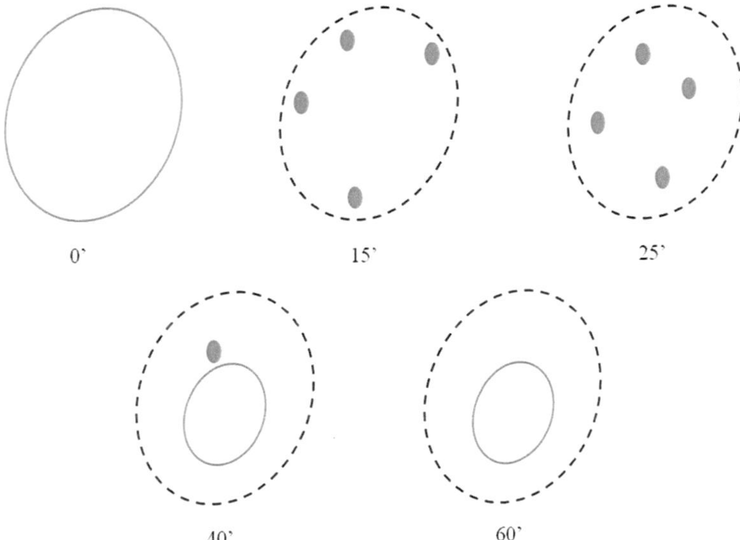

Fig. 3 Endocytic trafficking assay. A scheme of the typical subcellular localization of the endocytosed membrane fluorescent dye, FM4-64, during the time of the in vivo experiment in a wild-type yeast cell is shown (explained in Subheading 3.3.2). At time zero (0′) of culture temperature incubation, FM4-64 is localized at plasma membrane. Quickly, the dye is internalized; endosome compartments are labeled at 15 min (15′). At 25 min (25′), the dye is still at endosomes but soon it will begin to localize at the vacuole membrane. At 40 min (40′), few endosomes are still labeled and the vacuole membrane is fully labeled. At 60 min (60′), only the vacuole membrane is labeled by FM4-64. The plasma membrane is represented as a segmented-line *black circle*; endosomes are marked as inner *gray dots* and the vacuole membrane is shown as a continuous-line *gray inner circle*

3.4 Identification of Cellular Pathways Affected by Bioactive Chemicals

Once the hypersensitive and resistant strains are selected, the deleted genes in the corresponding strain can be easily identified in the indexed collection (http://sequence-www.stanford.edu/group/yeast_deletion_project/deletions3.html). Use FunCat [26] to classify the hypersensitive-conferred genes into functional categories relative to the *Saccharomyces* genome. This tool allows the analysis of the locational categories (CellLoc) of gene products of hypersensitive strains. Analysis for overrepresentation of found categories in the hypersensitive screening would suggest a relevant participation of the corresponding category. Of course, overrepresentation of a category in the hypersensitive-conferred genes has to be compared to the category representation over the whole genome. Statistical analysis of these data would support the dataset enrichment on the hypersensitive stains over the genome.

Strains that are resistant would probably be less abundant; therefore, their analysis could be done looking at gene function in the *Saccharomyces* Genome Database (SGD, www.yeastgenome.org).

To retrieve information about each deleted ORF, query directly in SGD for molecular function as well as Gene Ontology terms. Use either the GO Term Finder or the GO Slim Mapper at SGD to facilitate the identification of granular or general GO for all of the genes for which deletions cause resistance in *Saccharomyces cerevisiae*. Obviously, if there are a large number of hypersensitive strains, more analysis would be required.

1. Use FunCat and CellLoc classification systems from The Munich Information Center for Protein Sequences (MIPS, http://www.mips.gsf.de).
2. Go specifically to MIPS Functional Catalogue at projects section. Select functional distribution of gene list tool. *Saccharomyces cerevisiae* is the organism selected by default to perform the analysis.
3. Analyze the list of the gene products that are deleted in the hypersensitive mutants using their systematic name. Alternatively, the standard name can be used; however, this can be confusing for some genes with more than one name.
4. Perform the analysis to determine the functional categories and location categories. In both cases, the tool provides the representation in the hypersensitive list as well as in the genome with the corresponding *P* value.

4 Notes

1. The deletion strains in the collection were generated by swapping each ORF for the KanMX module that contains the selection marker KanMX and a bar code (unique tag with one or two 20mer sequences) [27, 28]. The deletions were checked by a PCR-based strategy. The mutant collection contains around fifty 96-well plates. Each plate is footprinted by leaving two blank wells (contain media, but no cells). In the MATα collection from Open Biosystems well H3 is always left blank. To facilitate the identification, each plate contains an extra blank well whose position is unique among the collection plates. For instance, plate number 101 has well A1 blank; plate 102 well A2, etc. *For this reason it is essential to respect the well order in each replicate plate through the screenings.* More information about the MATα or different collections available at http://sequence-www.stanford.edu/group/yeast_deletion_project/deletions3.html.
2. Tween-20® is highly viscous and very sticky, so it is difficult to manipulate with micropipette. It is recommended to make a 10 % v/v working solution which can be pipetted more easily.

3. Adjust the optimal dilution of your antibody solutions prior to primary screening. This will have to be optimized if different manufacturers or lots of antibody are used.
4. In hypersensitivity screening, the chemical compound is applied at the maximum concentration in which the wild-type strain does not show abnormal phenotype. In resistance screening, the bioactive chemical concentration corresponds to the minimal concentration that induces the abnormal phenotype in wild-type strain.
5. It is highly recommended to label systematically the nitrocellulose membranes to identify them later. Take into account that every master plate has a footprinting mark making its recognition easier for the operator (*see* **Note 1**).
6. Alternatively in this step, you can incubate the membrane overnight (16 h) at 4 °C.
7. In order to save both primary and secondary antibody, the antibody solutions could be applied in sealed plastic sheets to minimize their volume. In our experience, 2 mL antibody solution is enough for incubating a 96-well plate sized membrane. We do not recommend re-utilizing antibody solutions.
8. In a primary hypersensitive screening, the chemical concentration may not be able to trigger the phenotype. Therefore, most of the extracellular media from the strains will have no signal when strains are treated with a bioactive chemical or when treated in control condition. The hypersensitive hits are the ones where CPY is detected in the medium as a stained spot on the membrane from bioactive chemical treated strains (Fig. 1). The intensity of the signal could be different among hits. In this primary hypersensitive screening, it is expected that *vps* mutants would be a hit because their abnormal secretion vacuolar-protein phenotype [29]. This phenotype should be detected under control conditions. Then, *vps* mutants are considered the positive experimental controls. A *vps* mutant could be considered a hit when the spot in the chemical treatment is more intense than control.

   In a primary resistance screening, the majority of the spots are stained since CPY is present. Therefore, a hit is defined as an unstained spot which means that the strain is insensitive to the bioactive chemical concentration which is able to trigger the abnormal phenotype in wild-type as in most of the strains (Fig. 2). Unlike the primary hypersensitive screening, a resistance screening would not have a positive control among the mutant strains. The hypersensitive screening would identify genes that encode proteins that are either in the same pathway or in a pathway connected to the bioactive compound target. On the other hand, a resistance screen would identify the cognate target and proteins which participate in the chemical mode of action. Alternatively, a loss-of-function mutation can compensate for a bioactive compound effect. Therefore, it is expected to find fewer resistant strain hits than hypersensitive ones.

Once a primary screening hit is found, the identity of the mutant has to be searched on the master plate for information from indexed collection.

9. The final bioactive chemical concentration applied in both hypersensitive and resistance primary screening assays would be taken as a reference to verify mutant hits (*see* Fig. 2). Then, the secondary screening has to include both tested concentrations that will be used to confirm the result of the primary screening. The hypersensitive hits have to be challenged with a battery of lower bioactive chemical concentrations of what was used in the primary screening in order to determine the minimal concentration able to trigger the abnormal phenotype. Resistant hits have to be challenged with higher concentrations of what was used in the primary screening to test how insensitive the strains are. The best choice is to challenge mutants with a set of bioactive chemical concentrations to have a better profile of their sensitivity. In addition, this will be useful for comparison to the sensitivity of the parental strain.

   In all cases, we recommend preparing 100× chemical compound solubilized in 100 % DMSO for every final concentration used in the secondary screening to ensure the appropriate DMSO concentration in each final condition.

10. In a secondary screening, a hypersensitive hit is verified when extracellular CPY secretion from this mutant is detected at lower bioactive chemical concentration than the wild-type strain in a dose-dependent assay (*see* Fig. 2, C3 and D7 strains). In the case of a CPY secretor under control conditions, this is validated as a hypersensitive hit when extracellular CPY secretion is increased (stronger signal) at a lower chemical concentration that does not induce CPY secretion in wild-type strain (*see* Fig. 2, E12 strain). On the other hand, a mutant hit is classified as resistant when CPY secretion is detected at higher bioactive chemical concentration than a wild-type strain in this dose-dependent assay (*see* Fig. 2, A7 and D4 strains). The most extreme result for resistance would be the no CPY detection in any chemical concentration (*see* Fig. 2, D4 strain).

11. It is highly recommended to perform this assay by testing only one mutant strain with the wild-type strain as a comparative control at the same time. It is difficult to follow FM4-64 internalization for more than two samples because the incubation periods are short enough to overlap to each other. Additionally,

we recommend performing only one chemical concentration per test beginning with the primary screening concentration utilized before.

12. Images can be taken with a 40× water-objective. However it is preferable to use a 63× oil-objective because it resolves the dye intracellular localization with higher resolution, especially when dye is being trafficked through endosomes.

Acknowledgements

This work is supported by grant FONDECYT1120289 (LN, CRH) and National Science Foundation Grant MCB-0515963 (GRH).

References

1. Mishev K, Dejonghe W, Russinova E (2013) Small molecules for dissecting endomembrane trafficking: a cross-systems view. Chem Biol 20: 475–486
2. Zouhar J, Rojo E (2009) Plant vacuoles: where did they come from and where are they heading? Curr Opin Plant Biol 12:677–684
3. Xiang L, Etxeberria E, Van den Ende W (2013) Vacuolar protein sorting mechanisms in plants. FEBS J 280:979–993
4. Bassham D, Raikhel N (2000) Plant are not just green yeast. Plant Physiol 122:999–1001
5. Sanderfoot A, Assaad F, Raikhel N (2000) The Arabidopsis genome. an abundance of soluble N-ethylmaleimide-sensitive factor adaptor protein receptors. Plant Physiol 124:1558–1569
6. Rojo E, Zouhar J, Kovaleva V, Hong S, Raikhel NV (2003) The AtC–VPS protein complex is localized to the tonoplast and the prevacuolar compartment in Arabidopsis. Mol Biol Cell 14:361–369
7. Zouhar J, Hicks G, Raikhel N (2004) Sorting inhibitors (sortins): chemical compounds to study vacuolar sorting in Arabidopsis. Proc Natl Acad Sci U S A 101:9497–9501
8. Norambuena L, Zouhar J, Hicks G, Raikhel N (2008) Identification of cellular pathways affected by Sortin2, a synthetic compound that affects protein targeting to the vacuole in Saccharomyces cerevisiae. BMC Chem Biol 8:1
9. Lokey RS (2003) Forward chemical genetics: progress and obstacles on the path to a new pharmacopoeia. Curr Opin Chem Biol 7:91–96
10. Schreiber S (2005) Small molecules: the missing link in the central dogma. Nat Chem Biol 1:64–66
11. Hicks G, Raikhel N (2012) Small molecules present large opportunities in plant biology. Annu Rev Plant Biol 63:261 282
12. Eggert U (2013) The why and how of phenotypic small-molecule screens. Nat Chem Biol 9:206–209
13. Rojas-Pierce M, Titapiwatanakun B, Sohn E, Fang F, Larive C, Blakeslee J, Cheng Y, Cuttler S, Peer W, Murphy A, Raikhel N (2007) Arabidopsis P-glycoprotein19 participates in the inhibition of gravitropism by gravacin. Chem Biol 14:1366–1376
14. Robert S, Chary S, Drakakaki G, Li S, Yang Z, Raikhel N, Hicks G (2008) Endosidin1 defines a compartment involved in endocytosis of the brassinosteroid receptor BRI1 and the auxin transporters PIN2 and AUX1. Proc Natl Acad Sci U S A 105:8464–8469
15. Drakakaki G, Robert S, Szatmari A, Brown M, Nagawa S, van Damme D, Leonard M, Yang Z, Schmid S, Russinova E, Friml J, Raikhel N, Hicks G (2011) Clusters of bioactive compounds target dynamic endomembrane networks in vivo. Proc Natl Acad Sci U S A 108: 17850–17855
16. Rosado A, Hicks G, Norambuena L, Rogachev I, Meir S, Pourcel L, Zouhar J, Brown M, Boirsdore M, Puckrin R, Cutler S, Rojo E, Aharoni A, Raikhel N (2011) Sortin1-hypersensitive mutants link vacuolar-trafficking defects and flavonoid metabolism in Arabidopsis vegetative tissues. Chem Biol 18: 187–197
17. Park S, Fung P, Nishimura N, Jensen D, Fujii H, Zhao Y, Lumba S, Santiago J, Rodrigues A, Chow T, Alfred S, Bonetta D, Finkelstein R, Provart N, Desveaux D, Rodriguez P,

McCourt P, Zhu J, Schroeder J, Volkman B, Cutler S (2009) Abscisic acid inhibits type 2C protein phosphatases via the PYR/PYL family of START proteins. Science 324: 1068–1071

18. De Rybel B, Audenaert D, Vert G, Rozhon W, Mayerhofer J, Peelman F, Coutuer S, Denayer T, Jansen L, Nguyen L, Vanhoutte I, Beemster G, Vleminckx K, Jonak C, Chory J, Inzé D, Russinova E, Beeckman T (2009) Chemical inhibition of a subset of arabidopsis thaliana GSK3-like kinases activates brassinosteroid signaling. Chem Biol 16:594–604
19. Stearns T, Hoyt M, Botstein D (1990) Yeast mutants sensitive to antimicrotubule drugs define three genes that affect microtubule function. Gen 124:251–262
20. Chan T, Carvalho J, Riles L, Zheng X (2000) A chemical genomics approach toward understanding the global functions of the target of rapamycin protein (TOR). Proc Natl Acad Sci U S A 97:13227–13232
21. Butcher R, Schreiber S (2004) Identification of Ald6p as the target of a class of small-molecule suppressors of FK506 and their use in network dissection. Proc Natl Acad Sci U S A 101: 7868–7873
22. Giaever G, Flaherty P, Kumm J, Proctor M, Nislow C, Jaramillo D, Chu A, Jordan M, Arkin A, Davis R (2004) Chemogenomic profiling: identifying the functional interactions of small molecules in yeast. Proc Natl Acad Sci U S A 101:793–798
23. Hoon S, Onge R, Giaever G, Nislow C (2008) Yeast chemical genomics and drug discovery: an update. Trends Pharmacol Sci 29:499–504
24. Kemmer D, McHardy L, Hoon S, Delphine Rebérioux D, Giaever G, Nislow C, Roskelley C, Roberge M (2009) Combining chemical genomics screens in yeast to reveal spectrum of effects of chemical inhibition of sphingolipid biosynthesis. BMC Microbiol 9:9
25. Vida TA, Emr SD (1995) A new vital stain for visualizing vacuolar membrane dynamics and endocytosis in yeast. J Cell Biol 128:779–792
26. Ruepp A, Zollner A, Maier D, Albermann K, Hani J, Mokrejs M, Tetko I, Güldener U, Mannhaupt G, Münsterkötter M, Mewes HW (2004) The FunCat, a functional annotation scheme for systematic classification of proteins from whole genomes. Nucleic Acids Res 32:5539–5545
27. Baudin A, Ozier-Kalogeropoulos O, Denouel A, Lacroute F, Cullin C (1993) A simple and efficient method for direct gene deletion in *Saccharomyces cerevisiae*. Nucleic Acids Res 21:3329–3330
28. Wach A, Brachat A, Pöhlmann R, Philippsen P (1994) New heterologous modules for classical or PCR-based gene disruptions in *Saccharomyces cerevisiae*. Yeast 10:1793–1808
29. Robinson JS, Klionsky DJ, Banta LM, Emr SD (1988) Protein sorting in *Saccharomyces cerevisiae*: isolation of mutants defective in the delivery and processing of multiple vacuolar hydrolases. Mol Cell Biol 8:4936–4948

Chapter 20

Integrative Chemical Proteomics and Cell Biology Methods to Study Endocytosis and Vesicular Trafficking in Arabidopsis

Tomáš Takáč, Tibor Pechan, Olga Šamajová, and Jozef Šamaj

Abstract

We present a comprehensive approach combining proteomics and cell biology to study vesicular trafficking in plants. Within this approach, we exploit chemical compounds inhibiting particular vesicular trafficking events in plant cells. Treatment of plants with these relatively specific inhibitors results in intracellular accumulation of proteins being transported by vesicles as well as in a change in abundance of regulatory proteins. Such pharmacological inhibition allows for identification of key proteins, and for further detailed functional investigation using cell biological, molecular biological, and biochemical methods used for validation of proteomic results.

Key words Endocytosis, Vesicular trafficking, Proteomics, Plants, 2-D LC/MSMS, 2-D electrophoresis, Inhibitors, Brefeldin A, Wortmannin, LY294002

1 Introduction

Plant vesicular trafficking is a complex cellular process that can be studied by a wide range of cell biological, molecular biological, biochemical, and proteomic methods [1]. Its complexity, determined by intricate regulatory networks [2], requires the adoption of various independent but mutually complementary approaches. Several studies showed that integrative exploitation of proteomics and cell biological methods proved to be valuable in the elucidation of vesicular trafficking mechanisms in plants. Membrane dynamics and their regulation during vesicular trafficking is commonly investigated either by exploiting knockout mutants with altered vesicular trafficking, or by proteomic analysis of vesicles and endosomes isolated by subcellular fractionation and purified by immunoprecipitation [3]. Here, we describe an alternative to these approaches, which is based on the utilization of pharmaceutical compounds, such as brefeldin A [4], wortmannin [5], and

Marisa S. Otegui (ed.), *Plant Endosomes: Methods and Protocols*, Methods in Molecular Biology, vol. 1209, DOI 10.1007/978-1-4939-1420-3_20, © Springer Science+Business Media New York 2014

LY294002 [6], to inhibit vesicular transport processes. These inhibitors target either early endosomes (brefeldin A [7]) or late endosomes (wortmannin [8] and LY294002 [9]) and cause accumulation of cargo within these organelles. An important point is that molecular targets of such inhibitors and therefore the mechanisms of their action are known. For example, BFA targets selected adenosine diphosphate ribosylation factor-guanine nucleotide exchange factor (ARF-GEF) proteins, LY294002 targets phosphoinositide 3-kinase (PI3K), and wortmannin targets both PI3K and phosphoinositide 4-kinase. Consequently, inhibition of certain vesicular transport processes related to endosomes leads to the intracellular accumulation of proteins transported by these pathways, as well as the changed abundance of regulatory proteins. In this manner, the proteomic analysis of inhibitor-treated cells has a great potential to identify novel proteins involved in vesicular trafficking. Crucial to our approach is the evaluation and further validation of the proteomics data. The candidate proteins are selected on the basis of their known (or predicted) localization, and functional domains or signaling sequences. Furthermore, we apply independent methods for the validation of proteomic data, such as GFP-tagging, immunolocalization, or immunoblotting. Enzyme activity staining on native PAGE gels can be used for validation of proteomic data in the case that enzyme abundance is altered by the inhibitor treatment. As an example, we provide specific activity staining protocols for superoxide dismutase, catalase, and ascorbate peroxidase, antioxidant enzymes differentially regulated upon LY294002 treatment and connected with vesicular trafficking in PI3K-dependent manner [6]. Additionally, it was reported that the reactive oxygen species are produced in endosomes [10]. Genetic suppression of vesicle-associated membrane protein (VAMP) resulted in hydrogen peroxide accumulation in so-called hydrogen peroxide-containing megavesicles [10]. This should be connected with changed activities of above-mentioned enzymes involved in reactive oxygen species detoxification.

In addition to the validation, all these techniques may bring novel insights into the function and localization of candidate proteins and further specify their roles in vesicular trafficking.

2 Materials

2.1 Plant Material

1. Seeds of *Arabidopsis thaliana* ecotype Columbia.
2. Stable transformed lines expressing endocytic proteins (e.g., GFP-RabA1d [11], YFP-RabF2a, 2xFYVE-GFP [12]).

2.2 Equipment and Consumables

1. Cooled centrifuge capable of 13,000 × *g*.
2. Isoelectric focusing unit allowing the application of high voltage to gel strips.

3. Calibrated densitometer for gel documentation.
4. Vacuum dryer.
5. C18 cartridges: disposable cartridges with 130 mg sorbent per cartridge, 55–105 μm particle size (e.g., Sep-Pak C18 Plus Light Cartridge, Waters).
6. Electrospray ionization (ESI) mass spectrometer coupled with high pressure liquid chromatography apparatus, e.g., ProteomeX Workstation (Thermo Scientific) with a strong cation exchange column (e.g., SCX BioBasic 0.32 × 100 mm) and a reverse phase column (e.g., BioBasic C18, 0.18 × 100 mm, Thermo Scientific). It includes the Surveyor auto sampler and the Surveyor HPLC unit coupled directly in-line with a LCQ Deca XP Plus—ESI ion trap mass spectrometer, governed by XCALIBUR software (Thermo Scientific).
7. MALDI (matrix-assisted laser desorption/ionization) mass spectrometer (e.g., Ultraflex II, Bruker Daltonics).
8. Software for database search (e.g., TurboSEQUEST algorithm of the Bioworks Browser 3.2 EF2 software, Thermo Scientific).
9. Software for label-free quantification (the "in-house" made ProtQuant [13] as well as commercial software ProteoIQ 1.3.01, Bioinquire).

2.3 Solutions and Components of Proteomic Analyses

Prepare all solutions using ultrapure water (resistivity ≥ 18 MΩ cm at 25 °C) and analytical grade reagents.

1. Brefeldin A solution: 5 mg of brefeldin A in 1.78 mL dimethyl sulfoxide (DMSO). Add 50 μL of this solution to 10 mL of half strength Murashige–Skoog [14] media to make final concentration of 50 μM brefeldin A.
2. Wortmannin solution: 5 mg of wortmannin in 1.17 mL of DMSO. Add 33 μL of this solution to 10 mL of half strength Murashige–Skoog media to make final concentration of 33 μM wortmannin.
3. LY294002 solution: 5 mg of LY294002 in 1.46 mL of DMSO. Add 33 μL of this solution to 10 mL of half strength Murashige–Skoog media to make final concentration of 33 μM LY294002.
4. Control (mock) solutions: dilute DMSO in 10 mL half strength Murashige–Skoog media in the same concentration as for inhibitor treatments (50 μL of DMSO for brefeldin A, 33 μL DMSO for wortmannin and LY294002).
5. Phenol solution: Tris-buffered phenol (pH 8.0) (*see* **Note 1**).
6. Extraction solution: 0.9 M saccharose, 0.1 M Tris–HCl (pH 8.8), 10 mM EDTA, 100 mM KCl. Store at −20 °C. Add 2-mercaptoethanol (in fume hood) prior to use in final concentration of 0.4 % (v/v) (*see* **Note 2**).

7. Precipitation solutions: 80 % acetone and 70 % ethanol prepared with distilled water, 0.1 M ammonium acetate in anhydrous methanol. Store at −20 °C.
8. Ampholytes mixture of a pH range 3–10.
9. Rehydration buffer: 8 M urea, 2 M thiourea, 2 % (w/v) CHAPS, 2 % (v/v) Triton X-100, store at −20 °C for not more than 1 month. Prior to use, add 50 mM dithiothreitol DTT and 0.5 % (v/v) ampholytes. Add 0.002 % bromophenol blue following the protein concentration measurement.
10. SDS PAGE gel (10 %, 8 × 8 cm, 10 mL) components (second dimension of 2-D electrophoresis): 1.5 M Tris–HCl, pH 8.8 (2.5 mL), 40 % Acrylamide/Bis-acrylamide 37.5:1 (2.5 mL), distilled water (4.85 mL), 10 % (w/v) SDS (100 μL), 10 % (w/v) ammonium persulfate (50 μL), *N,N,N′,N′*-tetramethylethylenediamine (TEMED, 5 μL).
11. Running buffer for SDS PAGE: dissolve 3 g of Tris thoroughly in 800 mL of distilled water, add 14.4 g of glycine, dissolve, add 1 g of SDS, and make it up to 1 L with distilled water.
12. Agarose solution: 0.5 % (w/v) agarose in running buffer for SDS PAGE. Dissolve by boiling and add 0.002 % (w/v) bromophenol blue. Store at 4 °C. Thaw before use by heating at 100 °C (*see* **Note 3**).
13. Equilibration buffer: 6 M Urea, 50 mM Tris, pH 8.8, 2 % (w/v) SDS, 30 % (v/v) glycerol, 0.002 % (w/v) bromophenol blue. Store at −20 °C for not more than 1 month. Supplement with 1 % (w/v) DTT or 2.5 % (w/v) iodoacetamide (IAA) prior to use (*see* **Note 2**).
14. Gel spot picker with cutting diameter of 1.5 mm.
15. Colloidal Coomassie Blue protein stain solution: 0.08 % (w/v) Coomassie Brilliant Blue G-250, 8 % (w/v) ammonium sulfate, 1.6 % (v/v) ortho-phosphoric acid, and 20 % methanol (*see* **Note 4**).
16. Gel strips for isoelectric focusing of linear pH range 5–8, 7 cm long (*see* **Note 5**).
17. Protein lysis buffer for two-dimensional liquid chromatography coupled to tandem mass spectrometry (2-D LC/MSMS): 6 M Urea in 100 mM Tris–HCl, pH 6.8. Store at −20 °C for not more than 1 month.
18. Disulfide bond reduction solution: 50 mM DTT in 100 mM Tris–HCl pH 7.8 (prepare prior to use).
19. Thiol group alkylation solution: 100 mM IAA in 100 mM Tris–HCl pH 7.8 (prepare prior to use).

2.4 Solutions for Trypsin Digestion and Mass Spectrometry

1. Trypsin solution: 20 μg of sequencing grade trypsin in 200 μL of trypsin buffer provided by the manufacturer. For "in gel digestion" dilute this solution ten times with 50 mM ammonium bicarbonate (10 ng trypsin/μL).
2. Acetonitrile: all dilutions of acetonitrile have to be done with water of highest purity.
3. Trifluoroacetic acid (TFA) solutions: 0.1 % (v/v) and 0.5 % (v/v); all dilutions of TFA should be done with water of highest purity.
4. Peptide extraction solution: 60 % (v/v) acetonitrile, 1 % (v/v) formic acid.
5. α-Cyano-4-hydroxycinnamic acid solution: 4 mg/mL in 75 % acetonitrile and 0.2 % TFA.

2.5 Solutions for Isoenzyme Pattern Analysis Using Native PAGE

1. Extraction buffer for isoenzyme pattern analysis: 50 mM Na-phosphate buffer pH 7.8, 2 mM ascorbic acid, 10 % (v/v) glycerol. Add mercaptoethanol (1 %) prior to use.
2. The native PAGE gel components should be prepared as in Subheading 2.3, **step 10** without the addition of SDS. Substitute the volume of SDS with equal amount of distilled water.
3. Stacking gel (4 %, 5 mL) components: 0.5 M Tris–HCl, pH 6.8 (1.26 mL), 40 % Acrylamide 37.5:1 (0.5 mL), distilled water (3.68 mL), 10 % (w/v) ammonium persulfate (25 μL), TEMED (2.5 μL).
4. Running buffer for native PAGE: dissolve 3 g of Tris thoroughly in 800 mL of distilled water, add 14.4 g of glycine, and 0.352 g ascorbic acid, let dissolve and make it up to 1 L with distilled water. Cool down to 4 °C before use.

2.6 Solutions for Superoxide Dismutase Activity Visualization

1. KCN incubation solution: 2 mM KCN in 50 mM Na-phosphate buffer, pH 7.8 (*see* **Note 6**).
2. Hydrogen peroxide incubation solution: 0.05 % (v/v) hydrogen peroxide in 50 mM Na-phosphate buffer, pH 7.8 (prepare prior to use).
3. Solution A: 0.6 mM nitroblue tetrazolium chloride in 50 mM Na-phosphate buffer, pH 7.8 (*see* **Note 7**).
4. Solution B: 0.06 mM riboflavin, 0.25 % (v/v) TEMED. Dissolve in 50 mM Na-phosphate buffer, pH 7.8 (*see* **Note 7**).

2.7 Solutions for Ascorbate Peroxidase Activity Visualization

1. Solution C: 2 mM ascorbic acid in 50 mM Na-phosphate buffer, pH 7.0.
2. Solution D: 4 mM ascorbic acid, 2 mM hydrogen peroxide in 50 mM Na-phosphate buffer, pH 7.0.
3. Solution E: 2 mM nitroblue tetrazolium chloride, 28 mM TEMED in 50 mM Na-phosphate buffer, pH 7.8.

2.8 Solutions for Catalase Activity Visualization

1. Solution F: 0.006 % (v/v) hydrogen peroxide in distilled water.
2. Solution G: 1 % ferricyanide (w/v) and 1 % potassium ferric chloride (w/v) in distilled water.

2.9 Solutions for Live-Cell Microscopy

1. Liquid half-strength Murashige–Skoog culture medium (pH 5.7) containing 1 % (w/v) sucrose.
2. 4 μM FM 4-64 (*N*-(3-triethylammoniumpropyl)-4-(8-(4-(diethylamino)phenyl)hexatrienyl)pyridinium dibromide) solution: add 1 μL of FM4-64 stock solution (1.6 mM FM4-64 diluted in DMSO) into 0.4 mL of liquid half-strength Murashige–Skoog culture medium to make final concentration of 4 μM FM4-64.
3. Mock solution: add appropriate amount of DMSO in liquid half-strength Murashige–Skoog culture medium.

2.10 Solutions for Whole Mount Immunolocalization

1. Stabilizing buffer (MTSB): 50 mM PIPES (*see* **Note 8**), 5 mM MgSO4.7H_2O, and 5 mM EGTA, pH 6.9 (*see* **Note 9**).
2. Phosphate-buffered saline (PBS): 0.14 M NaCl, 2.7 mM KCl, 6.5 mM $Na_2HPO_4.2H_2O$, and 1.5 mM KH_2PO_4, pH 7.3 (*see* **Note 10**).
3. Reduction solution: 0.05 % (w/v) $NaBH_4$ in PBS (prepare prior to use and keep in dark).
4. Formaldehyde stock solution: prepare fresh 8 % paraformaldehyde stock solution by dissolving 0.8 g paraformaldehyde in 9 mL MTSB (using warm water bath of maximum 60 °C) and 10 N NaOH (5–10 drops) to clear solution. Remove solution from water bath and carefully adjust pH to 7.2–7.4 with NaOH. Bring to final volume 10 mL and keep at 4 °C not more than 1 week (*see* **Note 11**).
5. Glutaraldehyde stock solution: 25 % (v/v) glutaraldehyde in water (*see* **Note 12**).
6. Fixation solution: mixture of 1.5 % paraformaldehyde and 0.5 % glutaraldehyde in stabilizing buffer (MTSB). Glutaraldehyde should be added last.
7. Cell wall digestion cocktail: 1 % (w/v) meicelase, 1 % (w/v) macerozyme R10, and 1 % (w/v) cellulase R10 in PBS. Alternatively, use 2 % (w/v) driselase, 2 % (w/v) cellulase R10, and 1 % (w/v) pectolyase Y23 in PBS. Spin down by centrifugation before use (12,000 × *g*, 2 min).
8. Permeabilization solution: 10 % (v/v) DMSO, 2 % (v/v) Nonidet P-40 in PBS (*see* **Note 13**).
9. Blocking solution: 3 % (w/v) bovine serum albumin (BSA) in PBS.
10. Solutions with primary antibodies: Primary antibody is diluted in PBS containing 2 % BSA (e.g., antibody against 2S albumin is

diluted 1:200). Specific dilutions should follow manufacturer's instructions and be optimized on a personal experience basis.

11. Solutions with secondary antibodies: we routinely use anti-rabbit, anti-mouse, and anti-rat IgGs coupled to a wide range of Alexa Fluor dyes diluted 1:500 in PBS containing 2 % BSA.
12. 4,6-diamidino-2-phenylidone (DAPI) staining solution: 0.1 mg/mL (w/v) DAPI in DMSO, aliquoted and stored at −20 °C in the dark. Prepare working solution by diluting 1:1,000 with PBS.
13. Antifade mounting medium: 0.1 % (w/v) para-phenylene diamine (*see* **Note 14**) in 90 % (v/v) glycerol buffered with PBS or 1 M Tris–HCl at pH 8.0 (*see* **Note 15**). Store in the dark at −20 °C. Alternatively, use commercial antifade medium (e.g., Vectashield, Vector Laboratories).

3 Methods

3.1 Growth and Treatment of Arabidopsis Seedling Plants for Proteomics

Perform all treatments in at least three biological replicates.

1. Surface-sterilize seeds by initial incubation of seeds in 70 % (v/v) ethanol followed by incubation in 0.8 % v/v sodium hypochlorite supplemented with a drop of Tween 20 with occasional vortexing for 8 min. Wash the seeds five times in sterile distilled water.
2. Place the seeds on the surface of half-strength Murashige–Skoog culture medium (pH 5.7) containing 1 % (w/v) sucrose and 0.8 % (w/v) phytagel in a square Petri dish aligned in a row approximately 2 cm from the top. Place 20–30 seeds per Petri dish. To break dormancy store the plates at 4 °C for 48 h in the dark, and then keep vertically under 16 h light/8 h dark regime, at 22 °C, for 12 days.
3. Apply 10 mL of inhibitor solution by gently pipetting it close to the seedling roots (without touch) and incubate for 2 h. Seal the Petri dish with Parafilm and let it slowly agitate on variable speed rocker to prevent complete submergence of the roots in the liquid. Treat the control plants in the same way with mock solution.
4. Cut-off the roots quickly using a scalpel, and harvest the tissue for protein extraction.

3.2 Protein Extraction

1. Homogenize the roots to fine powder using mortar and pestle in the presence of liquid nitrogen.
2. Add 0.5 mL of extraction solution to 0.3 mg of homogenized material and incubate on ice for 10 min. Vortex.

3. Add 0.5 mL of phenol solution. Vortex and incubate at 4 °C for 30 min. Centrifuge at 8,000 × *g* and 4 °C for 5 min (*see* **Note 16**).
4. Collect phenol phase (top phase) and back-extract aqueous phase with 0.5 mL of phenol solution by vortexing. Centrifuge at 8,000 × *g* and 4 °C for 5 min. Collect phenol phase.
5. Combine the phenol phases and precipitate phenol extracted proteins by adding five volumes of 0.1 M ammonium acetate in anhydrous methanol (cooled down to −20 °C). Vortex the mixture thoroughly and incubate overnight at −20 °C (*see* **Note 17**).
6. Centrifuge at 12,000 × *g* and 4 °C for 30 min.
7. Discard the supernatant and wash the pellet by addition of 1 mL of ice-cold ammonium acetate precipitating solution. Resuspend the pellet thoroughly by pipetting and vortexing. Incubate at −20 °C for 15 min, and centrifuge at 12,000 × *g* and 4 °C for 10 min.
8. Discard the supernatant and wash the pellet by addition of 1 mL of ice-cold 80 % (v/v) acetone. Resuspend the pellet thoroughly by pipetting. Incubate at −20 °C for 15 min, and centrifuge at 12,000 × *g* and 4 °C for 10 min. Repeat this step once more.
9. Discard the supernatant, and wash the pellet by addition of 1 mL of ice-cold 70 % (v/v) ethanol. Resuspend the pellet thoroughly by pipetting. Incubate at −20 °C for 15 min, and centrifuge at 12,000 × *g* and 4 °C for 10 min.
10. Discard the supernatant, and wash the pellet by addition of 1 mL of ice-cold 80 % (v/v) acetone. Resuspend the pellet thoroughly by pipetting, and centrifuge at 12,000 × *g* and 4 °C for 10 min. Discard the supernatant.
11. Air-dry the protein pellet (leave the tube open upside down at RT in a fume hood for no longer than 10 min). Store at −80 °C, if not used immediately.

3.3 Two-Dimensional Electrophoresis

1. Resuspend the pellet in no more than 125 μL of rehydration buffer (*see* **Note 18**).
2. Spin down the non-dissolved particles by centrifugation at 10,000 × *g* and RT for 10 min.
3. Quantitate the protein concentration in the samples.
4. Use 50–80 μg of proteins per sample, and make the sample volume to 125 μL by rehydration buffer supplemented with bromophenol blue (0.002 % final concentration) (*see* **Note 19**).
5. Apply the sample to the focusing tray, place the gel strip into it according to the manufacturer's instruction and place the tray into isoelectric focusing unit. Rehydrate the strip actively

at 50 V for 12 h and perform the isoelectric focusing using the following parameters: step 1: 150 V—till 150 VH, step 2: 500 V—till 500 VH, step 3: 4,000 V—till 15,000 VH (*see* **Note 20**).

6. Wash the strips by immersing them in running buffer for SDS PAGE.
7. Equilibrate the strips by incubation in DTT-supplemented equilibration buffer (to reduce the disulfide bonds) for 15 min, followed by incubation in the equilibration buffer supplemented by IAA (to alkylate the thiol groups) at RT for additional 15 min (*see* **Note 21**).
8. Prepare 10 % PAGE gels by filling the gel slab up to 2 mm from its top. Overlay the monomer solution with 2-propanol. After gel polymerization, remove the 2-propanol (*see* **Note 22**), wash with distilled water, and dry remaining traces with filter paper (*see* **Note 23**). Add running buffer to the polymerized gel and apply the strip on the surface of the gel avoiding the generation of bubbles between the gel strip and the surface of the gel (*see* **Note 24**). To fix the strip in position, overlay the strip with agarose solution. Let it polymerize for 5 min.
9. Run SDS PAGE electrophoresis at 60 V for 10 min, allowing the movement of proteins from the gel strip to the PAGE gel. Afterwards, increase the voltage to 100 V. Do not let the protein front run out from the gel.
10. Wash the gel in distilled water three times for 5 min (*see* **Note 25**).
11. Stain the gels with colloidal Coomassie Blue solution for at least 2 h (overnight staining is possible) (*see* **Note 26**).
12. De-stain the gels by repeated washes in distilled water (overnight de-staining is possible) (*see* **Note 27**).
13. Scan the gels on calibrated densitometer according to the manufacturer's instructions. Set equal scanning area and resolution for each gel. Remove air bubbles from the scanned area.
14. Detect the protein spots and compare the spot densities on the gels using any of the commercially available software. Normalize the spot intensities according to total density in the gel images.
15. Cut the statistically significant, differentially abundant protein spots from the gels using a gel spot picker with cutting diameter of 1.5 mm, and place them on flat bottom 96-well plate. Store at 4 °C in running buffer for no longer than 1 week.

3.4 In-Gel Trypsin Digestion

Perform the trypsin digestion in 96-well plate in keratin and dust-free environment, preferably in the fume hood, wearing latex-free gloves.

1. De-stain the gel pieces with 100 μL of 50 % acetonitrile. Incubate for 5 min at RT on shaker. Discard the liquid from the gel pieces (*see* **Note 28**).

2. Add 100 μL of 50 % acetonitrile in 50 mM ammonium bicarbonate. Incubate at RT on shaker for 30 min. Discard the liquid.
3. Add 50 μL of 100 % acetonitrile. Incubate at RT on shaker for 15 min (*see* **Note 29**). Discard the acetonitrile, and dry the gel pieces in a fume hood overnight.
4. Rehydrate the gel pieces in 20 μL of trypsin solution of final concentration 10 ng/μL. Pre-incubate at RT (do not exceed 30 °C) for 2 h. The slices will rehydrate during this time. If the gel slices appear white or opaque, add an additional 10–20 μL of 50 mM ammonium bicarbonate, and incubate for another hour at RT. Store at −20 °C for no longer than a week.
5. Extract the peptides from gel plugs using 100 μL of peptide extraction solution.
6. Mix extracted peptides with matrix α-cyano-4-hydroxycinnamic acid solution (in equal amounts).
7. Spot 1 μL of the peptide/matrix mix onto an anchor chip target/MALDI plate.
8. Use available instrument for mass spectrometry (MS) analysis.
9. Set the MS-mode acquisition (1,000–4,000 Da) to 150 laser shots averaged from 5 sample positions. Use the top ten peaks from each full scan for subsequent MSMS analysis. Use collision-induced dissociation for peptide fragmentation and sum up 250 laser shots from 5 sample positions for each parent ion.
10. Use desired software for data processing.
11. The database search should be conducted using a MS tolerance of 100 ppm, and a MSMS tolerance of 0.2 Da. Allow one miscleavage. Choose carbamidomethylation as global modification, while oxidation of H, W, M residues and phosphorylation of S, T, Y residues should be set as variable modifications. Select the Arabidopsis database from any provider (e.g., Uniprot) to match the spectra. We chose standard scoring and a significance threshold of $p < 0.05$ for protein/peptide identification for Mascot results filtering.

3.5 In Solution Digestion

1. Resuspend the protein pellet in 50 μL of protein lysis buffer (*see* **Note 18**). Spin down the non-dissolved particles by centrifugation at 10,000 × *g* at RT for 10 min.
2. Measure the protein concentration. Take equal amount of proteins (100 μg) for all samples and dilute them to 50 μL with protein lysis buffer.
3. Add 10 μL of disulfide bonds reduction solution per sample. Mix gently and incubate at RT for 1 h.
4. Add 10 μL of thiol group alkylation solution per sample. Mix gently and incubate at RT for 1 h.

5. To consume any unreacted IAA, add 10 μL of disulfide bond reduction solution per sample.
6. Add 320 μL of distilled water per sample to dilute urea to less than 1 M to avoid inhibition of trypsin activity.
7. Add 20 μL of trypsin solution per sample (100 ng trypsin/μL). Incubate overnight in the dark at 37 °C.
8. Add 4 μL of acetic acid per sample to stop the digestion by lowering the pH to 4.
9. To remove precipitate, centrifuge the samples at 12,000 × *g* and 10 °C for 5 min. Collect the supernatant.

3.6 Desalting of the Peptide Digest Using C18 Cartridges

To load a solution on a cartridge, fill the syringe with desired solution and fix the cartridge thoroughly onto a syringe. Load any solution slowly, applying minimal pressure on a syringe. Avoid any air entering the cartridge.

1. Activate the cartridge with 1 mL of 50 % acetonitrile.
2. Equilibrate the cartridge with 1 mL of 0.1 % TFA.
3. Load the sample.
4. Wash with 0.5 % TFA acid in 5 % acetonitrile (1 mL).
5. Elute sample with 400 μL of 90 % acetonitrile.
6. Vacuum-dry samples, and store them at −80 °C.
7. Redissolve in 20 μL of solution containing 0.1 % formic acid and 5 % acetonitrile prior to the 2-D LC/MSMS analysis.

3.7 2-D LC/MSMS

1. Spike the sample with tryptic digest of BSA, horse myoglobin, and horse cytochrome C to final concentration of 0.5 pmol.L^{-1}.
2. The liquid chromatography—mass spectrometry analysis could be performed using any proper equipment. Our protocol is optimized for ProteomeX Workstation (Thermo Scientific). Set the flow rate to 3 μL/min for both strong cation-exchange and reverse phase columns.
3. For the strong cation-exchange column, use salt steps of 0, 10, 15, 25, 30, 35, 40, 45, 50, 57, 64, 90, and 700 mM ammonium acetate in 5 % acetonitrile and 0.1 % formic acid.
4. Elute the reverse phase column by acetonitrile gradient (in 0.1 % formic acid) as follows: 5–30 % for 30 min, 30–65 % for 9 min, 95 % for 5 min, 5 % for 15 min, for a total of 59 min elution and data collection for each of 13 salt steps.
5. Program the mass spectrometer to operate in the data-dependent mode with dynamic exclusion, and four scan events: one MS scan (m/z range: 300–1,700) and three MSMS scans of the three most intense ionized species detected in MS scan in real time.

3.8 Protein Identification and Label-Free Quantification

1. Use program of your choice for matching MS and MSMS data with the database (e.g., TurboSEQUEST algorithm of the Bioworks Browser 3.2 EF2 software, Thermo Scientific). Include cysteine carbamidomethylation and methionine oxidation in the search criteria as variable modifications. Match the data against both target and decoy databases. Select the taxonomically relevant and latest release of the database for a search.
2. There is a variety of programs available for data analysis beyond the primary protein identification. For label-free quantification, we used two independent software packages: the "in-house" made ProtQuant [13] and commercial software ProteoIQ 1.3.01 (Bioinquire). In the case of ProtQuant, the unfiltered .srf files were converted to .xmL files and imported to ProtQuant. Consider only proteins identified with at least three peptide hits (spectral count) for particular biological sample TurboSEQUEST cross correlation factors (*X*corr) of all the identified peptides from all three replicates for each protein were summed, and one way ANOVA ($p < 0.05$) was used to identify statistically significant differences in protein expression between biological samples, according to [15]. The results are validated by detecting correct relative amounts for known spiked proteins.
3. The ProteoIQ (Bioinquire) worked directly with unfiltered .srf files that were uploaded to the software, which uses spectral counting as a base for relative label-free quantification. The crucial parameters should be set as follows: minimum peptide length = 5 amino acids (AAs), maximum protein false discovery rate (FDR) = 1 %, minimum protein group probability = 95 %, normalization was based on total spectral count in triplicates of particular biological samples, and only the "Top" proteins (as defined by ProteoIQ—within a protein group, each and every respective peptide could be matched to the top protein) should be further considered. A minimum of three spectral counts per protein are required for quantitation. The outputs from the two software packages should be compared, and only concurring results should be accepted.

3.9 Validation of Proteins Identified by Proteomics: Immunoblotting

Perform the treatments in at least three biological replicates.

1. Grow and treat the Arabidopsis seedlings with the inhibitors as described in Subheading 3.1.
2. Harvest the roots, and homogenize them using mortar and pestle to a fine powder under liquid nitrogen.
3. Extract the proteins using phenol extraction method described above and dissolve the pellet with rehydration buffer. Do not add ampholytes to the rehydration solution. Measure the protein concentration.

4. Supplement the protein solution with 1/3 volumes of 4× Laemmli SDS buffer and centrifuge to remove undiluted particles. Collect the supernatant (*see* **Note 30**).
5. Load 15 μg of proteins per lane on 10 % SDS PAGE gels (*see* **Note 31**).
6. Perform the electrophoresis and immunoblotting according to the protocols for given instrumentation and detection method.
7. To validate the proteomic data, perform software-based band density quantification.

3.10 Validation of Proteins Identified by Proteomics: Isozyme Pattern Analysis

Perform the treatments in at least three biological replicates.

1. Grow and treat the Arabidopsis seedlings with the inhibitors as described in Subheading 3.1.
2. Harvest the roots and homogenize them using mortar and pestle to a fine powder under the liquid nitrogen.
3. Add extraction buffer for isoenzyme pattern analysis in a ratio 100 μL/0.1 g of homogenized material, vortex, and incubate on ice for 30 min.
4. Centrifuge at 12,000 × *g* and 4 °C for 15 min.
5. Collect the supernatant and measure the protein content.
6. Add bromophenol blue to the sample in final concentration of 0.002 % from the stock.
7. Prepare native PAGE resolving and stacking gels.
8. Load 40 μg of proteins per sample for superoxide dismutase, and ascorbate peroxidase isozyme pattern analysis. For superoxide dismutase, load samples on three separate gels in order to treat the individual gels with superoxide dismutase isozyme inhibitors. Load 10 μg of proteins for catalase isozyme pattern analysis.
9. Run electrophoresis at constant 20 mA/gel on 8 cm wide gels.
10. To differentiate activities of different SOD isoforms, incubate the first gel in 50 mM sodium phosphate buffer for 15 min. Incubate two remaining gels in KCN incubation solution (*see* **Note 32**), and the hydrogen peroxide incubation solution (*see* **Note 33**), respectively. Next, transfer the gels to solution A in the dark for 15 min, followed by additional 15 min dark incubation in solution B. Visualize the bands (achromatic bands on dark purple background) by exposing the gels to light for additional 5–10 min. Stop the reaction by transferring the gels into distilled water. Document the gels and evaluate the band densities using appropriate software (e.g., ImageJ).
11. For ascorbate peroxidase, incubate the gels two times for 15 min in solution C, next two times for 15 min in solution D. Wash the gels by brief immersion in 50 mM Na-phosphate buffer (pH 7.0) and visualize the bands (achromatic bands on dark

purple background) by incubation in solution E for 5–10 min. Stop the reaction by transferring the gels into distilled water. Document the gels and evaluate the band densities.

12. For catalase isozyme pattern analysis, incubate the gel in distilled water three times for 10 min. Consequently, incubate the gel in solution F for 10 min and visualize the bands (achromatic bands on dark purple background) by incubation in solution G for 2–5 min. Stop the reaction by transferring the gels into distilled water. Document the gels and evaluate the band densities using appropriate software.

3.11 Validation of Proteins Identified by Proteomics: GFP Technology and Live Imaging

Cloning and tagging with fluorescent protein such as GFP, YFP, or CFP are currently the most used methods for subcellular visualization of endocytic proteins. Endocytic proteins can be expressed under the control of constitutive cauliflower mosaic virus promoter 35S, the ubiquitin promoter UBQ10, or their own promoters [1]. Expression in stable transformed lines carrying fluorescently tagged molecular markers for endocytic compartments (e.g., GFP-RabA1d [11], YFP-RabF2a, and 2xFYVE-GFP [12]) is a very useful tool to study co-localization during endocytosis and vesicular trafficking.

1. Place surface sterilized seeds expressing fluorescently tagged candidate proteins identified by proteomics on half-strength MS culture medium (pH 5.7) containing 1 % (w/v) sucrose and 0.8 % (w/v) phytagel, closed with parafilm, and stratified at 4 °C for 2 days.
2. Move Petri dishes to environmental chamber (21 °C, 16 h light, 8 h dark) for 5–6 days.
3. Transfer 5–6 day old seedling plants very carefully to micro-chambers between microscopic slides and coverslips with one parafilm layer as a spacer. Fill chambers with liquid half-strength MS medium.
4. Perform microscopic analysis before treatment of seedlings with inhibitors using confocal laser scanning microscope. Acquire images using 40× or higher magnification objectives. Excite and detect the fluorophores with the adequate settings and filters.
5. Infiltrate inhibitor solution by perfusion into micro-chamber and incubate seedlings for 2 h.
6. As a control, use plant incubated with a mock solution containing half strength Murashige–Skoog medium.
7. Perform microscopic analysis 2 h after inhibitor treatment.
8. In the case of labeling with membrane dye FM 4-64, first apply FM 4-64 to seedlings by perfusion into micro-chambers in final concentration of 4 μM and incubate for 30 min.

9. Wash out the dye with liquid ½ MS medium containing inhibitor at appropriate concentration and incubate for additional 2 h.
10. Perform microscopic analysis (*see* **step 4**). Excite FM 4-64 at 488 nm and detect between 620 and 750 nm.
11. For post-processing of images, use appropriate software (e.g., Zeiss ZEN software) and Image J 1.38×, Photoshop 6.0/CS, and Microsoft PowerPoint applications.

3.12 Validation of Proteins Identified by Proteomics: Immunolocalization

Immunolocalization with specific antibodies raised against endocytic proteins can be used as a very faithful and simple technique for validation of proteomic data, especially in the case when fluorescently tagged proteins are not available [1]. In addition, immunolocalization can also support and strengthen localization results on endocytic proteins obtained by fluorescent GFP/YFP/RFP-tagging.

1. Use 3–6 days old seedlings of *Arabidopsis thaliana* grown as in Subheading 3.11. After inhibitor treatment (*see* Subheading 3.1) transfer them quickly but very carefully by using fine forceps or tweezers (EM grade) to fixation solution in cell culture plates or fixation vessels (*see* **Notes 34** and **35**).
2. Place the samples in vacuum desiccator and fix probes for 1 h (*see* **Note 36**).
3. Carefully remove fixative and wash samples with MTSB for two times 10 min and with PBS for additional two times 10 min.
4. Reduce residual aldehyde groups with $NaBH_4$ in MTSB for 3 × 5 min.
5. Wash probes with PBS 3 × 5 min. Fixation and reduction steps are done in the fume hood.
6. Incubate probes with enzyme cocktail digesting cell walls at 37 °C for 30 min.
7. Wash probes with PBS 4 × 5 min.
8. Pipette permeabilization solution to samples and incubate at RT for 1 h.
9. Wash probes with PBS 4 × 10 min.
10. To prevent unspecific binding incubate samples in blocking solution at RT for 1 h.
11. Remove blocking solution and add primary antibody appropriately diluted in PBS containing 2 % BSA. Incubate overnight at 4 °C.
12. Remove primary antibody solution and wash with blocking solution at least 6 × 10 min (*see* **Note 37**).

13. Add the appropriately diluted Alexa Fluor-conjugated secondary antibody and incubate at 37 °C for 1.5 h and at 25 °C for another 1.5 h.
14. Wash probes with PBS 4 × 10 min.
15. Pipette DAPI solution for staining of nuclei and chromosomes at RT for 10 min.
16. Wash probes with PBS for 10 min.
17. Transfer seedlings very carefully in a drop of mounting medium to microscope slide and cover with coverslip. By using filter paper remove excess of mounting medium. Seal with transparent nail polish. Store samples at −20 °C in the dark to preserve fluorescence signal.
18. Prepare adequate controls (*see* **Note 38**).
19. Microscopic slides with immunolabeled whole-mount probes are ready for observation in confocal laser scanning microscope using appropriate selection of fluorescence excitation and emission wavelengths.

4 Notes

1. Phenol is toxic. Follow all safety precautions.
2. Dissolve the components sequentially.
3. Cool down to 60 °C prior to use.
4. Other mass spectrometry compatible stains can be used as well.
5. Longer gel strips may be used as well. The procedure has to be modified accordingly.
6. KCN is highly toxic. Follow the manufacturer's safety precautions
7. Prepare prior to use and keep in dark.
8. Put solid KOH to dissolve PIPES first.
9. Prepare 0.5 M EGTA stock solution and titrate with solid KOH until the solution clears and adjust pH to 8.0.
10. We recommend preparing 10× PBS stock solution, supplemented with 0.01 % (w/v) NaN_3 to prevent contaminations, and storing at room temperature. Alternatively stock solution may be autoclaved and stored at room temperature.
11. Formaldehyde is highly toxic and carcinogenic. Work in fume hood. Dispose according to institutional safety regulations for toxic waste.
12. Glutaraldehyde is toxic and should be handled and disposed as formaldehyde. Use frozen stock aliquots (−20 °C).

13. DMSO is a developmental neurotoxin and should be handled with appropriate safety precautions.
14. Para-phenylene diamine is highly toxic, and should be handled and disposed accordingly following institutional safety regulations.
15. Weight 50 mg para-phenylene diamine, add 100–150 μL DMSO, swirl the glass vessel fast and vigorously until powder is dissolved, add 5 mL of 1 M Tris–HCl buffer (pH 8.4–8.8) or phosphate buffer (pH 8.2–8.6), by using glass rod mix gently but quickly with 45 mL of 100 % glycerol (DMSO is oxidant so it must be rapidly diluted with Tris–HCl and glycerol to prevent oxidation of para-phenylene diamine). Aliquot and store at −20 or −80 °C in dark, maximum 3–4 months (don't use if color turn to pink or brown).
16. Work in fume hood when handling the phenol.
17. Initial precipitation at −80 °C for 10 min has beneficial effects on precipitate yield.
18. We recommend incubating the pellet in this solution for 1 h to allow complete dissolution of the proteins. Keep the dissolved proteins in constant temperature.
19. The volume of the protein extract is limited by the maximum volume applied to the gel strip. The protein content applied to the strip is sample dependent.
20. Avoid and remove any bubbles under the gel strip.
21. This step might be performed in 15 mL falcon tubes or in commercially available specialized trays.
22. 2-propanol residues may disturb the protein separation.
23. Do not disturb the gel surface. Do not allow the gel dry.
24. Bubbles substantially alter the protein separation.
25. Thorough washing helps to eliminate unspecific background on the gel.
26. Overnight staining increases the unspecific background on the gel.
27. Prolonged washing does not hamper the staining intensity.
28. Repeat or prolong this step if necessary.
29. The gel plugs become opaque. Repeat or prolong this step if necessary.
30. Do not boil the protein extracts to avoid carbamylation of proteins.
31. The protein content loaded on the gel has to be optimized.
32. KCN inhibits CuZnSOD, but has no effect on MnSOD and FeSOD activities.

33. Hydrogen peroxide inhibits CuZnSOD and FeSOD, but does not affect the MnSOD isozyme.
34. Using 1 mL micropipette change solutions in cell culture plates or fixation vessels very carefully, avoid direct contact of micropipette with seedlings, especially with their root tips. For better preservation of probes and convenient handling we recommend using plastic baskets (diameter 14 mm) with nylon mesh grids (100 μm), which are placed to wells of 24 well cell culture plate. This pipetting-free procedure is minimizing sample damage.
35. Using forceps transfer plastic baskets in new wells of cell culture plate filled with fresh solutions during all steps in this protocol.
36. Place the plates in vacuum desiccator and use membrane pump; incubate samples at RT for 60 min.
37. This step is critical for minimizing of background fluorescence.
38. Control labeling with pre-immune serum determines specificity of primary antibody. Unspecific binding of the secondary antibody can be determined by omitting primary antibody in control sample.

Acknowledgements

This work was supported by the grant LO1204 from the National Program of Sustainability I to the Centre of the Region Haná for Biotechnological and Agricultural Research, Faculty of Science, Palacký University, Olomouc, by Czech Republic Operational Program Education for Competitiveness—European Social Fund (project CZ.1.07/2.3.00/20.0165), by Genomics for Southern Crop Stress and Disease, USDA CSREES 2009-34609-20222, and by grant No. NIH INBRE P20GM103476.

References

1. Šamajová O, Takáč T, von Wangenheim D, Stelzer E, Šamaj J (2012) Update on methods and techniques to study endocytosis in plants. In: Šamaj J (ed) Endocytosis in plants. Springer, Berlin, p 336
2. Otegui MS, Herder R, Schulze J, Jung R, Staehelin LA (2006) The proteolytic processing of seed storage proteins in Arabidopsis embryo cells starts in the multivesicular bodies. Plant Cell 18:2567–2581
3. Drakakaki G, van de Ven W, Pan S, Miao Y, Wang J, Keinath NF et al (2012) Isolation and proteomic analysis of the SYP61 compartment reveal its role in exocytic trafficking in Arabidopsis. Cell Res 22:413–424
4. Takáč T, Pechan T, Richter H, Müller J, Eck C, Böhm N et al (2011) Proteomics on brefeldin A-treated Arabidopsis roots reveals profilin 2 as a new protein involved in the cross-talk between vesicular trafficking and the actin cytoskeleton. J Proteome Res 10:488–501
5. Takáč T, Pechan T, Šamajová O, Ovečka M, Richter H, Eck C et al (2012) Wortmannin treatment induces changes in Arabidopsis root proteome and post-Golgi compartments. J Proteome Res 11:3127–3142

6. Takáč T, Pechan T, Šamajová O, Šamaj J (2013) Vesicular trafficking and stress response coupled to PI3K inhibition by LY294002 as revealed by proteomic and cell biological analysis. J Proteome Res 12:4435–4448
7. Nebenführ A, Ritzenthaler C, Robinson DG (2002) Brefeldin A: deciphering an enigmatic inhibitor of secretion. Plant Physiol 130:1102–1108
8. Wang J, Cai Y, Miao Y, Lam SK, Jiang L (2009) Wortmannin induces homotypic fusion of plant prevacuolar compartments. J Exp Bot 60: 3075–3083
9. Kim DH, Eu YJ, Yoo CM, Kim YW, Pih KT, Jin JB et al (2001) Trafficking of phosphatidylinositol 3-phosphate from the *trans*-Golgi network to the lumen of the central vacuole in plant cells. Plant Cell 13:287–301
10. Leshem Y, Melamed-Book N, Cagnac O, Ronen G, Nishri Y, Solomon M, Cohen G, Levine A (2006) Suppression of Arabidopsis vesicle-SNARE expression inhibited fusion of H_2O_2-containing vesicles with tonoplast and increased salt tolerance. Proc Natl Acad Sci U S A 103:18008–18013
11. Ovečka M, Berson T, Beck M, Derksen J, Šamaj J, Baluška F, Lichtscheidl IK (2010) Structural sterols are involved in both the initiation and tip growth of root hairs in Arabidopsis thaliana. Plant Cell 22:2999–3019
12. Voigt B, Timmers ACJ, Šamaj J, Hlavacka A, Ueda T, Preuss M et al (2005) Actin-based motility of endosomes is linked to the polar tip growth of root hairs. Eur J Cell Biol 84: 609–621
13. Bridges SM, Magee GB, Wang N, Williams WP, Burgess SC, Nanduri B (2007) ProtQuant: a tool for the label-free quantification of MudPIT proteomics data. BMC Bioinformatics 8(Suppl 7):S24
14. Murashige T, Skoog F (1962) A revised medium for rapid growth and bio assays with tobacco tissue cultures. Physiol Plant 15:473–497
15. Nanduri B, Lawrence ML, Vanguri S, Pechan T, Burgess SC (2005) Proteomic analysis using an unfinished bacterial genome: the effects of subminimum inhibitory concentrations of antibiotics on *Mannheimia haemolytica* virulence factor expression. Proteomics 5: 4852–4863

INDEX

Marisa S. Otegui (ed.), *Plant Endosomes: Methods and Protocols*, Methods in Molecular Biology, vol. 1209, DOI 10.1007/978-1-4939-1420-3, © Springer Science+Business Media New York 2014

Zeitfracht Medien GmbH
Ferdinand-Jühlke-Straße 7
99095 Erfurt, Deutschland
produktsicherheit@kolibri360.de